Barns of the Genesee Country

1790-1915

Publication of this book
has been made possible in part
with grants from
*
The Jones Chemical Company
&
The Bank of Castile

Barns of the Genesee

Including An Account of Settlement

JAMES BRUNNER, Publisher, Geneseo, New York

Country, 1790 - 1915
and Changes in Agricultural Practices

By DANIEL FINK

EDITOR: Claire Gavin
DESIGN CONSULTANT: Herbert Johnson
TECHNICAL ASSISTANCE: Robert Bretz

Except as otherwise noted the photographs in this book were taken by the author.

Publication of this volume has made possible, in part, with public funds from the New York State Council on the Arts through the Landmark Society of Western New York, Rochester, New York and the Association for the Preservation of Geneseo.

Library of Congress Catalog Card Number 86-72689

Library of Congress Cataloging in Publication Data

Fink, Daniel, 1937–
Barns of the Genesee Country, 1790–1915.
Bibliography: p.
Includes Index

Typesetting by Typographic Insight, Ann Arbor, Michigan.
Printed and Bound by Dai Nippon Printing Company, Ltd., Tokyo, Japan.

First Edition
Second Printing
1988

DEDICATED TO MY FATHER

Daniel Alfred Fink

WHO STARTED MY INTEREST
IN THE SUBJECT

AND TO MY WIFE

Mary Ann

WHO MAKES EVERYTHING WORTHWHILE

ACKNOWLEDGMENTS

It is well to acknowledge the generous assistance and sharing of personal knowledge, lifetime experiences and materials which contributed to the content of this volume. First, thanks go to the Association For the Preservation of Geneseo whose members, Jeanette McClellan and Nancy O'Dea, expressed the need for such a study and undertook its inauguration. Thanks also to Paul Malo whose initial study supported the worth of the project. A special debt is owed to William P. Wadsworth, farmer and descendent of the pioneer Wadsworths, who gave freely of his experience, personal research, and practiced eye. Without Mr. Wadsworth's assistance many important nuances of the evolution of local barns might have been missed. Special thanks to Paul Eshbaugh, whose wide knowledge, engineering experience and skill and hewing skill added much to the understanding of timber framed barns. Thanks to Dr. John Black, whose father ran a local dairy, and to Dr. Samuel Culbertson, ex-Mayor of Geneseo and son of a Wadsworth farm overseer, for their sharing of experiences and observations. Thanks to the Honorable J. Robert Houston, historian and supporter of the study whose knowledge and encouragement added an important dimension. Thanks to Bill Lane, Head Reference Librarian, historian and keeper of the invaluable collection of primary documents in the Genesee Valley Room of the Milne Library on the SUNY campus at Geneseo. Thanks to Stuart Bolger, Director of the Genesee Country Museum, Mumford, N.Y., whose enthusiasm for the study contributed mightily. Thanks to Henry McCartney, Executive Director of the Landmark Society of Western New York, whose faith in the project helped lead to financial support. Thanks to Austin Wadsworth, who flew me up and down the valley in order to obtain the aerial photographs. Thanks to the publisher, James Brunner, who in the formative stages of this book shared his collection of regional material and who often pushed me off dead center, sometimes kicking and screaming, in order to investigate some facet of barn evolution which had escaped my notice. Thanks go to those who read the drafts of the study and offered their editorial advice: J. Robert Houston, Nancy O'Dea, Paul Hepler and especially my wife, Mary Ann, whose editing skills were of tremendous value.

At the risk of excluding someone the following individuals have my enduring thanks for their significant contributions to the study:

Mrs. John Wooster, John Coe, Tom Cook, Dorothy Facer, Jean Melville, Ted Bartlett, Mr. & Mrs. Gordon Wingate, Mrs. W. P. Wadsworth, Clark Rice, Cathy Reinholtz, Anne Humphrey, Sanford Robinson, Dick and Vernice Platt, Walt Kingston, Richard Quick, Hans Tanner, Paul Bailor, Mike & Barb Mulligan, Mrs. Edward Mulligan, Lynne Belluscio, Ted Kinsey, Bob Blaker, Dave Nusbickel, Allan Turner, Andrew Macauley, Sue Conklin, Laurel Wemett, George Spanneut, Jim Kimball, Elsie Biery-Fink, Pat Schaap, William Siles, Melvin McNinch, Jean France, Paul Knickerbocker, Peg Parshall, Robert Moody, Kay Thompson, Ann Tanner, Virginia Gibbs, The Honorable Barber Conable. Jr. M.C., J. Howard Pratt, R. B. Bromeley, Pat Murphy, Alma Burner Creek, Karl Kabelac, Marian Bowman, Jean Richardson, James Wilson, Robert Halsted, Pauline & Louis Owen, Linda & Vincent Bruno, Norma Higgins, Peter Bruckle, Steven McNall, Robert Frasch, Joseph Kane, Charles Ennis, Marjory Allen Perez, Marilyn Strickland, David Best, Richard Galbraith, Tom Pulvino, Orville Gray, Ray Thomson, Shirley Whyman, Mary Menzie, Josh & Beth Bruner, Bruce Kelley, John W. Brown, Ted & Penny Woehr, the *American Agriculturist*, The Arts Council for Wyoming County, The Landmark Society of Western New York, The Visual Studies Workshop, The Genesee Country Museum, The International Salt Company, The Genesee Region of The New York State Office of Parks, Recreation, and Historic Preservation. The names of other contributors are mentioned, along with the materials which they contributed, in the body of the text.

CONTENTS

Acknowledgments VIII

Preface XV

Genesee Country Portfolio XVIII

CHAPTERS

I *The Fertile Land of the Genesee* 33

II *Log Barns and Cabins* 47

III *Experiments in Land Speculation* 61

Diary: Anonymous, West Bloomfield, 1814 67

Diary: Thayer Gauss, Bloomfield, 1826-1858 69

IV *Steamboats, Roads and the Triumph of Canals* 77

V *Hewn Timber-Framed English Barns* 95

Diary: Sheffield Peabody, Springwater, 1849-69 134

VI *Bank Barns and Other Enlarged Gable-Roofed Barns* 137

Diary:	*Some Winter Thoughts by L. E., 1859-60*	165
VII	*Mills and the Sawn Timber-Framed Barn*	167
VIII	*Cobblestone Construction*	189
Diary:	*Philo Hampton's Book, Ossian, 1853-77*	199
IX	*Greek, Gothic and Other Decorative Barns*	203
X	*A Crop Sampler*	233
XI	*Horticulture*	263
XII	*Railroads Change the Genesee Country*	277
XIII	*Nineteenth Century Agricultural Implement Manufacturing in the Genesee Country*	289
XIV	*Gambrel-Roofed Barns*	303
XV	*Agricultural Information in Almanacs, Periodicals, Manuals, at Colleges and Fairs*	331
XVI	*Livestock*	341
Diary:	*Jennie Mather, Livonia, 1867*	357
XVII	*Barn Arrangements*	361
XVIII	*The Black and White Farm Barn*	381
XIX	*Changing Sources of Power on the Farm*	391
XX	*Truss Framed Barns*	415

Diary: *Seneca Owen, Pulteney, 1908-1912* 431

XXI *Wells Barns* 451

XXII *Modern Materials in Barn Construction* 467

Endnotes 490

APPENDICES

I Notice to New Settlers 497

II Jennings Broadside 499

III Barn Evolution 501

IV Photographing Interiors 505

BIBLIOGRAPHY 508

INDEX 516

ADDENDUM

PREFACE

BARN, *n. A covered building for securing grain, hay, flax, and other productions of the earth. In* the Northern States of America, *the farmers generally use barns also for stabling their horses and cattle; so that, among them, a barn is both a corn-house, or grange, and a stable.*

Webster's American Dictionary of the English Language, 1860.

To every thing there is a
season, and a time to every
purpose under the heaven:
A time to be born, and a
time to die; a time to plant,
and a time to pluck up that
which is planted;
A time to kill, and a time
to heal; a time to break down,
and a time to build up; a time
to mourn, and a time to dance;
A time to cast away stones, and
a time to gather stones
together; a time to embrace,
and a time to refrain from
embracing; A time to get, and a
time to lose; a time to keep,
and a time to cast away; A time
to rend, and a time to sew; a
time to keep silence, and a
time to speak. A time to love,
and a time to hate; a time of
war, and a time of peace.

Ecclesiastes 3: 1–8

THERE was a season of wooden barns, barns that dominated the countryside in the Genesee Country. For almost 200 years barns were, almost without exception, the largest structures in the rural landscape of the Genesee Country.

To the pioneers of the 1790s all the land in New York State west of Geneva was Genesee Country. Here was a rich soil, soon to be wheatland, soon to be celebrated in the 1830s and '40s as the "breadbasket of the nation." As the wheatland expanded Genesee Country barns evolved from small log buildings to large structures of concrete and steel in just a little over 100 years. Elsewhere in the eastern United States, society had been evolving from wilderness settlements to colonial states through post-revolutionary growth to industrialization, all in a period of 200 years or more. The Genesee Country, however, experienced this transition in a rush of less than fifty years. Between 1790 and 1830 the population grew from 1081, representing just 205 families, to more than 400,000; as wave after wave of change affected lives and farming practices in the predominately agricultural society, barn designs changed as well.

Throughout the time period covered by this book, 1790–1915, periodicals, agricultural colleges, and word of mouth combined to produce and disseminate superior barn designs. As a result, in the latter half of the nineteenth and the early part of the twentieth centuries there is a greater consistency in barn design in spite of continued experimentation.

This book presents a sampling of appropriate, typical or significant barns, a small fraction of the more than 30,000 that once stood in the Genesee Country. The barns were selected after consultations with individuals interested in barn history, extensive road searches and examinations of the available literature. The principal counties included are Monroe, Livingston, Orleans, Genesee, Wyoming, Allegany, Steuben, Yates, Ontario and Wayne.

It is possible, of course, to look at barns without attempting to understand their forms or the changing conditions that led to their evolution, but to do so is to ignore the richness of the past. The historical context of each barn is the daily life of the people, and wherever possible the soldiers, pioneers, land speculators, merchants, industrialists, farmers and others tell their stories in words gleaned from diaries, early histories, newspapers and letters. Appropriate period maps, woodcuts (artistic and advertising), tintypes, stereographs and other photographs add visual support to the study.

Barns of red, white, yellow, grey and natural weathering still stand in the Genesee Country, quiet giants marking a productive land. Built of logs, hewn or sawn timber, English style or gambrel-roofed, the wooden barns of the nineteenth century reveal a history of the Genesee Country.

The barns are not static; they are constantly changing and, in fact, are fast disappearing. The season of the wooden barn is passing. The survivors are being winnowed away as rapid urban development and a changing economic base encourage the sale of farmland.

When a barn no longer has an active farm to serve, it may become redundant for an owner, an expensive luxury to maintain. The roof may be allowed to deteriorate, the weathered siding boards may blow off or be sold, letting in rain and melted snow to rot out the timbers and joints; slowly the structure settles upon itself. Or the end may come within minutes in a violent fire such as the conflagration that destroyed the "Wells" barn on the Brown farm in Scottsville (see figure P-1).

Some of the barns included in this study are already gone. Before the earth reclaims the rest, they deserve careful examination and preservation wherever possible. It is my hope that others will join in recording the significant structures, because the job is large and the time limited. Records and memories of barn building and use are too often lost because no one has taken the time to preserve and record this valuable information.

No single volume can treat all aspects of barn evolution, much less examine all the outstanding examples in such an extended geographical area. I have attempted to outline the evolution of barns, the forms to which additional specific examples can be compared. In addition, barns built in other regions which developed earlier and barns built in more recently settled regions can be compared to those of the Genesee Country.

Figure P-1. The fiery destruction in 1957 of the Wells barn on the Thomas Brown Farm in Scottsville, New York resulted from a tractor backfire which set fire to straw in the barn. Photograph by John Brown.

I like to have a man's knowledge comprehend more than one class of topics, one row of shelves. I like a man who likes to see a fine barn as well as a good tragedy.

—Emerson

GENESEE COUNTRY PORTFOLIO

Huge oak trees near Oneida Springs in Geneseo, New York are magnificent reminders of the region's past. The Genesee River Valley is rich in oak trees which date back well over 200 years to the era before settlement began in the late 1780s. Many of these survivors are more than four feet in diameter and seventy to 100 feet high. Tenant leases on the thousands of acres of Wadsworth lands forbade clear cutting and called for the preservation of one large shade tree to every two acres. It was the oak trees that for the most part were preserved.

A Genesee landscape looking west on the Nations Farm, Geneseo, toward York. The Culvert Farm basement barn in the center is fairly bursting with hay.

The Genesee Country Museum pioneer farmstead in Mumford, New York includes a typical single bay log barn, representative of the type built in the late 1790s and 1800s. In a land so rich in wood, it is not surprising that rail fences and log structures were common.

Building a barn into the side of a hill allowed wagons access to both the main floor and the basement. Adding sheds to the original frame, a common method of enlarging a barn, produces the saltbox silhouette of this barn on the Naples-Prattsburg Road.

The Westerly *stables of Piffard, New York were designed with the steep-pitched roof and board-and-batten siding associated with Gothic designs. The cupolas, which are Italianate in design, add an eclectic element to the stables.*

Barns sometimes were constructed with exterior characteristics matching those of the farm house. The steep- pitched, patterned, slate-covered roof matched that of the Gothic farmhouse on the John Coe farm in South Livonia, New York. Other Gothic details are the pierced and built-up detailing of the vergeboards and the pointed arches.

Special modifications to barn plans were made when these new designs could assist in farm operations. The horizontal doors on this tobacco barn could be opened or closed to adjust the rate of curing of the tobacco in the barn. Addison, Steuben County, New York.

The hewn timbers in the barn at Westerly *in Piffard, New York are good examples of this early practice. The scoring marks left by the hewer can be seen as well as the treenail that holds the joined post and beam in position, and the diagonal braces to prevent raking. Bark was often left on roof rafters and floor joists, which were hewn only on one side.*

For a time the octagon shape was espoused as the most efficient compromise between a round barn, the most economical in terms of materials, and the traditional rectangular plan. A few octagon barns remain in the Genesee Country, including this small barn on Coe Road in South Livonia, New York.

A barn complex on Jones Bridge Road on the Genesee River flats near Cuylerville, New York is reflected in the flood waters of 1972. Annual flooding used to replenish the soil of the Genesee Valley.

Silos for the storage of chopped corn stalks were relative late-comers to the farm landscape, appearing around 1885. This silo near Greenwood in Steuben County, New York is about all that remains of the barn complex. The silo was constructed of vertical staves and sheathed with horizontal boards.

During the 1860s the Ennis barn in Alloway, New York was enlarged. Based on the interior framing, it is clear that the main barn door with its transom light was the center of the original barn. The addition of the bay to the right, plus the building of an adjoining new barn (unseen to the rear of the new bay) added considerable space to the barn.

The vineyards on Seneca Owen's farm above Pulteney provide a view of the farmhouse, barn and a branch of Keuka Lake. Farming became increasingly specialized as the nineteenth century wore on.

The Jennings patent truss frame can be seen inside the Richard Galbraith barn in Alloway, New York. The canted timber purlin posts tie the purlins in with the barn sills, making a strong frame which nevertheless has an open barn center permitting the use of a hay track.

The Wells truss arch is formed of laminated 2x3 inch boards, often sawn on the farm where the barn was built. The uniform appearance of the interior of the James Guthrie barn in Mumford, New York is the result of a large supply of uniform lumber. The frame provided an open center for efficient use of a hay fork.

The gambrel roofed barn of the Mulligan Farm barn in Avon, New York was modified through the addition of decorative brackets, dormers and a cupola. Similar cupolas on adjacent barns helped unify the appearance of the barn complex.

Between 1882 and 1884 the Black and White Farm barn was built along the Mount Morris/Dansville Road near Sonyea. Almost 200 feet on a side, it was considered the largest barn in New York State at the time.

The gable ends of Wells truss-framed barns provide an external sign that these gambrel roofed barns contain the patented frame. The hood molding is shaped into a "lazy W" with the outer arms of the "W" horizontal. Wells barns were constructed between 1889 and 1942. Thomson/Mulligan farm, North Avon Rd., Avon, New York.

The interior framing of the Ennis barn in Alloway, New York was modified in the 1860s to make efficient use of a hay track, which can be seen as a thin line beneath the ridge. The purlin posts were canted out so the purlins would provide proper support without the need for standard crossbeams.

Gambrel roofed barns became the dominant exterior barn shape in the Genesee Country, as seen in this view near Pittsford, New York. Depending on the angles of the roof surfaces, the gambrel roof offered 50% more storage capacity than the gable roof.

This impressively large concrete barn, 347x50 feet, in Cuba, New York is known as the "Block Barn." William Simpson used the new 1907 technology of decorative concrete blocks, poured concrete and terra cotta roof tiles to produce a fireproof stable.

CHAPTER I

The Fertile Land of the Genesee

BEFORE THE BARNS, before the white settlers, even before the Seneca Indians, there was the Genesee Valley, carved by the Genesee River, the only significant north-flowing river in the northeastern United States. Approximately 160 miles long, the often strongly meandering Genesee flows from the foothills of the Alleghenies in northern Pennsylvania through a spectacular gorge, over a series of falls to a flat valley bottom, tumbling powerfully over the falls at Rochester before emptying into Lake Ontario.

The river lies just west of Conesus Lake, a last minor digit of the Finger Lakes (figure 1-1, a 1799 map, shows "the position of the Genesee Country"). Topographic features of the valley have been carved out by the glaciers and cut by the river. Before the 1952 construction of the dam at Mount Morris, the river regularly flooded flat portions of the valley floor, producing a rich topsoil.

The Genesee Valley, "the beautiful valley," was named by the Senecas, Keepers of the Western Door of the Iroquois Confederacy. The Senecas were excellent farmers. Early descriptions of their agricultural practices exist from various encounters with Europeans, including the French explorer Jacques Cartier in 1534 and the Jesuit priest Father Fremin, who took up residence with them in 1668. From these and other accounts of attempts either to reform or to destroy the Iroquois, much can be learned of their farming. For example, in the seventeenth century, after Louis XIV appointed the Marquis Denonville governor of New France, the marquis decided to put the Iroquois in their place and "save their souls." Landing east of the mouth of the Genesee River in 1687, the marquis, after some early reversals, proceeded to burn four Seneca villages, including the standing crops as well as the corn already stored for winter. The extent of cultivation can be roughly measured by the fact that it required some seven days to complete the destruction. Despite periodic European raids, the Senecas added beets, onions, peas, cabbages, apples and peaches to their earlier crops of beans, squash and corn, planting seeds obtained from other European visitors with intentions less deadly than those of Denonville.

Seneca agriculture shared some problems in common with Europeans and settlers in this country, particularly those of soil exhaustion and forest clearing. As the land became less productive through constant use, the Senecas moved their villages every ten years or so. With the men of the tribe involved chiefly in hunting and warring, the women were in charge of the land, often working together on some of the harder tasks of farming. For example, to clear new land before metal axes were available the Seneca women girdled trees with fire, then chopped them away with stone axes. The remaining stumps, as

Opposite: The Genesee River meandering through Letchworth Gorge.

always, proved to be very difficult to remove. Fortunately, much of the valley bottom was meadowland and required much less clearing.

Europeans needed barns for storing grain crops and housing cattle in the winter. The Senecas, who kept no cattle and did not raise oats, hay or wheat, had no need for barns or similar structures. Corn, however, was an important staple, preserved and stored in a variety of ways. In fact, there is considerable reason to believe that present-day silos evolved from the Indian practice of storing corn in underground silos or corn cellars. The Iroquois dried their corn on the ear for short-term storage, but for longer keeping green

FIGURE 1-1. 1799 map showing the Genesee Country east of the Genesee River, divided into township squares in relation to the "Middle States." The Genesee Valley Historical Collection, Milne Library, State University of New York at Geneseo.

corn was charred, placed in bark-lined storage pits, and covered with a bark roof. Some early European visitors to Seneca villages complained of the smell of food prepared from this fermented corn. One of the earliest records of such caches occurs in Denonville's description of the 1687 raid: "All that time we spent in destroying the corn, which was in great abundance, . . . including old corn which was in cache which we burnt . . . ."[1]

Throughout history, farming societies have paid particular attention to the spiritual forces that control the success or failure of their crops. Corn Husk Masks, worn by dancers representing a mythical farming society, were an important part of the Iroquois Mid-Winter Ceremonies, held in January or February to predict the size of the crops in the coming season (see figure 1-2). These ceremonies were second in importance only to the late summer Green Corn Festival, which was a three day period of thanksgiving for the renewal of the corn crop.

The Senecas, renowned for their crops and admired and feared as warriors, proved to be troublesome to the rebellious colonists during the Revolutionary War, in which they fought as British allies. The resulting reaction by the Colonial Army was in large measure responsible for spreading the word about the quality of the land in the Genesee Country. In 1778 the Senecas and Cayugas raided Wyoming, Pennsylvania and Cherry Valley, New York, reinforcing their reputation for fierceness in warfare. Hundreds were killed. Some eight pallisaded forts and 1,000 dwellings, mills and other buildings were destroyed during July, 1778 in the Wyoming area alone. In addition to their warring, the Senecas also provided the Tory forces with much-needed foodstuffs.

Obviously General Washington could not permit the raids and logistical support to continue. The Continental Congress authorized him to conduct a raid against the Tories and Iroquois, and on June 18, 1779, General John Sullivan led more than 3,000 troops from Easton, Pennsylvania on an expedition into the Iroquois country along the upper east branch of the Susquehanna River through Wyoming, Wyalusing, Tioga, Waverly, Seneca Castle (Geneva) and Canandaigua, finally stopping at Genesee Castle (Little Beard's Town near Cuylerville). Joining forces with those of General James Clinton, brother of New York governor George Clinton, who had been in action against the Onondagas northeast of Otsego Lake, the Sullivan expedition finally numbered some 4,000, fully one third of the Continental Army. (These were Continentals, not just militia.)

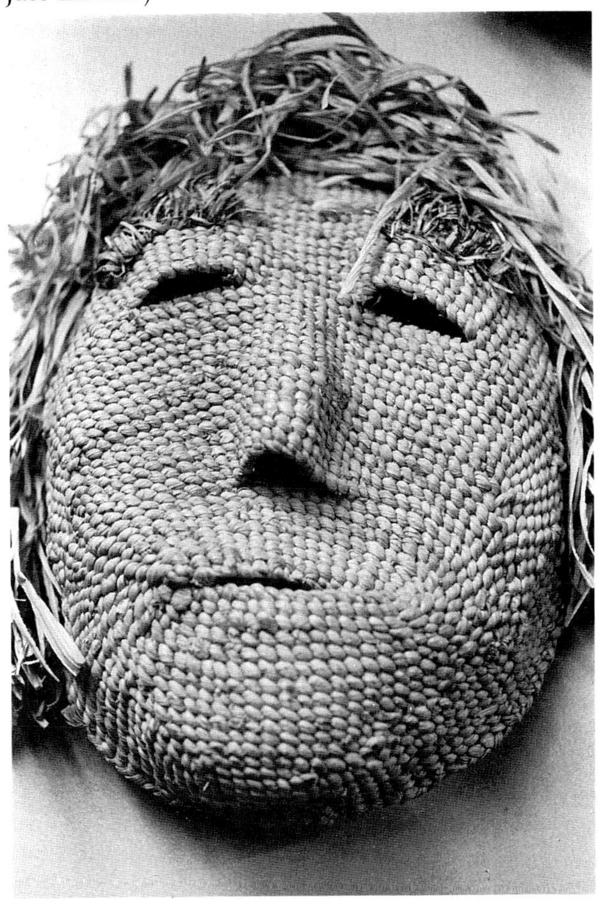

FIGURE 1-2. Seneca corn husk masks were worn during the Mid-Winter Ceremony. The ceremony was performed to elicit spiritual help in the production of corn and to ensure the birth of many healthy babies. Dancing and ritual movements were accompanied by puffing noises but no spoken words, a characteristic of Husk Face Society rituals.

The aims of the force were several: to end the Iroquois' and the Tory Rangers' ability to wage war, reduce the supply of foodstuffs to the Tories, capture major forts such as Niagara, and lastly—and perhaps most interestingly—expand the territory that would be won with a general surrender (see figure 1-3). On August 29th, 1779 at Newtown, near Elmira, the Iroquois were engaged with cannon and bayonets. They retreated, and fought no other significant battles with the Americans during the Revolutionary War. Sullivan followed his orders and continued toward the Genesee Country.

Freed from constant military pressure after Newtown, with only a few additional skirmishes, Sullivan's forces had time to eye an amazingly fertile country as they entered the Genesee Valley. Inasmuch as seventy percent of the colonial population was involved in some aspect of farming, the Continentals included a significant number of farmers-turned-soldier. In those late summer days, when the crops were ripening and the fruit trees hung full, the quality of the land was not lost on these practiced eyes. Lieutenant Erkuries Beatty described in his journal the Genesee Valley near Geneseo:

> Tuesday, September 14, 1779, . . . about 12 o'clock crossed over the branch of the Jinasee River and came upon a very beautiful flat of great extent growing up with wild Grass higher in some places than our heads.[2]

Dr. Jabez Campfield, on the same day, is a bit more reserved, describing the grass as growing . . . "as high as a man's head."[3] Rev. David Craft, who accompanied the troops, describes the arrival of the forces at Little Beard's Town:

> The castle [the name for a village] consisted of one hundred and twenty-eight houses, of which the most were large and elegant, and was surrounded by about two hundred acres of cornfields and gardens, filled with all kinds of vegetables. It was the western door of the Long House to which the Iroquois were accustomed to liken their confederacy . . . at 6 o'clock in the morning of the 15th of September, the whole army was turned out to destroy the crops, orchards, houses and gardens of the place. The corn was piled up in the houses and burned with them, or consumed on log heaps. It was estimated that from fifteen thousand to twenty thousand bushels were destroyed at this place. It was the largest corn the troops had ever seen, some of the ears being twenty-two inches in length . . .[4]

Now many knew that the Genesee Country was fair and fertile.

The Revolutionary war ended in 1783 without provision in the Treaty of Paris for the Iroquois nation in western New York. Because the Iroquois had aligned themselves with the losing side in the conflict, their lands were to be divided. However, both New York State and the Commonwealth of Massachusetts claimed the Iroquois lands which extended from Seneca Lake west to the present western New York border. Massachusetts and New York worked out a compromise at Hartford, Connecticut in 1786, in which New York acknowledged the right of Massachusetts to the preemption of the soil; that is, Massachusetts had the right to purchase the approximately six million acres from the Indians. Figure 1-5 shows the preemption line, which runs from Pennsylvania to Seneca Lake to Sodus Bay. At the same time Massachusetts recognized the political sovereignty of New York over the same land. In 1788 Massachusetts sold all of its land east and west of the Genesee River to a group of investors represented by Oliver Phelps and Nathaniel Gorham for about three cents an acre.[5] They agreed to pay the total sale price in three yearly installments.

Before the Phelps and Gorham land could be sold to others the rights of the Senecas had to be extinguished. In 1788 the Senecas agreed to give up the approximately 2.5 million acres east of the Genesee River and about 200,000 acres west of the river. No agreement with the Indians was made for the rest of the land west of the Genesee

River. In 1789 Phelps opened the first regular land office in the United States in Canandaigua. However, in spite of their best efforts in land sales Phelps and Gorham could not pay more than the first third of the purchase price, and the land west of the Genesee River was turned back to Massachusetts in 1790. The 1798 map of Ontario and Steuben counties shows the Phelps and Gorham division of the land (see figure 1-4).

The general situation for Phelps and Gorham did not improve and they were forced to dispose of all their remaining land except for two townships. In 1791 Robert Morris, signer of the Declaration of Independence, financier of the Revolution, and would-be land speculator, bought the remaining Phelps and Gorham purchase east of the Genesee* (see figure 1-5). Morris himself was beset by financial difficulties and in 1792 was forced to sell most of his land; the land east of the Genesee River was sold to Sir

*including the land sold to John Butler, located on the west side of the Phelps and Gorham land east of the Genesee River, commonly referred to thereafter as the "Morris Reserve" even after its inclusion in the sale to the Pulteney Estate.

FIGURE 1-3. Butler's Rangers (here enacted by members of The Brigade of the American Revolution), allied with the Senecas to create a formidable Tory force in the Genesee Country.

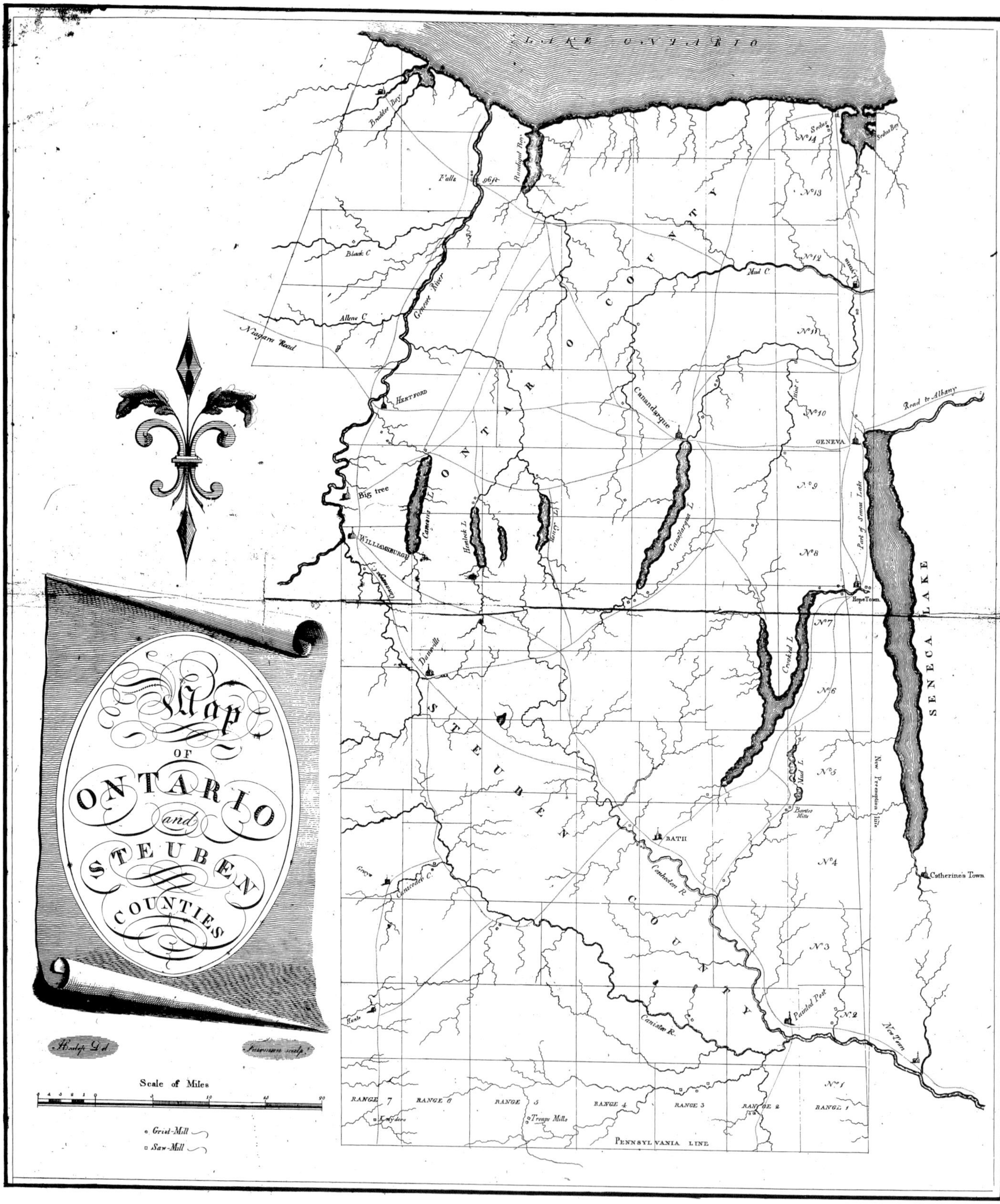

FIGURE 1-4. The original Phelps and Gorham lands ran from the second preemption line to the western border of the state. The original purchase was soon reduced in size through sale, inability to obtain complete Indian extinguishment of title and return of two-thirds of the land to Massachusetts. This 1798 Map of Ontario and Steuben Counties depicts the land remaining after the return of the two-thirds to Massachusetts. Structure symbols indicate the first villages. Monroe and Livingston Counties were formed from Genesee and Ontario Counties in 1821. The Genesee Valley Historical Collection, Milne Library, State University of New York at Geneseo.

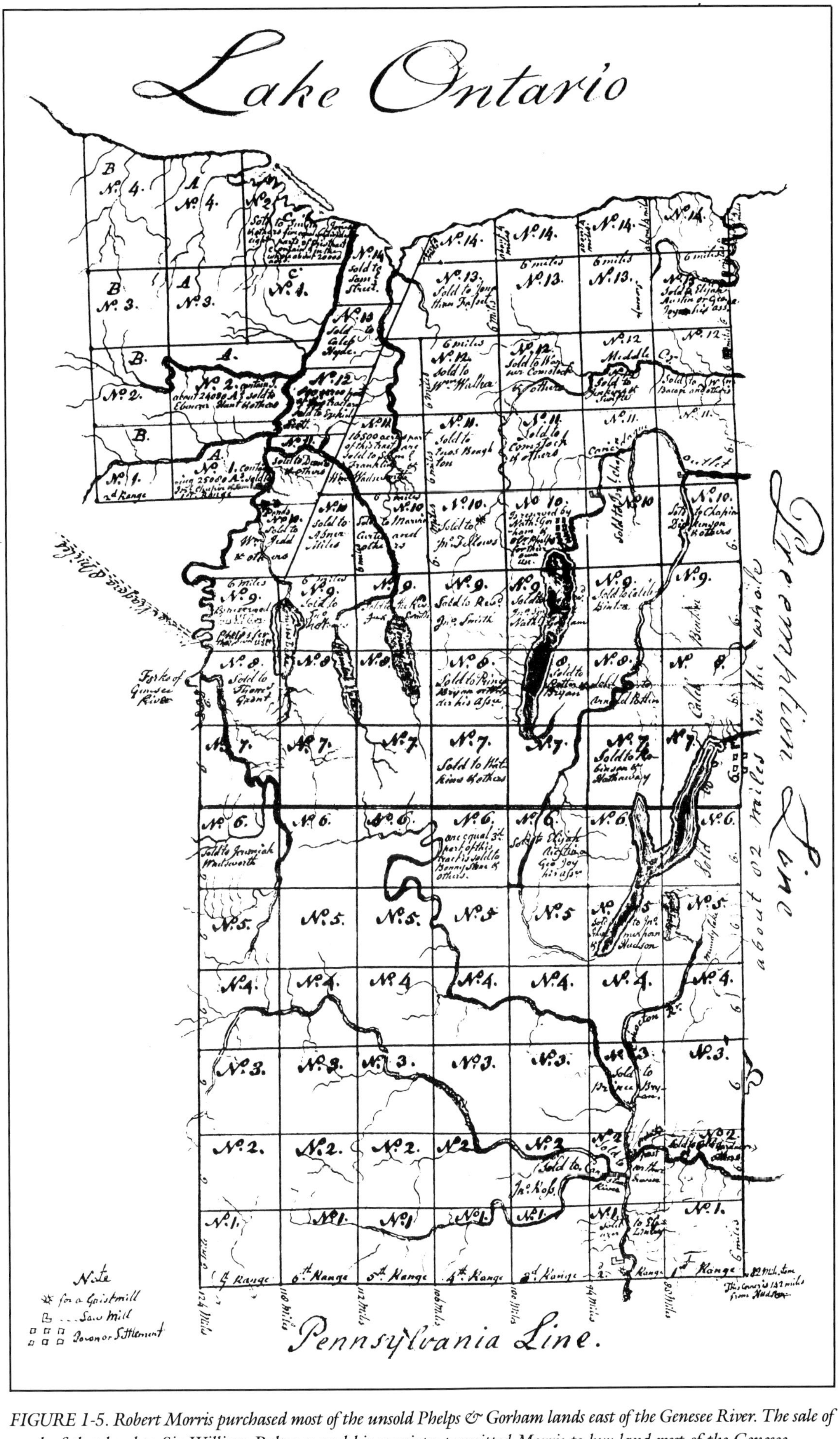

FIGURE 1-5. Robert Morris purchased most of the unsold Phelps & Gorham lands east of the Genesee River. The sale of much of that land to Sir William Pulteney and his associates permitted Morris to buy land west of the Genesee.

William Pulteney and his associates, a London-based group. The Pulteney estate included most of the original land east of the Genesee River except for that which had already been sold (about one-third). The profit from the sale permitted Morris to purchase the land west of the Genesee River which had been returned to Massachusetts. By 1793 he had arranged sale of most of this land to Theophile Cazenove, representative for a group of Netherlandish land speculators, pending the extinguishing of the Seneca rights. The 3.3 million acres involved became the holdings of the Holland Land Company, formed in 1795; this included most of western New York west of the Genesee River with some minor exceptions and, after 1797, the Indian reservations (see figure 1-6).

Because the Indians' rights had not been extinguished west of the river, only a portion of the total sale price was paid to Morris. The problem of "rights" was complicated by the fact that the Senecas had not been included in the Treaty of Paris following the Revolutionary War. Agreements were reached with the signing of the Pickering Treaty in 1794 between the United States and the Six Nations of the Iroquois. The Pickering Treaty limited Seneca lands to western New York and recognized the reservations granted to the other Iroquois nations through treaties with New York State. Facing financial ruin, Morris pressed for a conference with the Senecas to settle the question of his rights versus their rights, and the Treaty of Big Tree was signed on September 15, 1797. The Senecas retained less than 200,000 acres in reservation from the original four million acres. The treaty relegated the Iroquois Senecas to these reservations and the rest of the Genesee Country was opened to settlement. At this point Morris was legally able to transfer the land to the Holland Land Company and the rest of the purchase price was paid. The Holland Land Company established a land office in Batavia and subsequently built roads, bridges and mills to entice potential settlers to the western portion of the Genesee Country.

Many speculated in the lands of the Genesee Country with hopes of a tidy profit, but considerable effort was required in most cases to realize any gain. One person who was part of the Phelps and Gorham land speculation, Jeremiah Wadsworth of Hartford, Connecticut, Commissary General of the Continental army and friend of Washington, traveled in 1788 to the Genesee Valley to see the land in which he had a part interest. Wadsworth was generally delighted with the quality of his 200,000 acres. However, the land was rough and needed supervision if it was to be sold, so Jeremiah turned to two nephews, James and William Wadsworth.

Jeremiah offered James and William some 2,000 acres of the best land in the valley at a low price if they would live on the land, improve it, and help promote the settlement of the region. The brothers agreed to come, purchasing additional land upon their arrival, and worked hard to live up to their part of the bargain. Their efforts to interest others in the Genesee Country can be seen in two of their many broadsides, the first (figure 1-7) printed by James Wadsworth in 1809 and the second (figure 1-8) an 1823 advertisement for over 17,000 acres. The difficulties of attracting settlers many be gauged in this letter:

> Note—In a letter to Mr. Tronp [Troup] dated January, 1807, Mr. Wadsworth says:—When I commenced inviting settlement to West Pulteney, it was literally a wilderness without a road passing through it. It had been for sale ten years, and not a settler had gone upon the tract. Sales had been embarrassed by the cheap lands of the Holland Company; and yet, notwithstanding these obstacles, it has become the most respectable settlement west of the Genesee river. In a letter from same, to same, in May following, it is remarked: Mr. Mead has erected a saw-mill on Black Creek; nine barns have been erected in West Pulteney. There is not three frame barns in Caledonia.[6]

In retrospect, this early speculation seems attractive, since the land was at once so rich and so inexpensive, but that attitude tends to ignore the many problems associated with the venture, including competition and uncertainty. Only for those who knew the quality of the land was the end result certain.

Another person attracted to the Genesee Country was Henry Brewster, one of the pioneers of Riga, who describes how he and Samuel Baldwin came to the area, first to inspect and then to settle down and farm. Brewster's experiments in marketing are quite interesting.

> In October, 1805, Mr. Samuel Baldwin, a neighbor of mine, and myself, mounted our horses and came to see the Genesee Country,

FIGURE 1-6. The land west of the Genesee became the Holland Purchase once the Indian rights were extinguished. Rush Rhees Rare Book Collection, University of Rochester, Rochester, New York.

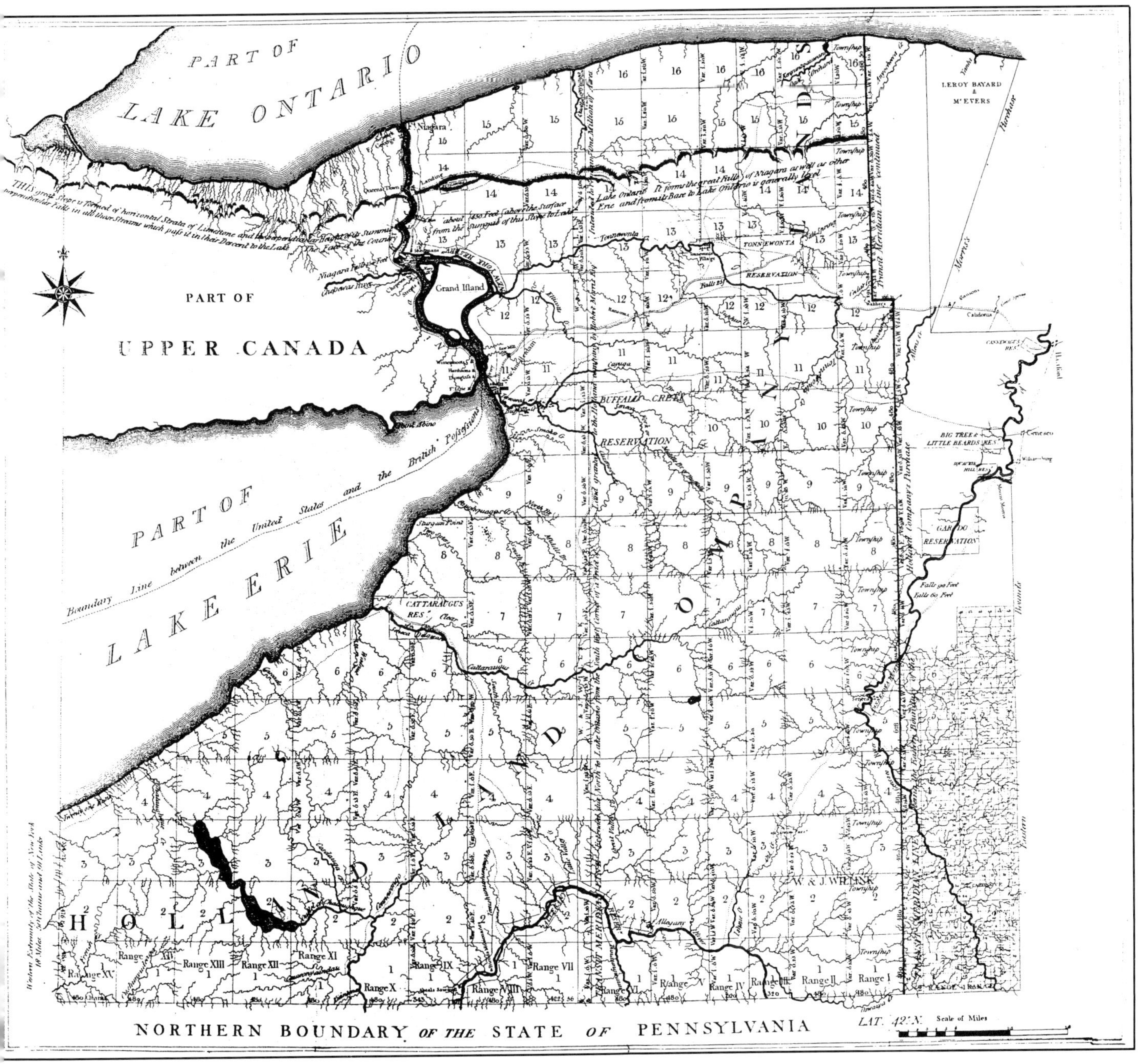

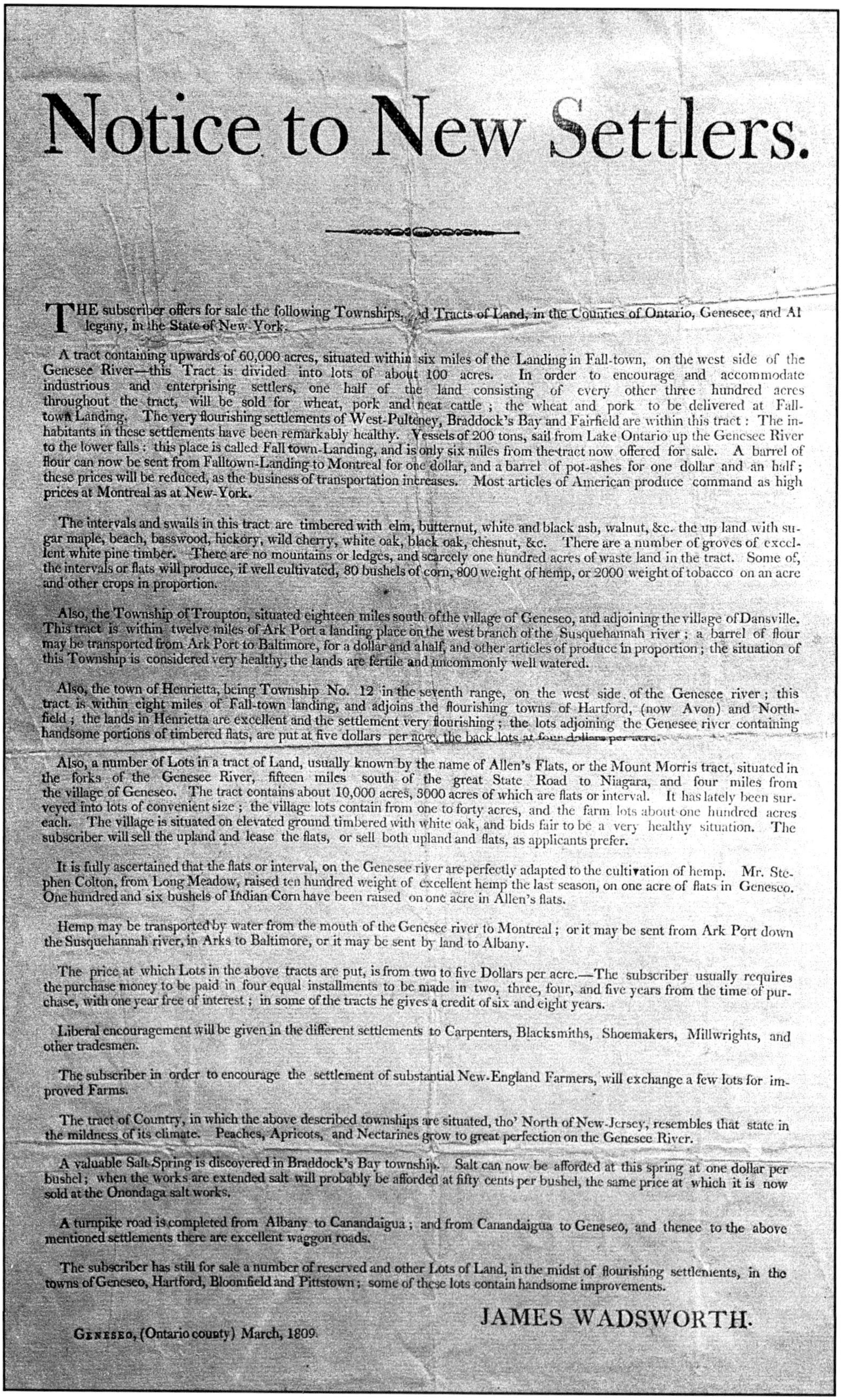

Notice to New Settlers.

THE subscriber offers for sale the following Townships, and Tracts of Land, in the Counties of Ontario, Genesee, and Allegany, in the State of New-York.

A tract containing upwards of 60,000 acres, situated within six miles of the Landing in Fall-town, on the west side of the Genesee River—this Tract is divided into lots of about 100 acres. In order to encourage and accommodate industrious and enterprising settlers, one half of the land consisting of every other three hundred acres throughout the tract, will be sold for wheat, pork and neat cattle; the wheat and pork to be delivered at Fall-town Landing. The very flourishing settlements of West-Pulteney, Braddock's Bay and Fairfield are within this tract: The inhabitants in these settlements have been remarkably healthy. Vessels of 200 tons, sail from Lake Ontario up the Genesee River to the lower falls: this place is called Fall town-Landing, and is only six miles from the tract now offered for sale. A barrel of flour can now be sent from Falltown-Landing to Montreal for one dollar, and a barrel of pot-ashes for one dollar and an half; these prices will be reduced, as the business of transportation increases. Most articles of American produce command as high prices at Montreal as at New-York.

The intervals and swails in this tract are timbered with elm, butternut, white and black ash, walnut, &c. the up land with sugar maple, beach, basswood, hickory, wild cherry, white oak, black oak, chesnut, &c. There are a number of groves of excellent white pine timber. There are no mountains or ledges, and scarcely one hundred acres of waste land in the tract. Some of the intervals or flats will produce, if well cultivated, 80 bushels of corn, 800 weight of hemp, or 2000 weight of tobacco on an acre and other crops in proportion.

Also, the Township of Troupton, situated eighteen miles south of the village of Geneseo, and adjoining the village of Dansville. This tract is within twelve miles of Ark Port a landing place on the west branch of the Susquehannah river; a barrel of flour may be transported from Ark Port to Baltimore, for a dollar and a half, and other articles of produce in proportion; the situation of this Township is considered very healthy, the lands are fertile and uncommonly well watered.

Also, the town of Henrietta, being Township No. 12 in the seventh range, on the west side of the Genesee river; this tract is within eight miles of Fall-town landing, and adjoins the flourishing towns of Hartford, (now Avon) and North-field; the lands in Henrietta are excellent and the settlement very flourishing; the lots adjoining the Genesee river containing handsome portions of timbered flats, are put at five dollars per acre, the back lots at four dollars per acre.

Also, a number of Lots in a tract of Land, usually known by the name of Allen's Flats, or the Mount Morris tract, situated in the forks of the Genesee River, fifteen miles south of the great State Road to Niagara, and four miles from the village of Geneseo. The tract contains about 10,000 acres, 3000 acres of which are flats or interval. It has lately been surveyed into lots of convenient size; the village lots contain from one to forty acres, and the farm lots about one hundred acres each. The village is situated on elevated ground timbered with white oak, and bids fair to be a very healthy situation. The subscriber will sell the upland and lease the flats, or sell both upland and flats, as applicants prefer.

It is fully ascertained that the flats or interval, on the Genesee river are perfectly adapted to the cultivation of hemp. Mr. Stephen Colton, from Long Meadow, raised ten hundred weight of excellent hemp the last season, on one acre of flats in Geneseo. One hundred and six bushels of Indian Corn have been raised on one acre in Allen's flats.

Hemp may be transported by water from the mouth of the Genesee river to Montreal; or it may be sent from Ark Port down the Susquehannah river, in Arks to Baltimore, or it may be sent by land to Albany.

The price at which Lots in the above tracts are put, is from two to five Dollars per acre.—The subscriber usually requires the purchase money to be paid in four equal installments to be made in two, three, four, and five years from the time of purchase, with one year free of interest; in some of the tracts he gives a credit of six and eight years.

Liberal encouragement will be given in the different settlements to Carpenters, Blacksmiths, Shoemakers, Millwrights, and other tradesmen.

The subscriber in order to encourage the settlement of substantial New-England Farmers, will exchange a few lots for improved Farms.

The tract of Country, in which the above described townships are situated, tho' North of New-Jersey, resembles that state in the mildness of its climate. Peaches, Apricots, and Nectarines grow to great perfection on the Genesee River.

A valuable Salt Spring is discovered in Braddock's Bay township. Salt can now be afforded at this spring at one dollar per bushel; when the works are extended salt will probably be afforded at fifty cents per bushel, the same price at which it is now sold at the Onondaga salt works.

A turnpike road is completed from Albany to Canandaigua; and from Canandaigua to Geneseo, and thence to the above mentioned settlements there are excellent waggon roads.

The subscriber has still for sale a number of reserved and other Lots of Land, in the midst of flourishing settlements, in the towns of Geneseo, Hartford, Bloomfield and Pittstown; some of these lots contain handsome improvements.

JAMES WADSWORTH.

GENESEO, (Ontario county) March, 1809.

FIGURE 1-7. James Wadsworth's Notice to New Settlers, 1809, presents an enticing account of the advantages to be gained in the Genesee Country. Livingston County Historical Society, Geneseo, New York. For a transcript see Appendix I.

and especially West Pulteney. Arriving at Avon, a guide had been provided by Mr. Wadsworth to conduct us to our destination. Reaching the "Hanover settlement," in East Pulteney, we went through the woods to the surveyors' cabin in West Pulteney, where we were lodged, fed, and provided with maps and a guide, while we made a pretty thorough exploration of the township. We found that several of our neighbors from Berkshire had been in . . . Liking the country, and especially the land we were viewing, Mr. Baldwin and myself selected 80 acres each, the quantity which the appraised value of our farms entitled us to. After this, we visited the mouth of the river, and ascending it, viewed the Falls, the Rapids and the present site of Rochester. All was a dreary wilderness, in which there was no opening, save that made by the river, and a small one immediately about the old Allan mill. There was a narrow and crooked wagon path on the east side of the river, and such it remained for several years after, during which I wagoned many loads

Indian Lands

FOR SALE.

10,000 Acres.

THE Subscribers having just purchased part of the *Gardeau Reservation*, (extending two miles on the Genesee River.) commonly called the "White Woman's Land," now offer for sale, to actual settlers, the most valuable tract of LAND in this part of the State. 7,400 acres of this Land is situated on the east side of the Genesee River, in the town of Mount-Morris, Livingston county, and through which there are three good roads, one of which is the State road to Angelica—2,600 are on the west side of the River, in the county of Genesee. This Land is two miles from the village of Perry, seven miles from Moscow, and ten miles from Geneseo, all of which are very respectable villages; the latter is the county town, and in mercantile operations, ranks with the most respectable and important villages in the western part of this State, and affords a good cash market, at all seasons of the year, for the surplus produce of that part of the country.—*Terms of Sale*, Seven Dollars per acre, one dollar per acre on the day of sale, and the balance payable in six annual instalments, with annual interest.

H. B. GIBSON, *Canandaigua*,
MICAH BROOKS, *Bloomfield*,
JELLIS CLUTE, *Moscow*.

***ALSO*---For sale by the subscriber, a tract of 2,900 acres of good LAND, eligibly situated for** farming purposes, in the town of Nunda, Allegany county, and through which passes the great Allegany road. *Terms*, Three Dollars per acre, fifty cents cash in hand, and the balance payable in six annual instalments, with annual interest.

***ALSO*---The remaining unsold lots of a tract of 4,450 acres of superior Wheat LAND, in** the town of Leicester, Livingston county, situated within three miles of the village of Moscow, and nine miles from the village of Geneseo. For water privileges, for mills, and other machinery, this tract is not surpassed by any other now for sale in this State. *Terms*, from Six to Seven Dollars per acre, seventy-five cents per acre cash, and the balance payable in ten annual instalments, (from the first day of May last,) with annual interest.

H. B. GIBSON.

Canandaigua, September 10, 1823.

BEMIS & CO. *PRINTERS.*

FIGURE 1-8. 1823 notice of available land once belonging to Mary Jemison, a white captive of the Senecas who had been granted lands by the Big Tree Treaty. Livingston County Historical Society, Geneseo, New York.

of pot-ash over it to the mouth of the river, made from the timber of my lands in West Pulteney.

I moved my family from Berkshire to the then new region of the Genesee country, in May, 1807. The town of Riga had a rapid and permanent settlement, the population being, with few exceptions, from New England. We saw, perhaps, less of the harsher features of pioneer life, than most of new settlers. We were tolerably well accomodated with a grist and saw mill; the substantial necessaries of life were obtained at convenient distances, and at fair prices; the lack of a market was a serious drawback. Before the completion of the Erie Canal, in one year, I raised three thousand bushels of wheat. After harvest, the nominal price was from 31 to 37½ cents per bushel. I tried the experiment of transporting flour to Northampton, Conn., by sledding. For this purpose, I had seventy barrels manufactured from the best quality of wheat. Purchasing six yoke of oxen, I put them upon two sleds, and two spans of horses, each upon a sleigh. With the four teams, I transported my 70 barrels of flour; was on the road twenty days; sold my flour at $6 a barrel, and my oxen at a profit; all for cash in hand. My teamsters cost me nothing but their board going and coming, as they wished to visit New England; and that was a part of my own object;—upon the whole, the experiment succeeded pretty well. We were about twenty days on the road, going down. I sold the balance of my crop of wheat the next June, for 56 cents per bushel. It went to the Canada market.[7]

The transportation problems endured by Brewster would eventually be alleviated by new roads and the Erie Canal.

It would be naive to think that all who came to the Genesee Country found it to their liking. Some stayed a short time and moved west to less expensive land, while others returned east. Benjamin West and John Trumbull, both artists of renown, were caught up in Genesee Valley land speculation. Benjamin's son, Raphael, came to live in the Valley and oversee the land his father had purchased from James Wadsworth in 1800. Young West came to Big Tree (Geneseo), but—unlike the Wadsworths, who thrived in the valley—West soon left this wilderness to return to his London home. He was an artist, not a farmer, and to emphasize that point he claimed that once while he was drawing a bear came to his window, as if to take a lesson.[8]

In essence the valley offered tremendous promise, but before the pioneers could realize that promise, the wilderness that Raphael West complained of had to be converted to thriving farms (see figure 1-9). The way of life the Senecas had known for so long was gone, and few could imagine the changes the next fifty years would bring to the land.

FIGURE 1-9. Oaks on the Genesee River Flats at Geneseo.

FIGURE 1-10. Fall Brook flows to the Genesee.

CHAPTER II

Log Barns and Cabins

NEW SETTLERS of a region usually give high priority to establishing permanent shelter of some sort. That shelter might be an adaptation of the type already in use by the natives of the region; in the Genesee Country that was the Iroquois long house, the multi-family structure of an elm pole frame covered with seasoned bark. But settlers in the Genesee Country did not adopt the long house, nor did they, at first, reproduce the single-family, framed houses common in their home regions of New England, Pennsylvania and Europe. The first structures were of log construction; in fact, the Senecas themselves began to build log cabins soon after the iron tools needed to produce them were made available.

In the late 1780s and early '90s, when the first settlers were buying land from Phelps & Gorham, Pulteney, the Wadsworths and the Holland Land Company, most of the Genesee Country was rich in both soft and hardwoods. Only the highly desirable—and already purchased—alluvial flat land was virtually treeless; the rest had to be cleared of most of its standing trees before it could be farmed.

Patrick Campbell, a Scotsman sent to America in 1792 to find an appropriate parcel of land for his countrymen, records his observations of the Genesee River valley, observations which include a list of potential building materials:

> The west side of the river Genesee is still possessed by the Indians, so that the water only divides them and Mr. Pulteney's settlers." . . . on each side of the river are large and extensive flats without a tree or shrub upon them, and of the richest soil I think I ever saw, but is every spring flooded by the freshes. The up land on each side of these flats is thinly timbered, with small crabs of *black and white oak, chestnut, and poplar,* the soil thin and sandy. From a hill which commanded a great prospect as far as my sight could carry on the Indian side of the river, the lands were flat, all covered with *stately hard wood* . . . [1]

There was thus a superabundance of wood for building houses and barns. Frames might have been hewn from the felled trees by those with the skill or the time, but timber frames require sheathing boards, as a skeleton requires a skin, and in the beginning of settlement many lacked the necessary cash to purchase mill-sawn boards. Historian William Siles points out that the high cost of settling in the Genesee Country—including moving expenses and the cost of land, tools, and materials—made the entire operation a financial risk, especially for a family just starting out in life together.[2] Even if he could afford to buy sheathing boards, a settler might find that the mills themselves simply were not convenient. The settlers turned to using their own axes and free materials to produce their first shelters of log cabins and log barns. As Capt. Basil Hall commented in 1827, "A settler, especially from the New England States, often begins the world in that coun-

Opposite: Log barn at the Genesee Country Museum, Mumford, New York.

try with no other fortune than a stout heart and a good axe. With these he has no fears, and sets merrily forward in his attack upon the wilderness."[3]

Again and again the references to original structures built by the new settlers in western New York are to a log cabin or log house. For some, these early attempts to provide shelter were only stopgap measures, and they planned to replace the log structures as quickly as possible with what they termed "proper framed buildings." Indeed, construction of the first frame buildings was cause for announcement and celebration.

Contrary to current popular legend, which associates the general use of log structures with the very beginnings of colonization in the new world, it actually wasn't until the eighteenth century, along the expanding frontier, that log houses and cabins became common.[4] In most of the eastern U.S. since before the Revolution, buildings were timber-framed, sometimes with brick or stone load-bearing walls.

A settler who wanted such a framed structure but lacked the cash to purchase the boards might have turned to pit sawing of his own sheathing boards; however, this process was laborious and required special tools and skills, not the thing for every person (see figure 2-1).[5] In a log structure, the frame and sheathing are one and the same, a considerable time-saver. Year after year, log structures answered the need for immediate shelter, as each new pioneer family arrived at the wilderness homesite, following the struggle of traveling the trails.

FIGURE 2-1. Pit sawing, as illustrated in the February, 1878 issue of the American Agriculturist, p. 58. As the term suggests, at times a pit was dug, but in other operations an above ground frame was constructed. In this particular instance a frame is built against a bank—a very good design. The top person is the sawyer, the bottom person the "pitman." Straight planks and boards depended on the sawyer's ability to guide the saw. The process can be slow, requiring hours to saw each 2 inch plank from a 24 inch wide green white-oak log some twenty feet long.

FIGURE 2-2, Opposite. The pioneer homestead is depicted in Turner's Pioneer History of The Holland Purchase; note the many improvements: the cabin has a chimney, the fencing keeps the animals out of the garden, and a cooperative logging/clearing "bee" is taking place in the clearing to the left of the cabin. O. Turner, Pioneer History of the Holland Purchase, "Second Sketch of the Pioneer," page 64.

Some Early Log Structures

After arriving at their Geneseo homesite on June 9, 1790, the pioneer Wadsworth brothers are said to have completed construction of their log cabin in three days, with the help of five or six people and using oak logs from the woods where the cabin stood. Ethan Lanphear describes the building of another log cabin, in the 1820s near Wellsville:

> There was not a frame building in the town, and but a few settlers. They cut a place for a door through the logs, laid a floor of split basswood flat side up, leaving an offset to build a chimney at one end sometime. The gables were not closed as they had no boards made in the town . . . I remember our first meal eaten in the new house. We had no chimney yet, gables to the house were all open, a quilt was hung up for the door. A fire was built on the ground, where my mother cooked the first meal.[6]

Mrs. John Young, the wife of the owner of the first Holland Company deed, came from Virginia in the fall of 1804 to a homesite just north of Batavia. Mrs. Young describes the building of their home:

> Mr. Clark was kind enough to give us shelter for a few days until my husband built a shanty. It was about ten feet square, flat roofed, covered with split ash shingles; the floor was made of the halves of split basswood; no chimney; a blanket answered the purpose of a door for a while, until my husband got the time to make a door of split plank. We needed no window; the light came in where the smoke went out . . . For chairs, we had benches made by splitting logs, and setting the sections upon legs. A bedstead was made by boring holes in the side of the shanty, inserting pieces of timber, which rested on two upright posts in front; a side piece completing the struc-

FIGURE 2-3. The Settler's first home and cattle shed. Many such early illustrations show rude lean-tos, sheds, or very small barns. Boyd, History of the Town of Conesus, 1887.

ture; pealed basswood bark, answering the place of cord. We bought a cotton bag of Mr. Brisbane, and stuffing it with cat-tail, it was far better than no bed. Buying a little iron ware, crockery, and a few knives and forks, we were soon under way, house, or shanty keeping (see figure 2-2).[7]

Frances Wright, an Englishwoman visiting the Genesee Country, describes log structures in an August, 1819 letter from Geneseo:

> Mr. Hopkins [who lived near Avon] . . . has . . . cleared the forest around his dwelling in such a manner as to give it the air of a magnificent park [Note: she calls this 'ornamental clearing'] . . . those who abhore the trees call them 'long weeds'. . . . During the summer nights, a log hut often presents a very singular appearance. It is not unusual, when the hot months set in, to clear away the mud which stops the interstices between the logs, . . . so as to allow a free passage to the external air. In the darkness of the forest, the light streaming through these crevices, gives to the cabin the appearance of being either illuminated or on fire.[8]

FIGURE 2-4. Log buildings at the Genesee Country Museum. The barn on the extreme left is typical of early log barns in the region, but is from Picton, Ontario, Canada (none could be located by the museum in the Genesee Country). The cabin is from Scottsville, Monroe County, New York, ca. 1809.

Henry Conklin remembers, "after they got the house fixed they put up a sort of barn or shed of logs for the team and a place for hay, and during the winter father and John used to go over on Blenheim Hill and draw their hay over on a sleigh for the teams."[9] Building the house first and a barn second seems to have been the usual order, except in those cases where animals and family shared the same structure, a system practiced in Europe. Figure 2-3 shows a small cattle shed of the sort that might be associated with early settlement.

Many, however, did not feel a strong need to house their animals in winter at all. As late as 1853, the Genesee Farmer complains of ignorance in the wintering of stock, claiming that most barns were still too open in winter and that unhoused animals consumed a third more fodder.[10] The pioneer Joseph Sibley describes the wintering of cattle along Honeoye Creek in the Town of Rush, at a place called "Rush Bottom," where cattle were brought from Lima, Bloomfield and Victor to feed on the rushes. "Cattle would winter well and thrive on the rushes . . . The rushes finally run out by being repeatedly fed down."[11] The log barn furnished a more permanent system of storing feed and housing cattle.

By the time of this writing, nearly all the early log structures in the Genesee Country have disappeared and there are, of course, no early photographs, since photography hadn't yet been invented. However, the Genesee Country Museum, a museum of original regional architecture at Mumford, includes a complex of original log structures (see figure 2-4).

Building

All the log builder really needed was skill with an axe. Alex Bealer, in his book, *The Log Cabin,* claims: "Almost any adult backwoodsman could use an axe with uncanny precision and as a true extension of himself, cutting as deeply as he wanted, at whatever angle he desired, cutting thin shavings or chips the size of dinner plates."

After the logs were cut to length they had to be shaped for joining at the corners. Variously shaped joining cuts were made, depending on the skill and background of the builder. As the students researching log construction for the *Foxfire* books of the 1970s soon realized, there is no one "right" way of joining; each building had its own peculiarities with many variables, each producing at once similar and yet dissimilar results. The log barn at the Genesee Country Museum illustrates the saddle notch, perhaps the most common and easiest form of joining (see figure 2-5).

The saddle notch is named after the "saddle" cut into the upper side of the log, sometimes requiring a special "gutter" adze for the purpose (see figure 2-6). The advantage of the saddle notch was that the joining surface of the upper log did not have to be shaped; its major fault was its tendency to retain water, making the joint susceptible to rot. Later, more complicated joints were evolved to permit drainage of rainwater. A close cousin of the saddle notch was the sharp notch, in which both joining surfaces were shaped with an axe. The upper side of the joint was cut to a "V" and the lower side notched to receive the "V" from the log below. No matter which style of joining was used, the corner joints were the weakest part of the construction. The three-bay log barn at the Farmer's Museum in Cooperstown (from about 1795 in South Berlin) is vertically sheathed with sawn boards, a process that extended the life of many a log structure because of the protection it afforded the corner joints.

If the logs were hewn square, allowing a tighter structure, the chamfer and notch or the square notch might be used. With a square notch a wooden pin known as a treenail (pronounced "trunnel") was needed to tie the timbers together. The other tools required for this joint included a broad axe for hewing, a felling axe and a crosscut saw. Perhaps the most complicated notches in terms of fitting were the half and full dovetail notches. Figure 2-7 shows a full dovetail notch on a smokehouse at the Genesee Country Museum. These dovetail joints are characterized by their complicated axe-shaped planes, which are self-draining and resemble the spread tail feathers of a dove.

FIGURE 2-5. The saddle notch. Genesee Country Museum, Mumford, New York.

FIGURE 2-6. Head of a "gutter" adze sometimes used for making the saddle notch.

FIGURE 2-7. Dovetail notch. In this joining the logs are hewn square. The dovetail notch sheds water to protect the joint. Genesee Country Museum, Mumford, New York.

Raising

After the logs were cut and notched on the ground, it was time for the "raising". Alexander Patterson tells of Asel Clark's raising near Turkey Hill in the town of Conesus, in 1814:

> I had the pleasure of being invited by Mr. Asel Clark, to assist him in raising . . . as soon as Mr. Clark had got ready to raise . . . he asked his neighbors to help him, while his wife, invited the ladies to aid her in quilting at his father's house . . . on my arrival I found the men assembled in the morning, and through the day had raised the house. When I arrived I found the only entrance to the building, was through a hole in the side, which they had cut for a window . . . having not the time to cut a hole for a door . . .[12]

By working together, Mr. Clark's neighbors completed the raising in less than a day. Unless a family was large and the children well grown, a raising was too much for one family. In a 1795 raising in Conesus, Hector McKay was helped by some Indians from Squakie Hill.[13] Extra hands not only made raisings less dangerous, but provided good fellowship as well.

The raising was accomplished in a number of ways. With the strong arms and backs of those assembled, logs could be lifted into place until the walls reached shoulder height. Above shoulder height, wooden rails were placed against the top log at roughly a forty-five degree angle and the new logs were rolled up these rails into place— a dangerous practice for those who stood inside the rails. When the walls were higher than a man could reach, pike poles were used to roll

FIGURE 2-8. Genesee Country Museum log barn, 14 by 18½ feet (the door is 3 by 5⅔ feet). The interior is open except for mangers at one end and a loft. The saddle-notched logs are chinked with split logs, typical of log barns because log chinking allowed proper ventilation. According to the usual practice, the roof is shingled. The gable ends (the triangular end portion beneath a double pitched roof) are closed with vertical boards. Due to the absolute scarcity of log barns in the Genesee Country this barn had to be obtained by the museum from Canada.

the logs. Also gins made of poles, the legs set in a tripod arrangement, were used with a block and tackle to raise the topmost logs into position.

Roof rafters of small saplings hewn flat on one side and half-notched at the top were held together at the ridge with a treenail, but usually no ridge pole was used. Finally, wide strips of bark, hand-riven shingles—usually of spruce, oak, or cedar—or the longer shakes were placed over the roof boards. Many builders prided themselves on not using a single nail or metal fastener in an entire structure, including hinges or latches.

Figure 2-8 shows a typical log barn, an original structure, not a reproduction. This log barn is part of the pioneer farm arrangement at the Genesee Country Museum, Mumford. Although once numerous, log barns have apparently entirely disappeared in the region, probably because their small size made them of limited use to increasingly productive farmers.

Stumps

After the house and barn were built, the land had to be cleared for farming, and this was slow work. Clearing consisted of two phases, the felling of trees and the removal of stumps. An illustration in Patrick Campbell's Travels of 1792 shows two farm structures surrounded by a field of stumps (see figure 2-9).

A Scottish farmer and writer, Patrick Shirreff,

FIGURE 2-9. A field of stumps from Patrick Campbell's Travels, 1792—"American New Cleared Farm," University of Rochester, Rush Rhees Rare Book Collection, Rochester, New York.

visited the Genesee Country in 1827 and noted "the fields . . . thickly covered with black stumps overtopping the wheat crops; and felling and burning of trees going on in all directions."[14] Captain Hall's account of the same period describes "subduing" the land, a local term:

> . . . the tops of the stumps were seen poking their black snouts above the young grain, like the shoal of seals. At other places we came upon ploughs, always drawn by oxen, making their sturdy way amongst the stumps, like a ship navigating through coral reefs, a difficult and tiresome operation.[15]

Figures 2-9 and 2-10 depict the sort of stump-filled fields described by Hall.

The stumps were eventually removed, some pulled for stump fences, some burned, and others left to rot until they could be pulled. Fences made from stumps pulled from the earth have been a part of the Genesee Country from the time of the earliest pioneer settlements. Figure 2-11 shows the stump fence on the Schumaker (formerly Gelser) property near Dalton. Built in the 1830s of virgin hemlock and white pine, the fence has survived virtually intact. The butts of the trunks are three to four feet in diameter and the height of the fence varies from seven to twelve feet. Because hemlock and white pine are shallow-rooted, tipping over the stumps to make a fence was not as difficult as it might have been with trees having tap roots, especially if the job was done when the ground was saturated with water. Even so, it is hard work to remove a stump and sometimes a machine was used. One of the many patented stump pullers, Traxler's 1857 model, looked much like a medieval war machine (see figure 2-12). Indeed, the battle against the stumps was a war for many settlers. Even after they pulled the stumps from the ground, the settlers had to drag

FIGURE 2-10. Stumps and cabins—Rochesterville in 1812. The engraving, which was made for O'Reilly Sketches of Rochester . . ., 1838, depicts the barn as vertical-boarded; in reality not many barns (1790-1812) in the region were thus boarded.

them into place if they were to have fencing. Miles of stump fences can still be seen in the area, their continued existence proof of the durability of the stump roots.

Blasting stumps was a more dangerous method of removal (see figure 2-13). The forests of the Genesee Country were a considerable nuisance to the farmer and there seemed to be no limit to the number of ideas on how to eliminate them. Eight to ten acres cleared and fenced in a year was considered good progress. In 1821 Henry Conklin regarded ten cleared acres as good work for a summer. Mrs. John Young tells that her husband cleared four acres in their sec-

FIGURE 2-11. 1830s original stump fence near Dalton, Livingston County, New York. Gelser/Schumaker farm.

ond year near Batavia. Thirty to forty acres cleared after ten years might have been a reasonable accomplishment.[16] Sometimes the settlers arranged logging "bees" to help each other out with the clearing (see again figure 2-2).

In addition to using the logs for buildings, splitting them for rail fences and making stump fences, some settlers burned wood to make potash, which could be sold for cash. Pot ash (potash or black salts) was used as fertilizer, in the manufacture of soap, in bleaching, glass manufacture, and baking in place of yeast. Potash got its name from the large cast iron kettle, an absolutely essential requirement, in which water was boiled after percolating through hardwood ashes. Potash was a relatively lightweight, easily transportable cash "crop." If desired, a second boiling would reduce the impurities, producing pearl ash. Canal records show that pot and pearl ash were important products throughout the first half of the nineteenth century (see figure 2-14).

Clearing the land, fencing, building a house, barn and other outbuildings constituted "improving the land." The value of such work can be measured by comparing the 1804 price of unimproved land east of the Genesee, which sold for two to four dollars an acre, with that of improved land: a farm of about 100 acres, including log house and barn with twenty to thirty cleared acres, sold for six to twenty dollars an acre.[17] See figure 2-15 for 1820s advertisements of farm land available in the Genesee Country.

The newspaper advertisements in figure 2-16 provide an insight into some of the concerns in the region during the period following the completion of the Erie Canal in 1825.

In the early years of settlement, when sawmills were distant or too expensive and when there was land to be cleared, log barns filled a critical need in the Genesee Country. Eventually, however, the successful farmer began to produce beyond his own needs and to require more storage capacity than the small log barns could supply. The timber-framed barn, which could be made much larger, was the only practical solution. Most settlers from Pennsylvania and New England had always wanted timber-framed barns anyway, and within a surprisingly short time the original log barns gave way to the "English" or "Connecticut" barn, which first appeared in the late 1790s. By the 1820s, most new barns were of post-and-beam construction.

FIGURE 2-12. Traxler's 1857 Patent Stump Puller. Peter Traxler was from Scottsburg, Livingston County, New York.

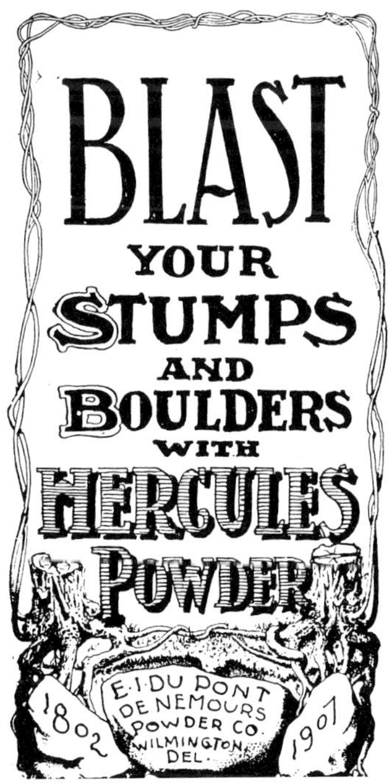

FIGURE 2-13. 1907 DuPont advertisement.

NONE SO CHEAP,

AS

BOSTWICK, DUNNING & CO.

THEY have on hand, an elegant assortment of

FANCY DRY GOODS

of every description, and of the newest fashions.

Their assortment of

Staple DRY GOODS,

is complete. Their

Groceries, Crockery, Glassware, and Hardware,

are unparalleled in their quality and prices.

They have just received a quantity of the first quality

SALT.

They continue to receive CASH and most kinds of country Produce in payment for GOODS, and will pay the highest market price in CASH for

Wheat, Pot & Pearl-Ashes.

Geneseo, Dec. 30, 1824 39

FIGURE 2-14. 1824 newspaper advertisement offering cash for pot and pearl ashes.

Tuscarora Lands

FOR SALE.

THE subscriber offers for sale, 8000 acres of the above tract of Land, in the towns of Mount-Morris and Nunda, commonly called the *Widow's Tract.* The greater part of this Tract is first rate Land, and well watered. There are a number of improved Farms, and valuable mill seats.

☞For the terms, apply to the subscriber, living in the town of Groveland, and county of Livingston.

C. H. CARROLL.

Groveland, April 23, 1827. 73tf

A FARM

For Sale.

THE subscriber wishes to sell his **DISTILLERY** together with a small

FARM

on which he resides: or he will sell either separate. The Distillery has attached to it a STABLE 80 by 24 feet the whole in excellent order and now in *blast.* There is no better land in the county of Livingston than the farm here offered for sale. On it are two good BARNS two excellent wells of water that have never run dry, a small DWELLING HOUSE, and a TURNING SHOP with 2 LATHES now in operation by water power. *He that has got* MONEY *can buy* CHEAP.

WM. FINLEY.

Geneseo, March 20, 1830. 51tf.

FOR SALE.

THE subscriber is authorised by Thomas Beals, Esq. of Canandaigua, to sell that valuable

FARM,

situate in the town of Groveland, known by the name of the William McNair Farm, containing about one hundred and fifty acres of land. The Farm lies on the main road leading from Geneseo to Dansville, a good proportion of which is on the Caneseraga flats; it is well adapted to grazing and growing of Grain. For terms (which are very liberal) make application to the subscriber in Dansville Village. J. FAULKNER.

May, 1828. 5tf

FARM FOR SALE.

FOR SALE, the *Farm* in Leicester occupied by Jesse Wadhams, containing about one hundred and fifteen acres, under good improvement. On the premises are two good *frame Barns* and a comfortable *log House*—also, an Orchard of 170 bearing apple trees—young and very thrifty, with a variety of other fruit trees; the farm is well watered, about seventy acres of the land may be sowed with wheat next fall. For further particulars enquire of Mr. Wadhams on the premises, or of the subcriber near Moscow.

JELLIS CLUTE.

Feb. 15th, 1830. *5w48.

FIGURE 2-15. Sample newspaper advertisements of 1820s farm land available in the Genesee Country.

VALUABLE FARM FOR SALE.

THE subscriber will sell the FARM, on which he formeriy resided, on the west bank of the Genesee River, being Lots No. 111, 112 and 113, in the town of Leicester, Livingston county, containing 310 acres.

The improvements are, an excellent two story framed house, with convenient out buildings, and a well of good water within a few feet of it—A first rate barn 40 by 50 feet, with a basement stable of 16 by 50 feet; 70 acres of land under high cultivation, and 160 acres enclosed by a good and durable fence.

The proposed route of the Canal through the valley of the Genesee, passes within a few rods of the barn, which, with the River, will afford two direct communications by water to Rochester, from the property.

As a country residence for a gentlemen, or a field for an experimental and industrious agriculturalist, there are but few locations more deserving of notice; and aside from the fine and extensive prospect from it, the soil will insure a rich and valuable return to the labors of the husbandman.

The tenant will attend any person who may wish to view the farm and buildings.

One half of the purchase money will be required in hand, the other half may be secured by Bond and Mortgage on the property, for a term of years.

JOHN A. GRANGER.

Canandaigua, Jan. 9, 1827. 2m42

CLOVER SEED.

THE subscriber has received a quantity of first-rate CLOVER SEED, which will be sold very low for CASH.

A. K. VAN RENSSELAER.

Geneseo, Feb. 16th, 1830. 46tf

NOTICE.

THE subscriber has on hand a quantity of

PLOUGH SHARES

of Wood and Swan's Patent, for sale.

JAMES KONKEY.

Mt. Morris, March 24, 1828. 4w51

To Haymakers.

THE subscriber wishes to contract for Mowing and securing in Barracks, the Grass on 200 acres of meadow flats for which payment will be made in cash

A house, convenient for a family, near the meadows, is now erecting.

WM. WADSWORTH.

Geneseo, May 25, 1826. 126tf

DRUGGIST STORE.

(Front of Bristol-Street.)

H. WARNER,

HAS, in addition to his former very complete assortment of

Medicines,
Paints,
Oils,
Dye Stuffs,
Confections, &c. &c.

JUST RECEIVED,

7 casks LAMP OIL;
6 boxes SPERM CANDLES.

☞ ***Cash for Flax Seed.***

Canandaigua, Nov. 29, 1826. 35tf

Hats, Window Glass, and Salt.

60 Boxes WINDOW GLASS.
500 Barrels SALT,

And a large quantity Eastern Manufactured HATS, of a *superior quality*, for sale cheap, by A. STEWART.

Geneseo, Dec. 4, 1829. 36tf

Lumber! Lumber!!

ANY quantity of LUMBER can be had of the Subscriber, and of all descriptions, *CHEAP.* Those wishing to purchase will do well to call.

I. EVERTSEN.

Aug. 30, 1826. *23tf*

FIGURE 2-16. 1820s newspaper advertisements from the Genesee Country.

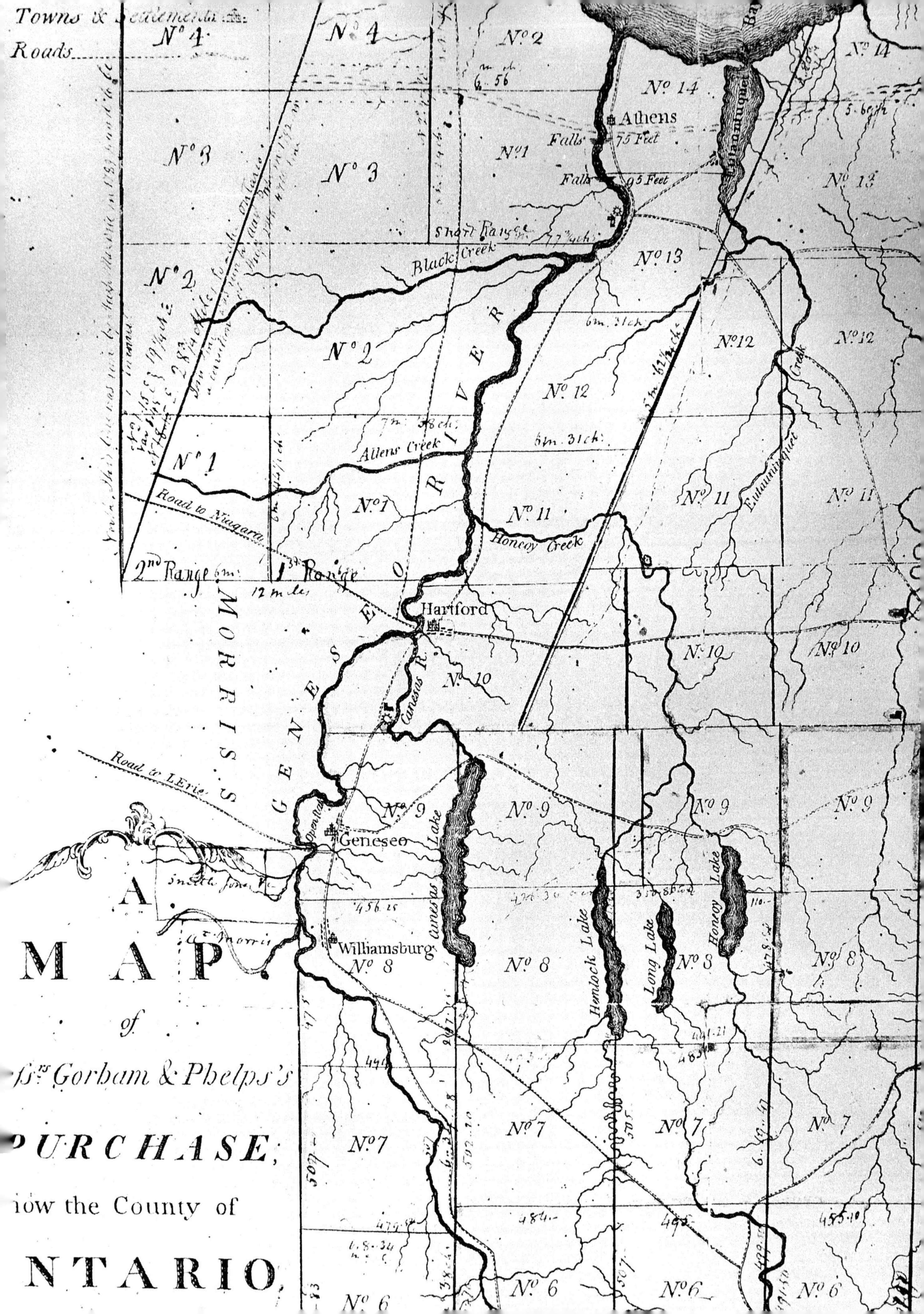

A
MAP
of
...s.rs Gorham & Phelps's
...PURCHASE,
...now the County of
...NTARIO,
Towns & Settlements
Roads
Athens
Falls 75 Feet
Falls 95 Feet
Short Range
Black Creek
Allens Creek
Road to Niagara
2nd Range 6m.
1st Range
12 miles
MORRIS'S
GENESEO RIVER
Honeoy Creek
Hartford
Canesius R.
Road to L.Erie
Geneseo
Williamsburg
Canesius Lake
Hemlock Lake
Long Lake
Honeoy Lake
Creek
Bay
No 1
No 2
No 3
No 4
No 6
No 7
No 8
No 9
No 10
No 11
No 12
No 13
No 14

CHAPTER III

Experiments in Land Speculation

THE LAND in the Genesee Country had been pronounced fertile, but it remained for the new owners to encourage settlers to come to the region. The rate of settlement varied considerably throughout the region, in part because different experiments in land speculation had widely differing results. Other influences included the War of 1812, which greatly retarded settlement, and the completion of the Erie Canal in 1825, which greatly stimulated it.

Some who came to the Genesee Country were former line soldiers from the Revolution, who had been given land (at least 600 acres each, depending on rank) by Congress and the State of New York. These allotments, collectively known as the Military Tract, were located east of the Genesee Country in what is now Cayuga, Cortland, Onondaga, Seneca and parts of Oswego, Schuyler, Tompkins and Wayne Counties (see again figure 1-2).

While many soldiers sold their allotments immediately for the cash, others came to work the land. Among the latter were some who, after a stint on the Military Tract, sold their allotments to buy land in Genesee Country from Phelps & Gorham, Pulteney, or the Holland Land Company. The rumor in the literature is that some soldiers sold their land in the Military Tract again and again—according to Maude, an English observer, perhaps as often as once a week.

Williamsburg

As discussed in Chapter I, the Phelps & Gorham land on the east side of the Genesee River was sold to Robert Morris in 1791; he in turn sold the land in 1792 to Sir William Pulteney and his partners. Pulteney's land was bounded by Lake Ontario to the north, Pennsylvania to the south, and the preemption line to the east. On the west, the boundary line extended from the Genesee River south of Geneseo northwest to Caledonia, and up to Lake Ontario (see figure 3-1). Of course the Pulteney land did not include portions already sold to others during the Phelps & Gorham ownership.

Charles Williamson, a countryman of Pulteney's, was selected to manage the Genesee lands, which were put in his name following his naturalization, since foreigners were not permitted to own land. Williamson established his land office in Bath and proceeded to sell the land in his own unique fashion. Williamson advertised land for a dollar an acre to those who would actually settle on it.

Opposite: 1794 Phelps & Gorham Map detail locating the Wadsworth land in the ninth range at Geneseo and the Pulteney Land in the eighth range at Williamsburg. Collection of the Genesee Country Museum, Mumford, New York.

FIGURE 3-1. The Geneseo Flats in fall, showing the gentle slope of the western side of the valley.

Williamson did some extravagant things to promote settlement. At one point he even considered offering ten pounds to anyone willing to settle on some of the less desirable land in the area. Early on, Williamson decided that to encourage settlement a road must be built from what is now Williamsport, Pennsylvania, through Tioga to Painted Post, ending at Dansville. The road was begun in 1792 and made passable the following year. Additional branches of this road were to serve other areas to the north. To the south, the Susquehanna River would permit travel between Williamsport and Harrisburg, Baltimore, and other towns in the East. O'Reilly reports:

> . . . in preference to the richer lands in the northerly part of the tract—The truth of the case was, that Capt. Williamson saw very clearly, on his first visit to the country, that the Susquehannah, and not the Mohawk, would ultimately be its best friend. Even now it has proved so; for at this day (1800) a bushel of wheat is better worth one hundred cents at Bath than sixty cents at Geneva. This difference will grow wider every year; for little, if any, additional improvement can be made in the water communication with New-York, while that to Baltimore will admit of very extensive and advantageous ones, . . .[1]

After surveying for a road, Williamson decided to build a "city" of some sort for settlers, to make the area appear less like an intolerable wilderness. Towns were begun or expanded at Bath, Lyons, Sodus Point, Geneva, and—perhaps his most ambitious undertaking—Williamsburg(h). Located near the mouth of the Canaseraga, where it meets the Genesee south of Geneseo, the village of Williamsburg was established and named for Sir William Pulteney. In 1792, some seventy German families worked their way to Williamsburg building Williamson's road, having been promised a house and fifty acres of land for each family. The village itself comprised about thirty acres, with houses on about forty of the 100 lots, as well as a tavern with a ballroom on the second floor, a dry goods store, blacksmith, grocery, distillery, and grain warehouse. In 1796, pioneer and historian Samuel Magee described an early barn at Williamson's own farm, Hermitage: "some two hundred feet long, say some of the early settlers, built to accommodate horses that came to the races . . . "[2] Church services were also apparently held in the barn. No drawings or plans seem to have survived; if the barn was indeed as large as Magee says, it was unusually enormous for the area and period, but then Williamson seemed to do everything on a large scale.

As a further inducement to settlement, for example, Williamson held a series of fairs and races beginning in 1793. He reasoned that in the process of coming to the fair, people would see what a fine country he had to offer and settle there. The fairground, some 1,000 acres on the Genesee

Flats, included a mile track and grandstands. Williamson advertised widely and, even though the roads were bad, expected a crowd of 2,000. Records indicate that by the time of the races in mid-September more than 3,000 people had arrived, from as far away as Virginia and Maryland. The main prize at the race was 1,000 pounds (see figure 3-2). The fair lasted into October, taking advantage of the Genesee Valley's fine fall weather. Later some of these fair-goers did return to settle, and land values jumped. Pulteney was pleased with most of what his land agent had accomplished.

Historian Lockwood L. Doty describes three Pennsylvanians and their encounter with Williamsburgh:

> Early in September, 1794, Daniel Kelly, John Jones, and John Harrison, all of whom were afterward notable farmers of Groveland, and all became deacons in the same church—left their Pennsylvania home, on the north branch of the Susquehannah, for a visit to the Genesee country by way of the Williamson road, "which was without bridges over creeks, or crossways in bad places, the underbrush and logs being removed a rod wide," says Mr. Harrison. William Ryans was also of the party. They had two horses between the four, riding and walking in couples by turns. The party on horseback would trot on far ahead, and hitching the horses beside the road, would start forward on foot, leaving their companions to come up and resume the saddle. A journey of eight days brought them to Williamsburgh, where, on the 13th of September, 1794, they put up at William Lemon's tavern, a small frame house, and the first frame house built in the town of Groveland. . . . The three others went to Geneseo to purchase lands of the Wadsworths, who were then laying the cellar wall of their homestead. James Wadsworth at once saw that they were good judges of farming lands, and advised them to look at some lots lying along the road leading to the foot of Conesus Lake . . . Mr. Harrison says that Williamsburgh, at this, his first visit, contained, besides the frame tavern and a house occupied by John Ewart, some five or six log houses built by Captain Williamson. On their way home to Pennsylvania, where they stayed until the following spring, they met several people driving cattle to the Williamsburgh fair, and heard frequent mention of the upcoming races, in which 14 speed horses were entered for the 50-pound purse.[3]

For all its great promise, Williamsburg itself was doomed; those first "imported" settlers were ill-suited to the frontier and for some unexplained reason others could not be found to populate the village.[4] Many believed that the lack of continued direct attention by Williamson himself caused the final demise of Williamsburg. Others believed that Williamson was both too idealistic and too ruthless in his management to be successful; he seemed to promise more than he could deliver.[5] At any rate, Williamson died of yellow fever in 1807 on a return trip to England, and it is about that time that Williamsburg itself ceased to be a living community. Today not a single building remains; only the cemetery can still be seen.

FIGURE 3-2. Williamsburgh Races advertisement describes less sumptuous prizes than those offered for the first fair. Ontario Gazette, September 8, 1797.

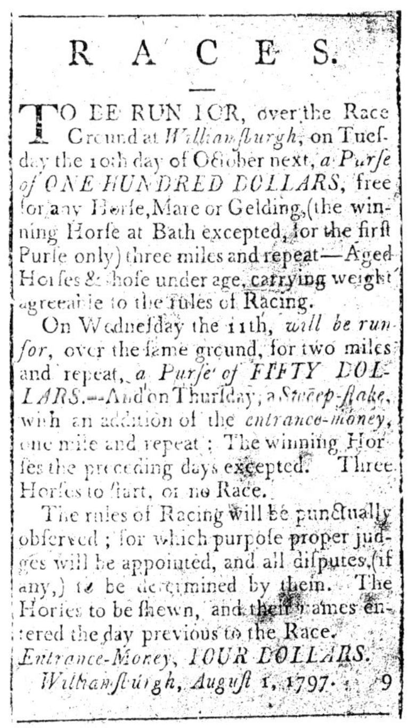

RACES.

TO BE RUN FOR, over the Race Ground at *Williamſburgh*, on Tueſday the 10th day of October next, *a Purſe of ONE HUNDRED DOLLARS*, free for any Horſe, Mare or Gelding, (the winning Horſe at Bath excepted, for the firſt Purſe only) three miles and repeat—Aged Horſes & thoſe under age, carrying weight agreeable to the rules of Racing.

On Wedneſday the 11th, *will be run for*, over the ſame ground, for two miles and repeat, *a Purſe of FIFTY DOLLARS*.—And on Thurſday, a *Sweep-ſtake*, with an addition of the *entrance-money*, one mile and repeat: The winning Horſes the preceding days excepted. Three Horſes to ſtart, or no Race.

The rules of Racing will be punctually obſerved; for which purpoſe proper judges will be appointed, and all diſputes, (if any,) to be determined by them. The Horſes to be ſhewn, and their [illegible] names entered the day previous to the Race.

Entrance-Money, FOUR DOLLARS.

Williamſburgh, Auguſt 1, 1797. 9

Leasehold Tenancy as Practiced on the Wadsworth Lands

James and William Wadsworth experienced considerable difficulty in overseeing the Genesee lands of their uncle, Jeremiah. The main part of the Wadsworth estate extended from the Mount Morris flats to the Town of Henrietta in Monroe County. On paper the investment was impressive, and, as the Wadsworths reinvested, the estate became the largest in the valley, comprising some 70,000 acres at its peak in 1811. The initial idea was to sell most of the land for profit, but it proved difficult to find buyers with enough cash or credit to purchase land, despite the enticing "Notice to Settlers" (see again figure 1-6).

For the Wadsworths, leasing provided an alternative to selling when the land market was soft, and held out the possibility of a later sale, when the market improved. The first leases date from 1809 and were arranged for life, with the tenant agreeing to pay the rent in wheat and to pay all taxes. The tenant also agreed to clear, fence, and plant wheat on at least 10 acres within two years. Additional clauses cautioned the tenant to keep the land fertile through crop rotation and fertilization. It was forbidden to cut more trees than necessary to clear the land for farming. Some records show that rent was reduced or eliminated when the tenant constructed a barn or other farm buildings on the property.

There can be no doubt that the lack of ready cash was a serious impediment to land sales, but the following comments from the historian O'Reilly clearly show that the War of 1812 was also a major inhibiting factor:

> The settlement of Rochester, commenced almost simultaneously with the last war between this country and Great Britain, was almost wholly checked by the alarm created by the movements of the belligerents. The hostilities along the Niagara caused a concentration of troops there, which left defenseless this point, then comparatively unimportant. The mouth of the Genesee was therefore not unfrequently visited by the British fleet under Sir James Yeo, commander of the hostile forces on Lake Ontario. The apprehension of attack prevented many from settling here as they had designed, and even caused the removal to more secure places of some who had already located hereabout. The distress of which the war was productive in this region was vividly portrayed in 1814 by the "Committee of Safety and Relief" at Canandaigua in a communication to the mayor and other citizens of New-York:
>
> *Canandaigua, 8th January, 1814.*
> GENTLEMEN—*Niagara county, and the part of Genesee which lies west of Batavia, are completely depopulated. All the settlements, in a section of country forty miles square, and which contained more than twelve thousand souls, are effectually broken up. . . . Our roads are filled with people, many of whom have been reduced from a state of competence and good prospects to the last degree of want and sorrow. So sudden was blow by which they have been crushed, that no provision could be made either to elude or to meet it . . .*[6]

After the treaty ending the war was signed in December, 1814, settlement resumed, but crop production remained low through 1816. 1816 was "the year without a summer," when frosts from the 6th to the 12th of June destroyed summer crops and gardens alike. Almost all of the corn crop was ruined, and the wheat crop was both small and late following replanting. The extreme cold weather was apparently caused by the enormous, sun-shrouding dust cloud created by the violent 1815 eruption of the East Indian volcano Tamboro.

Crops were good the following season, and in that year of 1817 James Wadsworth was still working hard to sell or rent acreage on the flats at Mount Morris. Leasing alternatives included ten dollars an acre for 100-acre farms, payable over a four year period, or "two-life" leases with no rent during the first two years (four years if a log house was built). As farms were let, several problems were solved at once: taxes were paid, land was cleared, and both landlord and tenant derived income from the property. Furthermore, even when a tenant left a farm before the lease expired, if it had been maintained according to the lease the property was generally worth more to the Wadsworths on its next rental (see figure 3-3).

The Wadsworth land was too extensive to manage without overseers, and so over the years managers, agents, or clerks were engaged to see that the leases were adhered to properly. In July, 1832, James S. Wadsworth writes in a ledger to Gardner Scott, his outdoor clerk:

You will first examine and have in your mind the provisions of the lease.

Buildings — Examine minutely the state of the buildings. Are the sills of the barn safe from decay? Are there any barn boards off? Is the barn floor and granary in good repair? Is the manure carted out? Is the yard fence in good repair? Are there convenient gates . . .

Fences — Examine all the fences—are they eight rails high, well righted up and made of oak or ash timber?

Orchard — Count the number of apple trees. Has any tenant any ambition to set out peach and other fruit trees around his house? Are there any burdocks, Canada thistles or other noxious weeds growing on the farm? Is the tenant sensible of the importance for his own interest of sowing clover seed and using plaster [gypsum] freely?

Woodland — Examine the wood lot, estimate its quantity, and be sure to ascertain if trespasses of any kind have been committed upon it. The occupant has no right to cut any wood or timber from the reserved lot. Has the occupant reserved any wood land for his own use? and to keep the fences in repair?

FIGURE 3-3. The Genesee River Flats at Geneseo in winter.

> Fields Ascertain the number of fields, about the quantity of land in each field and thereby ascertain the quantity of improved land on the farm. Does the tenant appear judicious and sensible in his rotation or change of crops? In carefully applying his manure to the best advantage—in his management of live stock, and to the benefits resulting from improved breeds of cattle, sheep and swine? Is he neat and tidy round his house? Has he set out shade trees round his house? Has he left shade trees in his fields? Does his farm generally indicate good management? From a look or two round the house and in the garden, you can easily ascertain whether the occupant drinks bitters in the morning and whiskey with his dinner or pure water and nothing else—If he drinks bitters you will find his garden full of weeds. If any remarks occur to you which you think will be useful to the tenant, suggest them to him in a conciliating and respectful manner— After you have made your notes on the above enquiry you will read them to the occupant in possession and inform him that in all particulars where he is deficient, he must make amends by immediate repairs and that the provision of the lease in future must be strictly complied with—.[7]

Late in the 1830s, leasing arrangements were changed to an annual renewal, which provided better control. A split rent of cash and wheat was required, with amounts varying from farm to farm. Generally, leased farms increased in size throughout the nineteenth century. By 1860 the leases were very tightly specified in printed form, leaving little room for interpretation. In hard times, such as those created by the devastating insects that destroyed area wheat crops in the 1850s, grain rents were reduced or eliminated: better to have a good tenant another year than lose him to debt. Some Wadsworth lands have been leased by the same family for generations, providing evidence of the mutual benefits of such leasing arrangements.

There were, to be sure, other attempts to attract settlers to the Genesee Country, but the Williamson and Wadsworth attempts were two of the more interesting solutions to the problem, in the case of the Wadsworth brothers a rather enduring and successful solution.

DIARY

Anonymous, West Bloomfield, Ontario County, 1814

THE DAILY FARM LIFE of the period comes alive in the words of those who troubled to keep day books or diaries. All the selections included in this book retain their original spelling and punctuation, to preserve the flavor of the original diaries. As might be expected, in some places the handwriting is unclear or smudged, making transcription uncertain, but the essence of the entry is usually discernible. Notice that in addition to his crops of hay, corn, and oats, this farmer is already improving his pastures with grass seed.

1814

April 12:	*PM went to west part of the farm for ashes to make soap and forgot the things to get them with*
April 14:	*spread some dung—plowed with the horses*
April 16:	*sowed grass seed on the north and south meadows*
April 20:	*begun to plough the planting ground with oxen*
April 21:	*ploughed planting ground*
April 22:	*ploughed with both yoke of oxen and had Edward [horse] to drive*
May 5:	*carried a ploughshire to Mr. Anthonys to be nosed etc. and harrowed with horses*
May 11:	*AM begun to plant PM planted till drove off by rain*
June 15:	*Helped Father log old logs..burned chunks in the south lot*
July 8:	*mowed and raked hay for myself, raked, got in three loads of hay*
July 14:	*AM mowed and spread hay PM carted and stacked hay*
Oct 27:	*AM husked corn PM covered the cribs and husked*

Oct 28:	*AM helped move the corn and cleaned up the floor PM thrashed oats with the horses*
November 28:	*PM helped dress a sheep and went to Mr. Roses to raise a log barn*

From the collection of the Genesee Country Museum

Other diaries in this book will reflect some of the changes, in both farming and society, taking place throughout the nineteenth century and early twentieth century; for example, see the January, 1855 entry in Thayer Gauss's diary.

FIGURE G-1. Interior of Benjamin Gauss's 1794 English barn [now enlarged] at Holcomb [formerly Bloomfield], New York. View looks west from the drivethrough.

DIARY

Thayer Gauss, Bloomfield, Ontario County, 1826-1858

THAYER GAUSS, a young farmer of Bloomfield, New York, began a diary in 1826. Thayer's father was Benjamin Gauss, a Revolutionary War soldier, who came to Bloomfield in 1789 from Alford, Massachusetts. Benjamin's marriage to Sarah Codding in 1793 was the first wedding in the town and one of the first in the area.[1] Thayer's almost daily entries continued until 1858, producing an extraordinary account of his successes and failures and the evolution of farming during the period. These excerpts concentrate on the early years and Gauss' description of an 1837 barn fire and the subsequent rebuilding.[2]

1826

First entry, Bloomfield, New York, April 21, 1826: "This has been a very uncomfortably cold day. It has snowed for short spels all day. . . . We began to plow today for oats on the Wright Farm"

27 April:	*This day I am 28 years old*
28 April:	*Plowed with two teams, two yoke of oxen and three horses on the other..*
29 April:	*Plowed for corn.*
May 2:	*Finished sowing oats.*
May 3:	*Finished harrowing and manured garden [next day would work garden for first time this year].*
May 9:	*. . . started for Michigan today. The Michigan fever is very high in this country . . . I have almost catch the fever.*
May 15:	*. . . we plowed with four yoke of oxen which is none too much . . .*
May 20:	*It thundered some this morning. In the forenoon we drew some off from the meadow in the afternoon went to T. Bronson's raising.*

May 26:	*I am verry tired tonight. I have been working on the highway today. No rain yet . . .*
June 12:	*I plowed and hoed corn for Mr. Richards, his corn is small for the season . . . we planted our beans this day.*
June 22:	*In the afternoon I repaired the fences . . .*
June 23:	*We had a small sprinkle rain but not enough to do any good. I have been drawing logs to make a hog pen I believe it is about as cheap to build a framed one but I am not prepared to do that this year*
July 1:	*. . . finished sowing corn and my potatoes in the garden . . . work is done.*
July 6:	*We have been haying today Have had a pretty good hayday Our crop is verry light indeed it will not yield half a ton to the acre on an average.*
July 12:	*I went to the carding machine with some wool . . .*
July 19:	*I have been cradling. I wish that I was able to live without work when I pleased.*
July 20:	*. . . I have been cradling It is hard work to cradle.*
[Finishes wheat reaping end of July.]	
August 1:	*I have been haying We got five loads in to a stack near the barn.*
August 8:	*It has not been verry warm today . . . We have finished haying today and I am very glad of it.*
August 9:	*Mr. Richards and myself cradled our oats eight acres but small*
August 15:	*We have drawed in our oats*
August 16:	*We mowed what peas we had and drawed them down*
August 31:	*In the after drove the oxen to plow. The ground is very dry and hard*
September 1:	*I have been harrowing with horses*
September 5:	*I have finished plowing the oat stubble*
September 18:	*We have been cutting and drawing of corn*
October 6:	*It has been most as warm as summer came home . . . and I drove home a yoke of oxen In the afternoon finished bringing in the wheat*
October 14:	*We suffer much for rain my will has nearly failed, but we are blessed with a spring of water within a short distance of this house which furnishes an abundance of water*
October 21:	*. . . we have been gathering apples*
October 28:	*. . . we had a very severe frost last night the first of consequence that we have had, we have been gathering apples, ground a small cheese [of cider]*

October 30:	*I have been making cider drawed one load to Carters brought home a few winter apples.*
November 4:	*It has snowed most all day some of the time pretty fast. In the forenoon, S and myself drew two loads of stalk, in the after— drawed some wood. Amos and Benjamin finished making cider today.*
November 9:	*I finished digging potatoes*
November 11:	*We have been husking corn . . .*
November 17:	*Drawed one load of stalks in the morning and one at night. Plowed the rest of the day for wheat, this evening we had quite shower.*
November 23:	*The weather has not been verry cold today Tho snow fell last evening about six inches deep which remains on the ground, I have drawed manure and some wood.*
November 29:	*..I have been chopping*
November 30:	*We have had to fodder our cattle several days, I have been chopping [firewood].*
December 7:	*The weather continues warm and pleasant This has been Thanksgiving day*
December 15:	*I have drawed some wood to school house and some home mended my hearth*

[Rest of Year Missing]

1827

January 26:	*Not quite so cold today. It has snowed verry little. In the forenoon fan[n]ed 24 bushels of wheat the first our threshing this year. In the afternoon drawed some wood.*
January 29:	*I have not done much except to fix some sheet iron over my fireplace.*
January 31:	*This morning went to Mr. Quinns carryed my hams to be smoked..then went down to the barn threshed a flooring of wheat, it has not thawed any today*
February 2:	*..it has snowed hard most of the day and the wind has blowed it into drifts verry much. In the woods it is about 18 inches deep. I have not done much. In the afternoon went down to the barn and fan[n]ed about seven bushels of wheat.*
February 3:	*In the forenoon there shed [?] a flooring of wheat with the horse.*
February 4:	*This after went down to the barn and threshed a flooring*
February 6:	*thawed some, In the forenoon fanned 19 bushels wheat In the afternoon threshed some corn*

February 8:	*In the morning carryed my ashes to the ashery, drawed some wood the rest of the day.*
February 10:	*Not much alteration in the weather, Fan[n]ed 21 bushels of wheat and threshed ten shocks more*
February 11:	*In the forenoon sheared the wool off some sheep*
February 13:	*Last night the snow fell about four inches This morning it was warm but it has grown cold all day. This evening it is blustering. I have helped thresh a flooring of wheat with the horses and made preparation for starting for Buffalo*

[Buffalo Page Missing]

[Came home on the 21st]

February 24:	*The snow dissolves fast. It rained all day considerable fast. This evening it has rained very hard attended with thunder. We have fanned 28 and one half bushels of wheat, and threshed one flooring.*
March 8:	*Not quite so warm today..worked at gathering and boiling sap*
March 14:	*I have been sowing clover seed most of the day boiled some sap.*
March 15:	*The forepart of the day boiled sap in the after noon not do much, too much wind to boil sap.*
March 18:	*It has been rather unpleasant today rained most of the time. I have not been to church. In the afternoon went for doctor for one of our babes which is verry sick*
March 19:	*Some snow this morning, cloudy all day. Gathered and boiled sap. Our babe is no better, hardly think it will survive the night.*
March 24:	*Babe is getting better, we hope.*
March 26:	*This has been I think the warmest day we have had this spring. This evening I have heard the frogs sing for the first time.*
April 2:	*In the forenoon sowed clover seed. In the afternoon yoked a pare of steers and began to plow.*
April 11:	*The weather continues warm but not pleasant. some prospect of rain, In the forenoon went to Canandaigua after the newspapers, In even set out some fruit trees.*
December 18:	*The snow this morning is about eight inches deep*
December 19:	*Quite comfortable weather we have been to W. Coddings on a visit sleighing pretty good*
December 20:	*Weather comfortable Sleding pretty good. I have been drawing wood down to father*
December 24:	*This has been quite a comfortable day. Thawing some. In the morning*

	drew some loads of wood then went to the meeting house to attend the singing and meeting.
December 25:	*It has thawed most all day this evening rains considerable. I have been drawing wood. In the evening my wife and I went to the Sleigh is about spoiled.*
December 28:	*Weather much as yesterday rained most of the day have been threshing with the horses and oxen*
December 29:	*..have been fanning up wheat*
December 31:	*It has thawed very little today. I have been threshing wheat with the oxen in the Wheelers Barn*

1828

January 1828:	*This has been quite a pleasant warm day It has thawed considerable but the sleighing is pretty good yet I have been threshing*
April 3:	*In the forenoon harrowed on my corn ground in the meadow In the afternoon helped J. Darwin raise his barn.*
May 20:	*..finished leveling my barn*
May 26:	*Weather cool, In the afternoon verry little rain In forenoon split out some bars. In the after underpined my barn*
May 31:	*I have been making a bridge to my barn*
June 15:	*In the after got in my clover hay*
June 16:	*In the forepart of the day pleasant. In the after had a shower, mowed in the forenoon, In the afternoon got in my hay in the dooryard before the shower*
June 17:	*This has been a good hay day; mowed in the forenoon In the after got it to the cock, all that we had drawn*

1837

April 27:	*Thus I am forty years old; Time what an empty vapor Hrs and days how swift they are—*
August 29:	*. . . fixed my machine—thrashed 15 shocks and run it through the mill one time*
September 8:	*This has been quite a pleasant warm day In the forenoon finished crop; plowing my fallow In the after commensed raking my peas with a horse rake broke out several teeth and quit*
October 28:	*Not so pleasant as yesterday snowed a little I have been thrashing wheat with machine have not got along verry well-verry tired. most out of patience with thrashing machine-guess 'Bob' [3 letters] never had one*

December 5:	*..then went over to Mr. Mans after my barrel— fixed for butchering*
December 6:	*no material alteration in the weather-I have butchered my hogs*
December 19:	*The weather has been comfortable today In morning——a while broke my rope and quit come down and went threshing oats, about three discovered that my upper hay was on fire. It catched from the cilendar [difficult to read, probably fanning mill cylinder] Insured for $200. [APPARENTLY THE BARN BURNED DOWN!]*
December 26:	*In the morning drawed some wheat that we got out after my barn was burnt down in a damaged state*

1838

January 9:	*Last froze some-has been growing cold all day I have chopped down some trees for sawlogs-piled up some wood*
January 10:	*I have been to West Bloomfield to get my insurance money which I got without any trouble $200*
January 18:	*It thawed all night—all day. This afternoon had a hard shower-roads verry muddy-I had drawed my stack of wheat down to Mr. R's for the purpose of thrashing in his machine [Apparently Thayer Gauss's threshing machine had burned.]*
January 25:	*I have carryed my burnt wheat and some barley over to Shepards Mill*
January 30:	*In the forenoon I drawed three logs to mill..and got my horse shod*
January 31:	*I have drawed six logs to the saw mill*
February 9:	*Froze pretty hard last night today snow fell two inches—I have been to Springwater after boards—*
February 10:	*Weather comfortable snowed a little came home with my boards, Slaying pretty hard from Honeyoye Lake*
February 15:	*I have been to south Bristol after a load of boards*
Mar 29:	*..went down to the sawmill and drawed off my lumber come home by the Baptist meeting house to engage a carpenter*
April 4:	*The carpenters commensed frameing my building today*
April 5:	*..then went in to the woods to get the ramainder of my timber—I have not been so much fatigued in a great while.*
April 11:	*In the forenoon we got some polls for rafters, joices*
April 23:	*In the after went to Mill and engaged some hands to help me——tomorrow*
April 24:	*Last night the snow fell three or four inches. The m—st it has gone today. I have raised my hog pen, cow house and addition to my barn*

May 8: *..got three bunches of shingles*

May 9: *We have been shingling..and laying stable floor*

May 17: *In the forenoon I put the roof boards on my cow house*

May 29: *Sunday, The weather today has been quite unpleasant In the afternoon it rained most of the time. I went to church at five O'clock and heard Ichabod Codding deliver an antislavery lecture at the church I was highly gratifyied with his eloquence*

June 6: *We have been working at my granary; got along pretty slow*

Aug 8: *finished drawing my wheat in the after I raked stubble with a horse rake*

September 13: *In the afternoon Bramble came with his thrashing machine set it up and thrashed my Spring wheat and cleaned it up*

[Often mentions helping wife to do wash]

[In 1843 mentions the State Fair in Rochester in September]

January 18, 1848: *I have been thrashing in the barn with flails most of the day*

January 22, 1850: *..attended a Plank Road Meeting*

October, 1850: *[goes to New York City and Boston for ten days]*

January, 1855: *Attended Womans Rights meeting*

August 22, 1857: *went to help move Toll House*

The temptation is strong to include all of Thayer Gauss's entries but he wrote almost every day for thirty-two years.

CHAPTER IV

Steamboats, Roads, and the Triumph of Canals

FROM THE VERY BEGINNING of land speculation in the Genesee Country, broadsides had promised access, by either river or lake, to agricultural markets in Montreal, New York City, Philadelphia and Baltimore. Access to markets was essential if the region was to prosper. It was true that Lake Ontario provided a "path" to Montreal and the first leg of a trip to Albany and New York City, the temperamental Susquehanna River reached Harrisburg and Baltimore, and from Harrisburg even Philadelphia could be reached by wagon. But both the waterways and roads, those two earliest routes of transport, were at best difficult, seasonal and uncertain. Only when transportation improved and farmers gained reliable access to distant markets could they expand their operations, diversify their products—and ultimately need larger barns.

A measure of early transportation limitations comes from Joseph Sibley, who came to the region in 1806. As he told the historian Orsamus Turner, "In early years, there was none but a home market, and that was mostly barter:—It was so many bushels of wheat for a cow; so many bushels for a yoke of oxen, &c."[1] Fortunately this "home market" situation did not last very long or the region would not have prospered.

On the Genesee River, during this early period, boats could only operate from the falls at Rochester south to Mount Morris or from below the falls north to Lake Ontario. In 1820, some eight or ten rowed or horse-drawn boats operated on the Genesee, including barges and some shallow-draft Durham boats. Durham boats were about sixty feet long and eight feet wide, but with only a 24 to 30 inch draft so that even in times of low water they could get all the way to Mount Morris. The most important valley products shipped by water were wheat, lumber, and ashes—all bulk items. O'Reilly records exports from the Genesee River to Montreal in 1818, including 26,000 barrels of flour, 3,653 barrels pot and pearl ashes, 1173 barrels of pork, 190 barrels of whiskey, and a large quantity of barrel staves and other items.[2] In just two years shipments increased to 67,468 barrels flour, 5310 barrels pot and pearl ashes, 2643 barrels beef and pork, 709 barrels whiskey.

In addition to the boats already mentioned, one-way boats called "arks," or flat boats, carried Genesee Valley wheat down the Susquehanna to Baltimore. Wheat from loaded Genesee River boats was transferred via wagon to the flat boats at Arkport in Steuben County. Since the Sus-

Opposite: Detail of a sixteen passenger private Road Coach built by Healey & Co., New York City, circa 1900. Collection of The Granger Homestead Society, Inc. Carriage Museum, Canandaigua, New York.

quehanna did not permit upstream travel, the arks were broken up at their destination and sold for lumber.

Up to twenty of the one-way arks were constructed at Arkport each year. The number might not seem significant but it made Arkport one of the most active shipping centers in the region until the Erie Canal ended significant navigation on the Susquehanna. While arking lasted, wheat, cheese, butter, corn, cattle (William Wadsworth of Geneseo sent two arks of oxen in 1804), and of course the lumber of the ark itself were transported with economic success.

Henry Brewster's experiments in marketing have already been mentioned in Chapter I, but his experience seems to be rather straightforward compared to that of "a large farmer" mentioned in L. L. Doty's *History of Livingston County.*

> His wheat was ground at Wadsworth's mill near Geneseo. He then drew it to Avon, paid storage there, paid freight on the river and storage above the falls at Rochester, freight to Carthage (below Rochester) and storage there, freight to Ogdensburgh and storage there, freight to Montreal and storage there, commission for selling, and 'cooperage everywhere' on the line. After paying for a draft on New York he had eighteen pence per bushel left for his wheat, without counting the cost and labor of transportation to Wadsworth's mill, and thence to Avon bridge.[3]

Clearly, something more certain was required, especially in terms of access to distant markets.

Although many experimented with steam-powered boats, it was Robert Fulton's 1807 commercial success on the Hudson River with his steamboat, the *Clermont,* that whetted the appetite for speed in water transportation. A curious monopoly was granted to Fulton and Robert Livingston, Fulton's financial backer, limiting steam-powered craft on New York State waters to those of their manufacture or license. Under this monopoly, eight steamboats operated on the Hudson by 1810, and by 1818 there were others on Lake Ontario and the rest of the Great Lakes. These steamboats navigated up the Genesee River as far as the falls at Rochester (see figure 4-1). There were many court challenges to the Fulton steam monopoly, and it was finally declared illegal by the Supreme Court in 1824, in the same landmark decision that mandated congressional regulation of interstate commerce.

FIGURE 4-1. The Port of Rochester at the Ontario Steamboat Landing. O'Reilly, Sketches of Rochester . . . , 1838.

Before this decision, Fulton and Livingston had expanded their partnership to include Nicholas Roosevelt, with the express purpose of extending the steamboat monopoly to the western United States. However, the sidewheelers produced in the East were of too deep a draft to be practical in shallow waters like those of the upper Mississippi. This problem was solved by the inventive Henry Shreve, whose first steam vessel, the *Enterprise,* had been pressed into service by General Andrew Jackson at New Orleans in 1815. A year later, together with a West Virginia boatbuilder, Shreve built a shallow-draft sternwheeler and equipped it with Oliver Evans' high pressure (for the period), horizontal engines. The deck was close to the water line, creating the lines that were later to become commonplace among Mississippi River sternwheelers.[4]

Inevitably, questions were raised concerning steamboat navigation on the Genesee, which in late summer can become even shallower than the Mississippi. However, on July 28, 1824, the year the Fulton monopoly was ended, the steamboat *Erie Canal* made the first recorded steamer trip from the falls at Rochester up the Genesee River to Geneseo—a demonstration that the feat could be accomplished. The steamboat's Captain Battle gave the waiting celebrants in Geneseo a free trip on the river. The *Erie Canal* was 77 feet long, had a beam of about 15 feet and a draft of 11 inches exclusive of her keel; she could make four miles per hour when not impeded by snags, sand bars or bridges.[5] Encouraged by this successful experiment, a company was formed to operate a sternwheeler on the river. This steamer, the *Genesee,* carried passengers and, more importantly, towed up to two or three river barges, although an advertisement of the period suggests that towing was not a regular practice. In 1824 passage was scheduled upriver to Geneseo one day and downriver to Rochester the next, stopping at York, Fowlerville, Avon and Scottsville Bridge on the way north to Rochester. The fare, one way, was a dollar.

Several considerations suggest that the Genesee River steamers were constructed above the falls at Rochester. First, the considerable height of the falls themselves did not permit through travel upriver. Second, the Federal Registry of Steamboats does not list either the *Erie Canal* or the *Genesee.* Since this registry included only ocean-going vessels or those that plied interstate waters, these two were probably confined to the Genesee River.

Although not recorded, it is likely that the *Genesee* had a hinged stack, which could be lowered on approaching a low bridge, momentarily showering anyone on deck with cinders before the draft was restored by raising the stack after the bridge was cleared. Based upon contemporary paintings and drawings of similar vessels, a drawing has been made of the *Genesee* with one barge in tow (see figure 4-2).

Even though dams were built on the river to

FIGURE 4-2. Sternwheelers on the Genesee River were probably of very light construction, with an exposed boiler but a housed engine. These were the early days of steamboating and almost every new boat was experimental. The width and depth of the river figured in determining the exact size of the vessel; in addition its abrupt meandering would have limited the number of barges that could be taken in tow at any one time. Thomas Birch's 1821 painting at the Pennsylvania Academy of Fine Arts called The Fairmount Water Works *shows the working end of a "sternwheeler" in which two "side" wheels are mounted aft, creating in effect a stern-side wheeler. Birch's painting very well may depict a boat similar to the Genesee.*

help control the depth for navigation, steamboating on the Genesee River came to a halt after only two years. Both the waters and early steamboats may have proved uncertain. To most, the ideal solution to the problems of travel on the Genesee was a canal running the entire length of the valley. But little did anyone realize in the early 1820s that contracts for construction of the Genesee Valley Canal would not even be authorized until 1837.

Road Development

Rivers and lakes were the very first transportation routes, and later on canals were built to connect existing waterways. But for farmers, waterways were not enough; they needed roads, and roads required much greater effort to become practical and economical routes for transporting goods to distant markets.

In 1794, New York State budgeted the modest sum of $5,000 for a new state road (now Route 5), which traced the existing east-west Iroquois Indian trail. The Iroquois trail, which had been primarily suitable for foot travel, was widened to four rods (66 feet), improved to accommodate vehicles and officially named the Seneca Turnpike. The turnpike, nicknamed the "Genesee Pike," ran from Utica via Canandaigua to Hartford (now Avon), where the first bridge across the Genesee was begun in 1803 (see figure 4-3). Before this bridge was built, the river could be crossed only by rope ferry or on a "bridge" of winter ice.

Other early roads, as earlier noted, were built by Captain Charles Williamson, acting on behalf of the landowner Sir William Pulteney. The first, built in 1792-3, connected the southern portion of the Genesee Country with the Susquehanna River at Williamsport, Pennsylvania (see Chapter III). Within the Genesee Country itself, Williamson built many roads to connect the towns he was establishing to entice settlers. For example, in 1794 he built a road from Geneva through Lyons to Sodus Bay.

Meanwhile, roads were being constructed to connect major localities such as Geneseo and Batavia, the location of the Holland Purchase Office; another road, between Williamsburg and the Allegheny River, connected the region with Pittsburg. One especially interesting early road, appropriately called Ridge Road, was constructed on the natural gravel ridge that roughly parallels the shore of Lake Ontario. In a letter dated September, 1819, Frances Wright describes this road, which runs to the west between Rochester and Lewiston (near the falls of Niagara) and to the east between Rochester and Sodus Bay: "[It is] chiefly remarkable from its being, such as it is, the work of nature, a bed of gravel . . . run [ning] almost in a direct line."[7]

Early roads, like the Seneca Turnpike, tended to follow established trails, partly because generations of use had shown that these paths avoided many natural obstacles. The old paths, however, often did not suffice for oxen and horse-drawn vehicles. It wasn't merely a question of widening the right of way; when winter gave way to spring or when autumn rains saturated the ground, roads needed a substantial surface if wagon traffic was not to squoosh to a soggy halt. Those paths with the best natural gradients were improved. In places, logs were laid across the roadway, providing a teeth-jarring but passable corduroy road that was especially useful in marshy areas. Frances Wright describes her experiences on these corduroy "causeways" again and again, summing

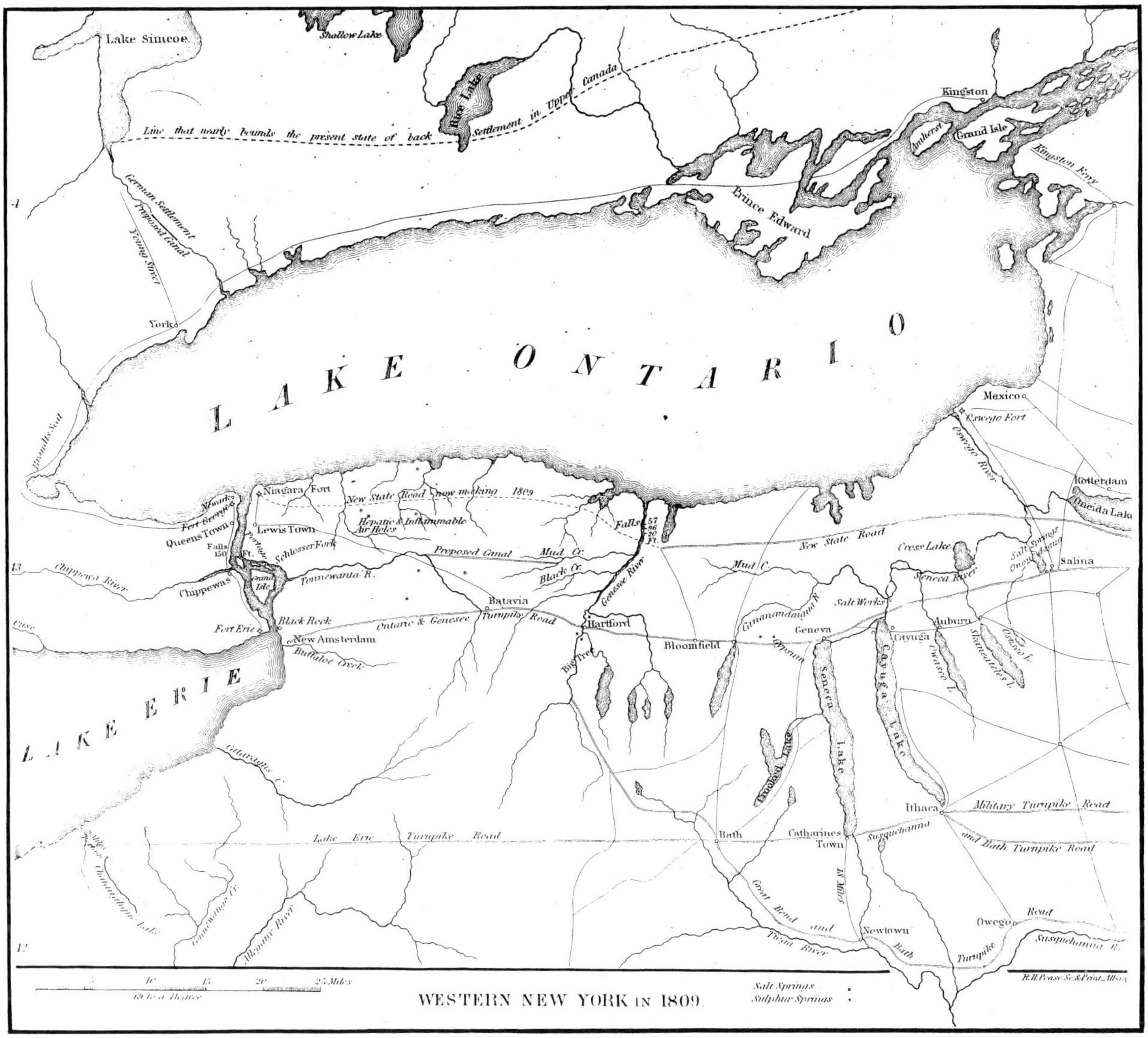

FIGURE 4-3. 1809 map of existing and proposed roads in the Genesee Country. The Genesee Valley Historical Collection, Milne Library, State University of New York at Geneseo.

them up this way: "the log causeway, through a deep swamp . . . is only to be crossed on foot; . . . a log road, or causeway, . . . is very grievous to the limbs."[8] Because smoothing the logs with an adze required considerable effort, it was seldom done. In the few instances where adzing appeared, the going improved temporarily.

Turnpike roads represented a big improvement over these early roads. In 1802 the legislature authorized a turnpike road from Canandaigua to Bath and, in 1806, from Bath to the Susquehanna River. This was essentially the same road that had been constructed by Williamson but a much better one. McIntosh outlines the contract conditions for a turnpike road:

> The road must be opened four rods wide, and be thirty-three foot between the ditches on each side; twenty-five feet thereof, when necessary, should be bedded with stone, gravel, sound wood, or other hard substance, well compacted together, and of sufficient depth to secure a good and solid foundation. The grades ascent should

> not exceed fourteen inches in any one rod. The bridges to be twenty-four feet wide, and covered with good white oak, white pine, or hemlock plank, clear of sap, and three inches thick. In the estimate, tollhouses were included, and ten miles was the extreme limit of a single contract.[9]

Work on such a state road could substitute for paying highway taxes (see Thayer Gauss's diary). Similarly some farmers were able to pay the rent or mortgage on their farms by working on private roads for the landowner. A highway tax deduction of twenty-five cents was permitted for each tree planted along the road that survived a year, and placing a watering trough by the highway exempted a person from three days' tax.[10] Turnpikes built by private stock companies were generally considered even better than state roads, but were expensive for moving products to market. The improvement represented by all the turnpikes meant that teams could draw loads one-third heavier than before.

A better solution than the corduroy road to the problem of all-season travel evolved with the building of wooden plank roads. These roads were first built in the Genesee Country about 1846. In 1847 a plank road was built from Rochester through Avon, Geneseo and Mount Morris to Dansville. The sills, or stringers, of a plank road and the planks themselves were usually made of mill-sawn hemlock, abundant in the region. Unlike turnpikes, plank roads were single width roads only about eight feet wide, except for sections with passing lanes. Traffic was governed by an 1801 law giving eastbound traffic the right of way under penalty of a fine. Caroline Richards writes in her 1852 diary of an outing with her grandfather, when they paid two cents for the pleasure of traveling the plank road from Canandaigua to Chapinville.[11] Plank roads turned out to be effective in spite of competition but they were somewhat expensive to maintain, with renewal needed about every five years. As competition from free roads increased and repairs were not kept up, the plank roads all but disappeared by the 1870s. In many instances the planking was replaced by gravel, making the road usable again but now without a charge to the user. Today only the plank road names remain, on road signs.

Even on improved roads, for much of the nineteenth century a farmer ordinarily used oxen rather than horses to pull a wagon; the tremendous pulling power of oxen was needed in the days before roads were graded (see figure 4-4). An 1804 issue of the *Albany Gazette* describes the transport of a wagonload of wheat from Bloomfield to Albany using four yoke of oxen and making the round trip in twenty days (underlining theirs). With the onset of frozen ground and snow cover, sleds could be used, and much more could be hauled in a single load. To make the going easier the snow was sometimes rolled and even regularly sprinkled with water to produce a frozen surface for the sled runners. Winter shoes with sharpened points were necessary for oxen and horses alike.

FIGURE 4-4. Oxen drawing a wagon. Anonymous photograph from the Letchworth Collection, The Genesee Region of The New York State Office of Parks, Recreation, and Historic Preservation.

Along all the roads, inns and blacksmith shops sprang up to offer service to the traveler, and as roads improved so did public transportation. James Mathers notes that, while the first mail on Ridge Road was carried by horseback, in 1816 daily stage coaches were established, numbering eight to ten loaded coaches daily at their peak before the opening of the Erie Canal. The law limited commercial road coaches to twelve passengers with fifty pounds of luggage per person.[12] Advertisements for various stage lines in the 1820s provide some idea of traveling times as well as fares: Up to forty miles a day over good roads might be expected in the summer and about half that in winter. Stage lines survived the competition from canals, but gave way to the railroads by the 1850s, except where they were needed to make connections (see figure 4-5).

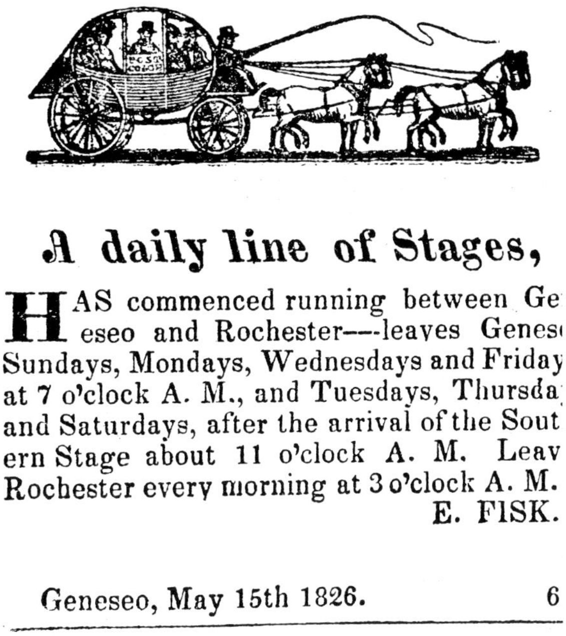

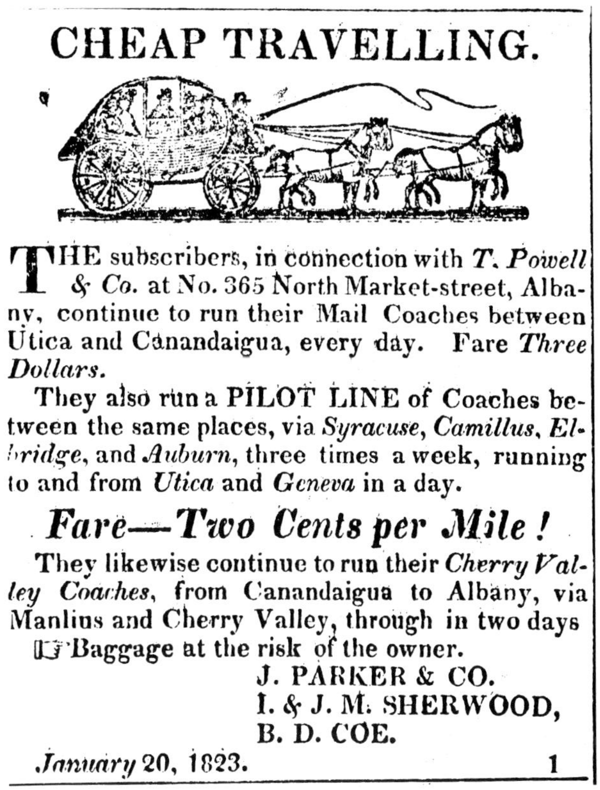

FIGURE 4-5. Stage line advertisements from the Ontario Republican, 1823, and the Ontario Freeman, 1826.

Canals

In the eighteenth century the cry was "canals!" They were in use throughout Europe and touted as the highways of the future. In 1784 General Washington reexamined earlier plans for a Potomac River waterway, complete with needed canals, which when finished would connect with the Ohio Valley. Many people, including New Yorker DeWitt Clinton, an Erie Canal commissioner, saw great potential benefit to the new nation from a canal connecting Lake Erie with the Hudson River, but not everyone agreed with them.

Travel and transport by water had been a part of America from Colonial days. Initial settlement had occurred along the banks of navigable rivers such as the Connecticut, James, Delaware and Hudson, with the water providing vital access inland from the coast. But these were major arteries. Elsewhere the rivers were less reliable, subject to summer lows, spring flooding and other problems. Sometimes they simply did not flow where settlers wanted to go. The construction of canals offered a solution.

Clinton visited the Genesee Valley in 1810 as

part of an Erie Canal feasibility study. O'Reilly reports Clinton's observations:

> There is a good sloop navigation from the lake to the Lower Falls [now called the Ontario Steamboat Landing in Rochester]. These falls, as also those of Niagara, and perhaps of Oswego, are made by the same ridge or slope of land. The Genesee River, in former times, may have been dammed up at these falls, and have formed a vast lake covering all the Genesee Flats forty miles up. The navigation above the ford is good for small boats to the Canaseraga Creek, and ten miles above it, making altogether fifty miles.
>
> We dined and slept at Hanford's tavern, who is also a merchant, and carries on a considerable trade with Canada. There is a great trade between this country and Montreal in staves, potash, and flour.
>
> I was informed by Mr. Hopkins, the officer of the customs here, that 1000 barrels of flour, 1000 barrels of pork, 1000 barrels of potash, and upward of 100,00 staves, had been already sent this season from here to Montreal; that staves now sold there for $140 per thousand, and had one time brought $400; that the expense of transporting 1000 staves from this place to Montreal is from $85 to $90; across the lake, from $45 to $50; that of a barrel of potash to Montreal, $2; pork $2; flour, $1.25; but that the cheapness of this article is owing to competition and is temporary.
>
> A ton of goods can be transported from Canandaigua to Utica by land for $25.00.
>
> Notwithstanding the rain, we visited in the afternoon the mouth of the river. On the left bank a village has been laid out by Col. Troup, the agent of the Pulteney Estate, and called Charlottesburgh, in compliment to his daughter. He

FIGURE 4-6. The Erie Canal Aqueduct over the Genesee River at Rochester. O'Reilly, Sketches of Rochester . . . , 1838.

FIGURE 4-7. A Clinton Line barge loaded with flour; the barrels arranged on their sides are visible through the door. The Erie Canal, Rochester, New York. O'Reilly, Sketches of Rochester . . . , 1838.

> has divided the land into one-acre lots. Each lot is sold at $10 per acre, on condition that the purchaser erects a house in a year. This place is in the town of Geneseo. The harbour here is good. The bar at the mouth varies from eight to eight and a half feet, and the channel is generally eleven feet. Here were four lake vessels in it.[13]

Clinton and others were delighted with the potential they saw for developing the lake and river arteries through the construction of more manageable connecting canals.

Over considerable opposition, the digging of "Clinton's Ditch," as detractors called it, was begun in 1817 from Albany and the Hudson to Rochester. The last link of the Erie Canal between Rochester and Buffalo was built between 1823 and 1825, finally connecting the waters of Lake Erie with those of the Hudson more than 360 miles away. The completion of the Erie Canal signaled a change in some fairly well-established transportation patterns: It reduced the importance of the Susquehanna River as a route to Harrisburg and Baltimore and virtually eliminated the established route of western migration that led through western New York down the Allegheny River to Pittsburgh.

With the completion of the Erie Canal, Rochester became a boom town with growth in every area, including manufacturing, warehouse storage, and milling capabilities, which were a boon to local agriculture. With these new operations the Genesee Country was bound to grow and grow it did. Soon people throughout the East and indeed even in Europe would hear of Genesee Wheat. Even before the expansion in agriculture, many had already heard of the marvelous 804-foot stone aqueduct that carried the waters of the Erie Canal over the Genesee River at

Rochester. The Erie Canal's aqueduct, with its round Roman arches, was touted as the largest bridge in the United States (see figure 4-6).

The canal quickly paid back its investment. O'Reilly[14] provides a sample of goods shipped from Rochester to Albany by the Erie Canal in 1836:

Domestic Spirits, gallons	44,978
Boards and scantling, feet	753,173
Timber, feet	9,500
Staves, lbs.	585,688
Flour, barrels	368,842
Wheat, bushels	151,714
Ashes, barrels	4,249
Clover & Grass seed, lbs.	491,976
Hops, lbs.	21,450
Cheese, lbs.	81,844
Tobacco, lbs.	15,805
Pig iron, lbs.	68,095
Iron ware, lbs.	383,097

A wood engraving from O'Reilly shows a typical, barrel-laden packet boat tied up at one of Rochester's many flour mills (see figure 4-7). Success soon demanded that the canal be widened to 70 feet and deepened to seven. Canal mania swept the United States. As might be expected, a number of canals feeding into the Erie were built; of these perhaps the one with the greatest potential was the Genesee Valley Canal.

By 1823 it was possible to ship goods on the Erie Canal from Rochester to Albany, and it became obvious that a feeder canal running the entire length of the Genesee Valley would benefit the area. In 1823 the citizens of the valley began a quest for permission to build just such a canal. After many delays, the canal was completed in 1840 from the Erie Canal at Rochester south to Mount Morris. Figure 4-8, an 1838 map, shows the connection of the two canals soon after the construction of the Genesee Valley Canal had begun.

By 1841, a branch extended from Shakers (Sonyea) to Dansville. However, great difficulties were experienced in building the main canal further south through the Genesee Gorge, due to slumping of the hillside on the east side of the gorge along the high banks above the middle Genesee River Falls. Charles Bierstadt's stereo view (half shown here) of the Genesee River gorge near Portage neatly illustrates some of the difficulties (see figure 4-9). At this point near the Middle Falls the canal clung to the high banks, an area where the rock formations are unstable and slides are common, causing high construction and maintenance costs. When the canal bed was finally "pinned" to the hillside, the canal was continued south, crossing above the Genesee on a forty-foot-high aqueduct at Portageville (see figure 4-10). At Cuba a maximum elevation of 978 feet above the Erie Canal at Rochester was reached. This main portion of the canal, from Rochester to its southern terminus at Mill Grove Pond, near Portville, was 107 miles long and had 132 locks; the Dansville branch increased the total mileage to 124 miles. Visionaries saw a canal that would connect with the Erie Canal at its northern terminus and then go south to connect with the Allegheny, Ohio and Mississippi Rivers!

An 1858 State Engineer and Surveyor map sketches the route of the Genesee Valley Canal from its connection at Rochester to its proposed southern connection at Mill Grove with the Allegheny River (see figure 4-11). In addition to passengers, the chief items transported on the Genesee Valley Canal were grain, lumber, unsawn wood, building stone, coal and gypsum. McNall, in his *Agricultural History of the Genesee Valley*, provides a view of the wheat transported by the Genesee Valley Canal: In 1844, the first year listed, some 276,000 bushels were delivered to Rochester. Wheat shipments peaked in 1852 and 1853, when 535,000 and 539,000 bushels were

FIGURE 4-8, Opposite. Detail of a map drawn by S. Cornell, showing the connection of the Erie and Genesee Valley Canals. Included in O'Reilly, Sketches of Rochester . . . , 1838.

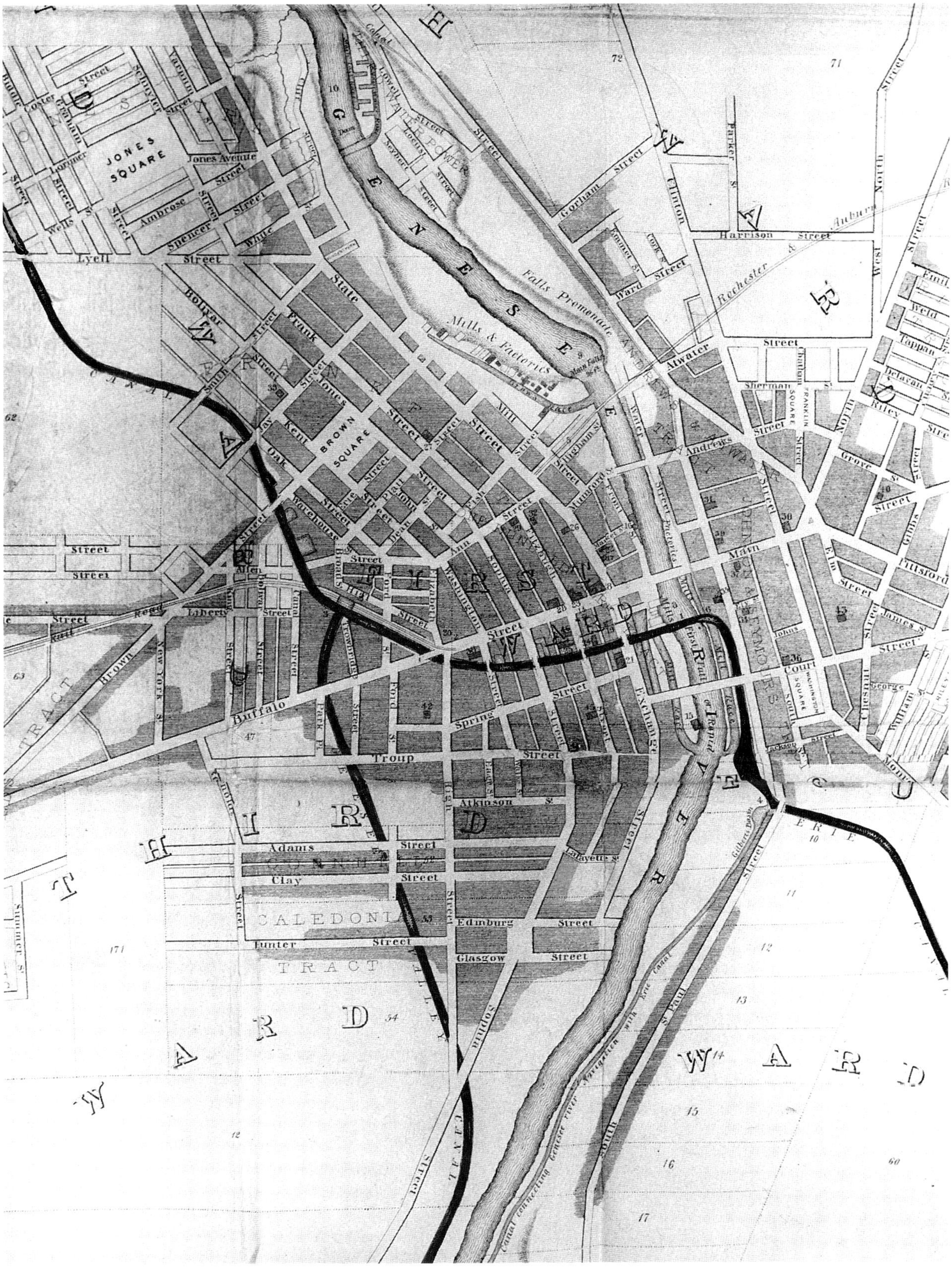
JONES SQUARE
Jones Avenue
Ambrose
Spencer
Lyell Street
BROWN SQUARE
State
Frank
Falls Promenade
Mills & Factories
Gorham Street
Clinton
Harrison Street
Rochester & Auburn
Atwater
Sherman
FRANKLIN SQUARE
Andrews
Main
Buffalo
Troup Street
Atkinson St.
Adams Street
Clay Street
CALEDONIA
TRACT
Edinburg Street
Glasgow Street
Hunter Street
Exchange
Court
Spring
Liberty
Brown
New York St.
Lafayette St.
Pittsford
James St.
Canal connecting Genesee river Navigation with Erie Canal
St. Paul
South
Sophia
Street
CANAL
WARD

FIGURE 4-9. The Genesee Valley Canal clings to the cliffside of the Genesee River near Portage, New York, showing the elevation of the canal above the river at this point. Appearing in the background of the stereograph is the Erie Railroad bridge at Portage, symbol of the competition which would spell the end for the canal. Stereo view by Charles Bierstadt.

tallied at Rochester. With the subsequent decline in wheat production, shipments slipped to 115,000 bushels by 1855.[15] The canal operated for thirty-eight years to Mount Morris, but finally closed along its entire length in 1878, when it had become clearly unprofitable.

The 124 miles of canal generated many stories, which can be gleaned from account books, early photographs, personal diaries, newspaper accounts and advertisements. In addition, careful inspection of newspaper advertisements provides information about running times, rates and connections (see figure 4-12).

The canals took products to market at a reasonable rate, returning an influx of materials and manufactured goods into the Genesee Country. Teams of horses or mules, usually two, three or four, were used to draw the barges and were changed about every twelve miles. Several examples of passage times might prove useful: The trip from Dansville to Rochester on the Genesee Valley Canal was about eleven hours, and a trip on the Erie Canal from Albany to Rochester lasted almost five days. In transporting products for which speed was not a major consideration the canals offered the advantage of being able to move large quantities of goods; by 1850 the average loaded canal boat on the Genesee Valley Canal weighed 100 tons.

FIGURE 4-10. The Genesee Valley Canal aqueduct at Portageville, New York. In this winter view from a stereo card, the spill of canal water has frozen into "arches". The aqueduct allowed the canal to retain the altitude it had gained without having to descend to river level for a crossing. The descent and ascent through the numerous locks which would have been required for a river level crossing would have been extremely wasteful and time-consuming. Stereo card from the James Dorr collection.

FIGURE 4-11. A detail of the 1858 State Engineer & Surveyor's map of New York State Canals, designed by Van Rensselaer Richmond, State Engineer & Surveyor, depicts the proposed route of the Genesee Valley Canal from Rochester to Mill Grove and the Allegheny River connection.

CANAL MEETING.

THE inhabitants of the county of Livingston friendly to the construction of a Canal through the valley of the Genesee River, are requested to meet at Hamilton's in Geneseo on the *18th day of July instant*, at 11 o'clock in the forenoon, for the purpose of choosing delegates to meet in Convention with those from other counties; and to transact such other business in relation to such Canal as may be deemed advisable.

Geneseo, July 4th, 1829.

John H. Jones,	Ph. C. Fuller,
E. Hill,	H. P. North,
Wm. A. Mills,	E. N. Buell,
Wm. McCartney,	R. Austin,
Wm. H. Spencer,	C. H. Bryan,
H. D. Mason,	C. H. Carroll,
Wm. Fitzhugh,	S. Frost,
O. P. Olmsted,	D. H. Fitzhugh.

Canal Transportation.

WASHINGTON LINE,

FOR FREIGHT AND PASSENGERS,

(Running Night and Day, with relays of Horses.)

THE proprietors of the above line, having fitted up their Boats in superior style, will on the opening of navigation, run two daily lines of Boats, with the greatest regularity, leaving Albany and Buffalo and passing the intermediate places, each way daily.

They have made arrangements with the Steam Navigation Companies on the North river, and any property transported by this line will be delivered as addressed, free of storage at Albany.

For the benefit of the shipper $16,000 is insured on all property bound upwards, and $8,000 on all downwards; at all times property will be delivered as directed by the owners. For freight or passage apply to

M'COLLUM & HULBERT.

Rochester, March 12, 1827. 50tf.

PACKET BOATS.

THE regular Line of PACKET BOATS between Schenectady and Buffalo, will be run the present season, by the Utica and Schenectady and the Erie canal Packet Boat Company, in connexion.

TWO DAILY LINES, will run between Schenectady and Utica.

THE MORNING LINE, will leave Utica at 8 o'clock A.M. on the arrival of the Western Boats, and Schenectady at 2 o'clock P.M

THE EVENING LINE, will leave Utica and Schenectady at 8 o'clock P.M.—all running through in 24 hours.

ONE DAILY LINE, will run between Utica and Buffalo, leaving Utica at 8 o'clock P.M on the arrival of the Eastern Boats, and Buffalo at 8 o'clock A.M.

The Packets will touch at all the principal villages on the canal to receive and land passengers.

STAGES will be always in readiness to receive and carry Passengers between Schenectady and Albany or Troy, on the arrival of the Steam and Packet Boats at those places respectively, and also at such points on the Canal as may accommodate passengers passing to and from the principal villages situate off the line of it.

Utica, April 15, 1827. 4w1

MERCHANTS' LINE,

For Freight and Passengers.

A BOAT belonging to the above line, well fitted up for Passengers, and running *night* and *day*, with relays of horses every 12 miles, will leave Rochester for Albany every morning (Sundays excepted) at 7 o'clock, and every evening at 8 o'clock, and will run through *in less than five days*, or at the rate of 60 miles every 24 hours.

FIGURE 4-12. Newspaper advertisements centered on the Erie and Genesee Valley Canals.

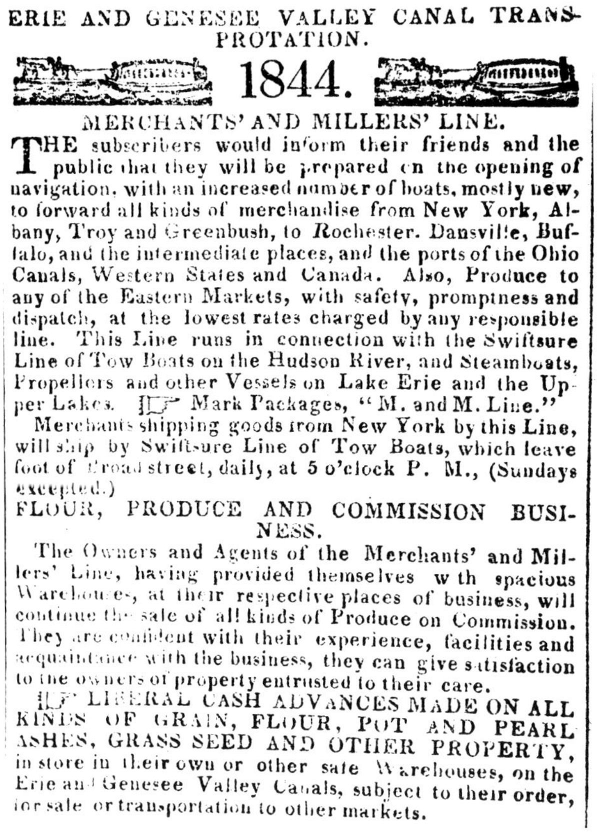

ERIE AND GENESEE VALLEY CANAL TRANSPROTATION.

1844.

MERCHANTS' AND MILLERS' LINE.

THE subscribers would inform their friends and the public that they will be prepared on the opening of navigation, with an increased number of boats, mostly new, to forward all kinds of merchandise from New York, Albany, Troy and Greenbush, to *Rochester*. Dansville, Buffalo, and the intermediate places, and the ports of the Ohio Canals, Western States and Canada. Also, Produce to any of the Eastern Markets, with safety, promptness and dispatch, at the lowest rates charged by any responsible line. This Line runs in connection with the Swiftsure Line of Tow Boats on the Hudson River, and Steamboats, Propellers and other Vessels on Lake Erie and the Upper Lakes. ☞ Mark Packages, "M. and M. Line."

Merchants shipping goods from New York by this Line, will ship by Swiftsure Line of Tow Boats, which leave foot of Broad street, daily, at 5 o'clock P. M., (Sundays excepted.)

FLOUR, PRODUCE AND COMMISSION BUSINESS.

The Owners and Agents of the Merchants' and Millers' Line, having provided themselves with spacious Warehouses, at their respective places of business, will continue the sale of all kinds of Produce on Commission. They are confident with their experience, facilities and acquaintance with the business, they can give satisfaction to the owners of property entrusted to their care.

☞ LIBERAL CASH ADVANCES MADE ON ALL KINDS OF GRAIN, FLOUR, POT AND PEARL ASHES, GRASS SEED AND OTHER PROPERTY, in store in their own or other safe Warehouses, on the Erie and Genesee Valley Canals, subject to their order, for sale or transportation to other markets.

FIGURE 4-12. Newspaper advertisements centered on the Erie and Genesee Valley Canals.

This copy of an 1844 packet boat schedule yields additional information about the various stations and traveling times encountered on the Genesee Valley Canal:

> DAILY DEPARTURES
> 8 AM Rochester
> 3 PM Cuylerville
> 4 PM Mt. Morris
> 5 PM Shaker Settlement
> 8 PM Dansville
> 6 AM Dansville
> 5 PM Rochester (In time for Packet Boat to Buffalo to Syracuse)

From the collection of the Livingston County Historical Society

Many wrote of their experiences on the Erie Canal: The trip seemed too long to Frances Trollope, Nathaniel Hawthorne and other writers, but generally pleasant to H. C. Blinn. In his diary of 1856, Blinn, a Shaker, recalls the trip from Shaker Settlement in the Town of Groveland to Rochester on the Genesee Valléy Canal:

> This was evidently the largest delegation of Shakers that ever met in the Genesee Valley. At 7 P.M. we are on board a canal boat, bound for Rochester, a distance of 40 miles, but David Parker had provided one of the best boats for passengers. For this special favor as well as for many others that we are receiving daily, we feel very much indebted to him. Our company on the boat was not quite as large as the one at the table. It now numbered 16 persons.
>
> Beside our company we had that of 20 other passengers, which was all that could be comfortably accommodated. While on the canal we were allowed the privilege of being on the deck or in the cabin. There were two rooms below used for sitting or sleeping rooms by the passengers. The partition between the two was made by stretching a curtain across the boat. At 9:30 P.M. all the passengers were ordered on deck, to give the boatmen a chance to make the beds. These were arranged on both sides of the boat, the whole length, in three tiers. The back part of the frame which made our bedstead was hasped to the side of the boat, while the outer or front side was fastened to a leather band that was attached to the roof. Over this frame was stretched a cloth, on which we lay. We also had a small piece of cloth to put over us in case we were inclined to be cold. Most of the evening our company remained on deck as we enjoyed the passing scenery.
>
> Even this pleasure was not unalloyed, as we were obliged to stoop very low while passing under the bridges. The highways over the canal were of frequent occurrence and at the time that we had arrived within a rod of one of these places the man at the helm would sing out, "Bridge," at which sound every head would drop, and all get on their hands and knees to avoid a bumping from the timbers overhead. During this passage we passed through ten locks. Our speed was about four miles an hour. Some of the time we were drawn by two horses and at other times by three. The riders took

FIGURE 4-13. Relief map of the Genesee Valley looking south from Lake Ontario. The deep gorge of Letchworth is located between Mount Morris and Portage while a broad flat valley extends southeast from Mount Morris to Dansville.

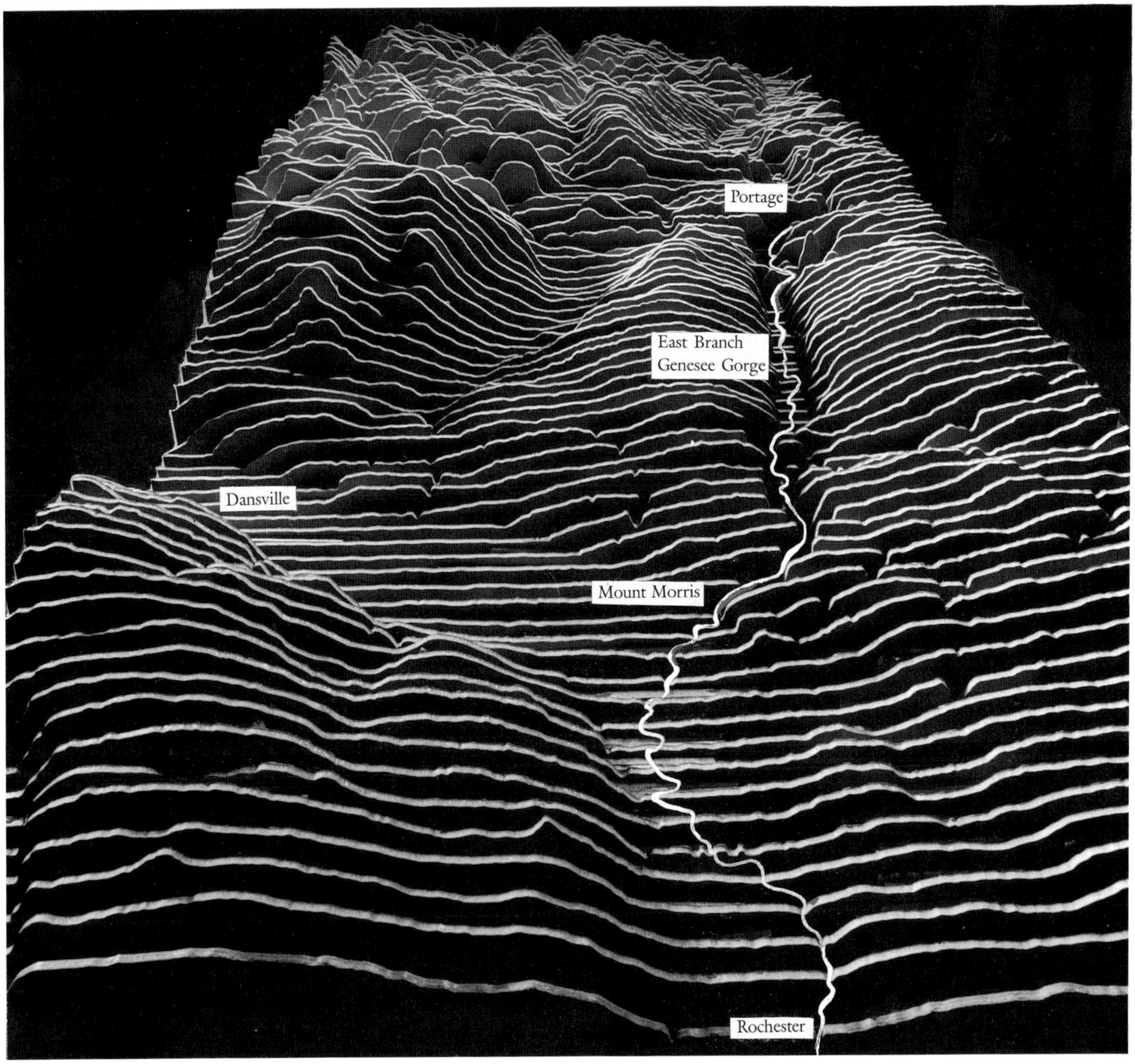

turns, and it was well that they did, for one would have been unable to use the lash so constantly as was the case with these drivers. They seemed to have but little mercy for the poor horses, and the whip fell heavily on their sides. At 10 o'clock we began to think of retiring, as it had grown quite dark and we were weary of looking. We accordingly packed ourselves away as best we could for the night, but the idea of sleeping in such a confined place and with so many persons, was not an agreeable thought. The writer had an upper berth and it proved to be a very uncomfortable place. In our cabin the men were stretched all over the room, on little or nothing for a bed. By the time that half of the night had passed, some of the company were up and on deck, thankfully inhaling some fresh air. Weary of this place we try the couch again but sleep had fled from our eyelids.

Our old boat moved steadily along and this novel ride was quite agreeable, outside of our offensive bed chamber, —the thumping while descending in the locks and the trouble of prostrating the body while going under the many bridges. At 5:30 A.M. we arrive at Rochester, and proceed immediately to the Clinton Hotel.

July 23

Soon after breakfast we all proceed to the station and enter the cars for Albany. Our friends from Groveland bid us adieu and will leave for their home in Sonyea. . ."[16]

Figure 4-13 shows a map of the Genesee Valley looking south from Lake Ontario toward the Pennsylvania border. A slight vertical exaggeration makes topographic features visible. The stark contrast between the Genesee Gorge from Mount Morris to Portage and the flatter land between Mount Morris and Dansville is emphasized on this map.

Agriculture and commerce improved with the building of roads and bridges and the construction of canals, in effect binding the Genesee Country to important markets and commercial centers in the East. Figure 4-14 shows the loading of bluestone on a Genesee Valley Canal barge. In agriculture, the small log barns of early settlement quickly gave way to large, more flexible and efficient framed "English" barns.

FIGURE 4-14. Loading bluestone used in building from the quarry approximately three miles south of Portageville, New York. This view of loading the "Harvey Tucker," taken from a stereo card, is one of the very few pictures of a canal boat on the Genesee Valley Canal. At this location the canal follows the curve of the Genesee River, seen to its right. The dark horses in the center provide the power for the loading gin. The workman in the rear is posed adjusting the ropes from the boom's block and tackle around a large piece of bluestone. From the Collection of Mrs. Lucille Frank.

CHAPTER V

Hewn Timber-Framed English Barns

ALTHOUGH the very earliest barns and houses were log structures, in which the frame and the sheathing were one, most barns in Europe and the eastern United States had hewn frames covered with exterior sheathing. The first timber-framed barns in the Genesee Country were one, two or three-bay structures with the most common being the English or three-bay style. The three-bay barns were also called Connecticut, hay, grain or common barns, but the design was originally English. Because many of the early settlers in the Genesee Country came from New England, where this design was common, it is not surprising that they built the English barn rather than the Dutch barn, familiar in eastern New York, or the German/Swiss barn of eastern Pennsylvania.[1] In England these three-bayed barns were used almost exclusively for grain and hay storage, but in America some included stall space for livestock. These barns were also essential threshing centers in a region of uncertain weather.

The first few English barns in the Genesee Country appeared at the end of the eighteenth century; small by modern standards, they were 20 to 30 feet long, the same size as many of the larger log barns in the region. Later examples were a bit larger, 30 by 40 feet. Size was limited by available materials and the difficulties of splicing timbers and handling large members. Because farmers' needs for storage and livestock housing did not change significantly during the 1820s and 1830s when the English barn was being constructed in large numbers, its size remained fairly constant.

English barn frames were hewn, but the sheathing was sawn; hewing boards is wasteful of both wood and time. Pit sawing by hand was possible (see Chapter II, figure 2-1), but water-powered saw mills were much more efficient. So important were sawn boards to framed construction that the establishment of saw mills was on the minds of the land speculators and settlers from the outset (see Chapter VII). Water-powered saw and grist mills were built along streams, rivers, and lake outlets very early, even in places where there were few other signs of habitation. L. L. Doty relates Samuel Magee's recollection of mills in the area south of Groveland Hill in 1795:

> Dansville, where not a building of any description had been erected . . . Groveland hill did not count a single settler . . . The road . . . led to the Williamson grist-mill and saw-mill, the latter

Opposite: The 9 by 12 inch, 30 foot swing beam of the Nations Farm English Barn. Geneseo, New York.

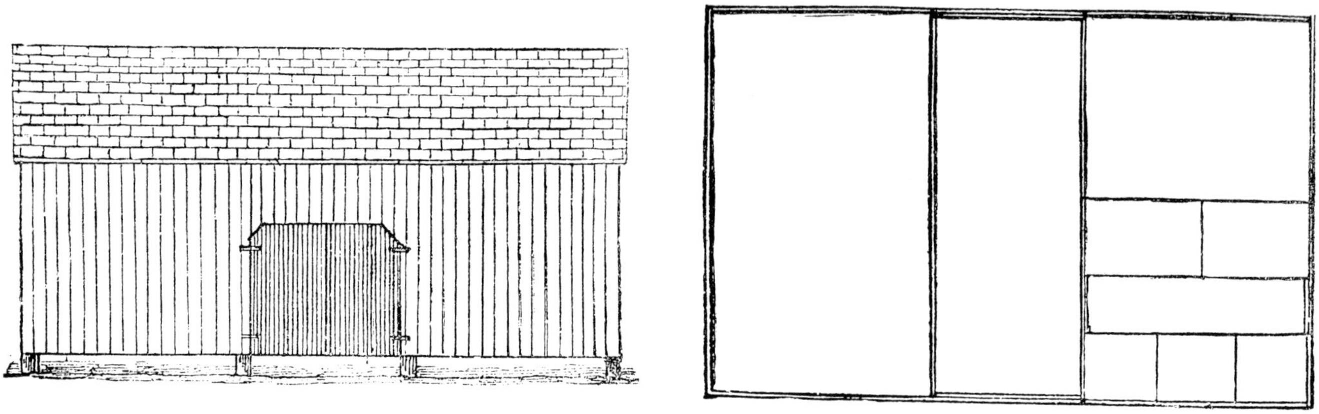

FIGURE 5-1. Common English three-bay barn. American Agriculturist, April, 1884.

> standing a few rods below the former, near Dansville . . . A mere wagon-track led to the mill, and to right and left 'the pine bush was so thick that a person could not possibly see one rod into it on either side.'[2]

The 1794 Phelps & Gorham survey map shows three saw mills in the Genesee Valley, including Ebenezer (Indian) Allan's on the west of the Genesee River Falls at what is now Rochester, the mill on the Conesus outlet below Avon, and the mill at Dansville. However crude and inefficient, Allan's mills were already operating in 1789, and by 1845, according to Hedrick, more than 7,000 saw mills were operating throughout the state.[3]

Dating Barns

The English barn remained the standard for framed barns in the Genesee Country for more than thirty years, and surprisingly many of them remain, probably because of their flexible design and durable construction. Unfortunately, individual barns are difficult to date based solely on physical evidence, such as nail or spike characteristics, because nailmaking techniques remained unchanged for a very long time (see Appendix II for a further discussion of barn dating). The designs of the hewn bents also remained remarkably similar and the act of hewing itself continued into the twentieth century. When one turns to contemporary sources for help in dating barns, disappointingly little is available.

The historian Orsamus Turner talked with many pioneers before the mid-nineteenth century and obtained some information on when barns were built, such as "Enos Boughton..built the first framed barn west of Seneca Lake," but no actual dates.[4] Turner quotes James Mathews, the first settler of Gaines in 1810, who built the first framed barn in Orleans County, "procuring my boards at Turner's mill on the Oak Orchard," but again with no definite date. Jehiel Kelsey remembered that "In '98 or '9, Peter Shaeffer put up a framed barn [near Avon]; it took all the men in the region— twenty, all told."[5] L. L. Doty describes the barn built by Horatio Jones in 1796 as "a frame barn, a little to the west of Jones's Bridge

FIGURE 5-2. Aerial view of The Nations Farm, town of Geneseo, Livingston County, New York. The 30½ by 40½ foot English barn on the lower right of the complex clearly shows the centered drive-through door. A standard three-bay English barn, it stands in a typical relationship to the other barns.

[approximately one mile south of the Village of Geneseo] . . . and soon he built a frame distillery at the Fort farm."[6] Besides this tantalizing information, no further details about the character of the barn are given; however, it was probably an English barn.

Since it was frequently the practice to build new, bigger barns next to the old, it may well be that in a barn complex the smallest barns are the oldest. Such is the case at the Wadsworth farm known as "The Nations Farm" (see figure 5-2).

The English barn was usually gable-ended with three interior bays, spaces enclosed by two framing bents; later versions sometimes included additional bays. Generally a central door led to the threshing floor or "driveway" (drive-through), with a bay or mow (rhymes with "cow") on either side. A team could draw a wagon loaded with loose hay or unthreshed grain sheaves into the center of the barn, the load often just clearing the door opening, and from this central position the load could be "mowed away"

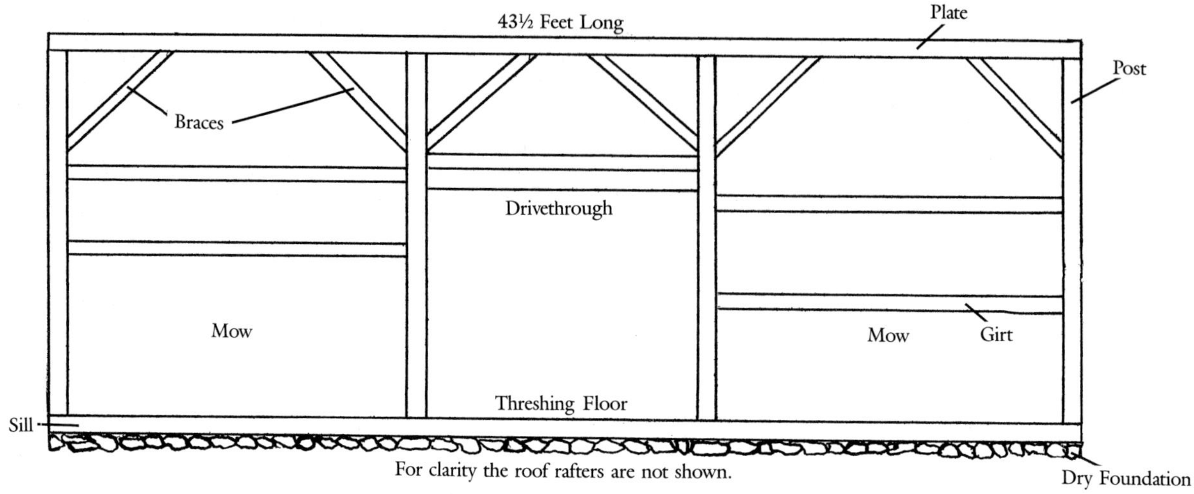

FIGURE 5-3. Elevation of the Nations English barn showing the mows and drive-through.

with pitchforks. The empty wagon was driven out the doors on the opposite side of the barn to fetch another load. The hay mows would be entirely full after the haying season and after the grain harvest the rest of the barn storage spaces would be filled with unthreshed sheaves. In some alternative interior configurations, one of the mows was divided into either a granary for the storage of threshed loose grain or into livestock stalls (see figure 5-3). One farmer wrote to *The Genesee Farmer* that he left space at the base of both mow and granary to "allow the cats room to carry on their useful avocations."

Flailing and Flooring

The initial design of a three-bay English barn made it the center both of grain storage and of threshing, in which the grain, or berry, is separated from the straw. The threshing floor included the drive-through and often a floored adjacent bay as well. In addition to providing storage space for unthreshed sheaves and space for threshing, the English barn was designed and oriented in relation to the prevailing wind to facilitate winnowing.

Threshing, whether with a jointed flail or with hoofed animals, was a slow process, often continuing through an entire winter until it was time to plow again; during this time the center bay or threshing floor had to be kept clear. A sheaf or bundle of wheat was placed on the threshing floor, untied, and then beaten with a jointed flail (see figures 5-4 and 5-5).

Threshing with horses and oxen was an age-old process which required considerable space in

FIGURE 5-4. Flail hanging on a barn door. Farmer's Museum, New York State Historical Association, Cooperstown, N.Y. The beater is about three feet long.

a barn (see figure 5-6). "Flooring," as it was called in the Genesee Country, was also conducted in the fields but that way more grain was lost. The heavy swing beam in an English barn permitted the circular treading of the animals without the obstruction of posts, and the wooden floor helped when it came time to collect the grain for winnowing.

FIGURE 5-5. Engraving of hand threshing.

Winnowing

After the berries had been separated from the straw, winnowing removed the chaff from the grain. The large hinged doors at either end of the drive-way were opened, allowing the prevailing winds to blow through. The strong blast of air can still be experienced on almost any day, simply by opening the doors. The winnower threw the grain, husks, chaff and straw from the basket up into the air or let the mixture fall to the floor (see figure 5-7). The funneling action of the wind allowed the lighter dust and chaff to be blown away while the grain settled back into the

FIGURE 5-6. The horses' hooves treading on the bundles separate the grain from the straw. The swing beam provides an open work area free of posts.

basket or onto a square of material on the barn floor. Repeated action was needed to clean the grain, and one was at the mercy of the elements, because a certain minimal wind velocity was needed for efficient winnowing. (It should be noted that three-bay barns built after 1825, when fanning mills came into use, might not display any orientation to the prevailing wind.)

Fanning mills were mechanical grain separators; the earliest versions were usually hand-cranked to produce a strong artificial draft from a rotary fan for the initial separation of the grain and chaff. The grain then moved over a series of shaking "riddles," or sieves, further separating the grains into various sizes which were then collected in sacks or buckets. Although fanning mills had been introduced to agriculture before the Revolution, they were seldom used in the Genesee Country until the 1820s (see figure 5-8). Fanning mills not only made winnowing independent of prevailing breezes but also easier and faster. In addition, it was felt that grain winnowed with a fanning mill was less apt to be damaged: the January 22, 1831 issue of *The Genesee Farmer* comments, "It is proper to observe, that wheat threshed by a machine, comes into market in bet-

FIGURE 5-7. Winnowing basket at the Genesee Country Museum, Mumford, New York.

ter condition than that threshed either by horses or by the flail."[7]

The winnowed grain was sacked or stored loose in the barn's granary, which was generally a separate room, located in one of the bays and characterized by a protective sheathing of tight-fitting, smooth boards on the walls as well as the floors.

FIGURE 5-8. Advertisement for a typical fanning mill. American Agriculturist, 1877.

Building An English Barn

Selection of the site for an early English barn was a major consideration. In the 1980s one thinks little of moving vast quantities of earth, but in the nineteenth century the selection of a proper natural seat was critical. If at all possible, land was selected that was properly drained and level. Later, sidehill barns were built against a bank, or low hill (see Chapter VI). If compatible with the major site conditions, a location central to the farm fields was also desirable.

While the log barn could be entirely crafted by the owner, the English barn required the special skills of a hewer. Hewers began coming to the Genesee Country in the late 1780s and early 1790s, soon after the treaties with the Iroquois permitted settlement.

An ancient craft, hewing is still practiced today. A hewer learns his craft by watching and imitating a master hewer. Paul Eshbaugh of Geneseo represented the third generation of hewers in his family. Eshbaugh spent most of his life as a civil engineer, but as a young man in the Pennsylvania oil fields he learned to hew beams for the oil derricks, which at that time were made of wood. The largest beams commonly hewed by Eshbaugh were oak sills for the drilling rigs, 16 by 18 inches and 36 feet long. He hewed in much the same way as his father and grandfather before him— in fact, he used his grandfather's broad axe, (see figure 5-9).

The following photographs reveal how the hewer of a barn frame squared up the logs into timbers and then notched and prepared the wood joints. Most of these photographs of Paul Eshbaugh were made in 1981.

In Figure 5-10, a short piece of ash (a mere stick to Eshbaugh, but weighing close to 300 pounds) is being "debarked" with a spud by Jerry Rowe of Avon. Eshbaugh is determining the size timber that can "be gotten out of the log." The log is marked out with plumb bob and square on each end and a chalk line is snapped to provide the hewing guide. A log house at the Genesee Country Museum provides the background for the photograph.

In Figure 5-11, Jerry Rowe is scoring the log with a double-bitted axe (the more uniform the depth of the scoring the smoother the timber will be), while Eshbaugh hews the log with the broad axe, removing the wood between the score marks. This heavy, ultra-sharp axe, flat-faced on the side toward the log, its handle angled to pro-

FIGURE 5-9. Basic hewing tools. Right: the slick, which, as its name indicates, is used for smoothing surfaces; it is, in effect, a two-handed plane. Left: Eshbaugh's grandfather's broad axe, with its angled handle.

FIGURE 5-10. Debarking and measuring the log.

FIGURE 5-11. Jerry Rowe (right) scores the log while Paul Eshbaugh hews away the scored section. The camera angle makes the two hewers seem closer together than they actually are. The smell of the newly hewn oak and the individual ringing of the axes biting into green wood are parts of hewing that, once experienced, are never to be forgotten.

tect the fingers on the downstroke, is capable of removing either large chunks or paper-thin slices. If, through repeated sharpening, the broad axe becomes beveled (angled) on the flat side, it is referred to as being "wood shy," meaning it will not cut as smoothly and must be faced flat again.

After the timbers are squared on all sides, they are joined without metal fasteners, usually with a mortise and tenon joint (see figure 5-12). The

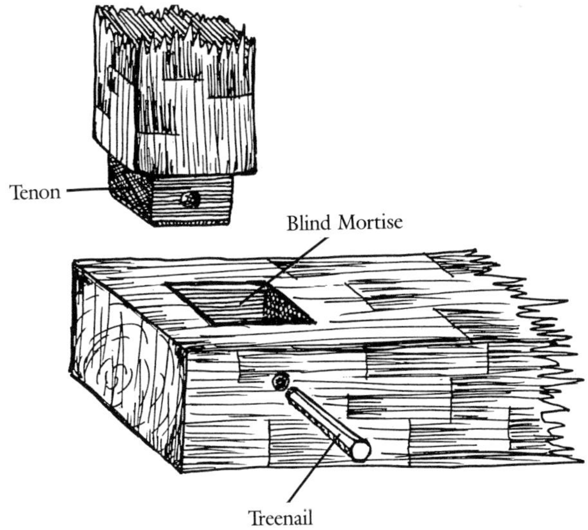

FIGURE 5-12. Mortise and Tenon joining. The tenon (tongue) fits tightly into the mortise and is held in place with the treenail.

FIGURE 5-13. Augering the mortise. The auger bit is slightly smaller than the final width of the mortise.

mortise (hole) is started with an auger of the proper diameter and depth. Two bored holes are required with a rectangular tenon (see figure 5-13). The wooden "bridge" left between the two bored holes is removed next, then the corners are squared with a corner chisel and the bottom is routed out to make it flat.

After the shoulder of the tenon is formed with a cross-cut saw, the axe is used to cut it to its final shape (see figure 5-14).

The hewer must be careful to create a tight fit between tenon and mortise. The two are joined together at assembly of the bent with a "persuader" such as a maul, beetle (a wooden mallet), or sledge. A hand-cut, many-sided treenail ("trunnel") keeps the joint from separating. Holes for the treenail are bored separately in the mortise and the tenon, with an offset of about

FIGURE 5-14. Shaping the tenon.

FIGURE 5-15. Augering the small-diameter offset hole for the treenail in the tenon.

1/8 to 1/4 inch to allow the joint to be drawn tight when the treenail is driven home (see figures 5-15 and 5-16). Some workers use augering or boring machines for the treenail holes, especially when an angle other than 90 degrees is needed. In this region the treenail was often made of locust wood, chosen for toughness and rot-resistance and dried so it would expand after being driven home in the green timber. Oak and hickory were alternatives to black locust. Expansion of the treenail, together with its undulating path, make removal difficult, as many have discovered after attempting to remove a treenail.

Any scoring marks that might interfere with a joint are removed by the hewer with a slick or an adz. In the Genesee Country the frame was smoothed only at the joints. The individual character of the scoring and hewing was and is pleasing to most eyes.

A minimum of fifty-eight mortise and tenon joints were necessary in a small English barn with four bents and a swing beam. When the hewing was completed, the various pieces were laid out on the ground and joined into bents (see figure 5-17). Next, the very important diagonal braces were "let into" the bent to provide structural stability: the triangles formed by the braces, posts and beams create an extremely rigid frame. The total time required for timber hewing and joining varied considerably, depending on the skill and number of the hewers, the character and seasoning of the wood (green is best for hewing) and, of course, the size of the structure. Figure 5-18 shows a typical swing beam bent in the Genesee Country.

FIGURE 5-16. Augering the small diameter hole in the beam.

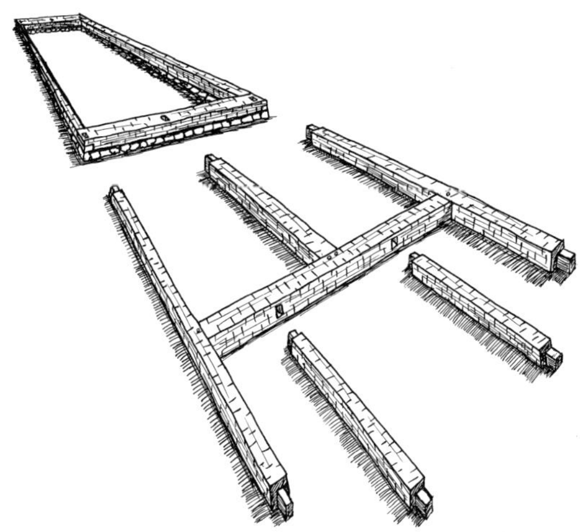

FIGURE 5-17. Assembling the bent. After the timbers are hewn and the mortise and tenons cut, the bent is assembled on the ground prior to raising.

Early contracts or articles of agreement for barn construction are difficult to locate. A rare 1816 article of agreement between Alexander Fraser and Joseph Walker for construction of a house in the Town of Caledonia provides some interesting proportions and an idea of the relative worth and difficulty of hewing. Fraser agreed to pay Walker $10 when the hewing of the beams for his house was completed and $40 when the frame was raised. The total cash cost was $235, with Fraser supplying the logs, boards and nails; he also provided room and board for the builders.[8]

While the timber was being hewed, other workers were laying the foundation, which in the late eighteenth and early nineteenth centuries was

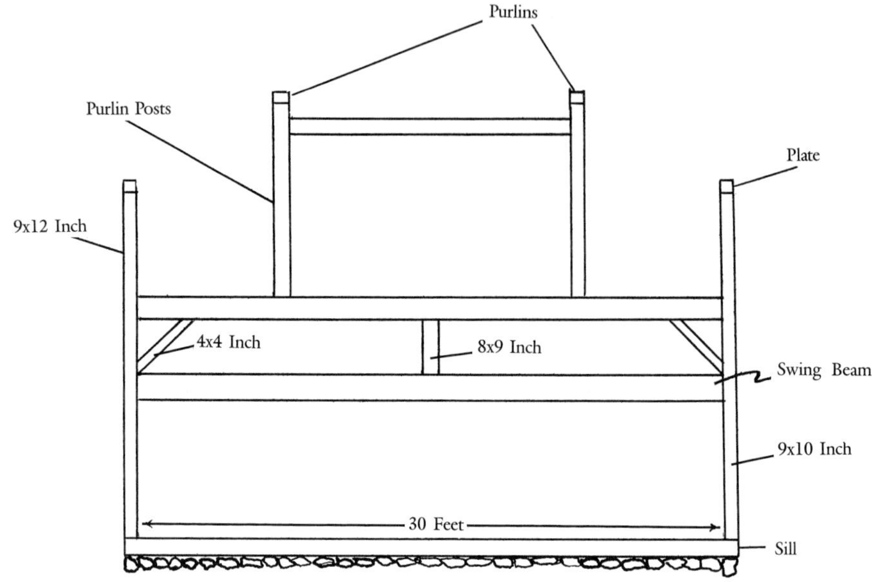

FIGURE 5-18. The swing beam bent from the Nations English barn, shown after raising.

FIGURE 5-19. Large rocks used in a typical dry foundation. Alloway, Wayne County, New York.

usually mortarless. This dry foundation often consisted of large rocks wrestled into place (see figure 5-19). Sills were laid upon the foundation stones and joined at the corners, then mortised to take the tenons of the bents. Depending on the lay of the land, early barns and houses were not elevated much by the foundation, just enough to keep the sills from rotting through contact with the soil. After each winter the barn owner, then and now, checks the foundation to correct for frost heaving and loosening of the foundation stones. If the foundation was high, the floor boards were often set in place without nailing, to give the men a surface to stand on during the raising.

Raising the Barn

To date, not one nineteenth-century photograph has surfaced of a hewn timber barn-raising in the Genesee Country. The difficulty may be due in part to the fact that photography was not invented until 1839 and was not used widely by amateurs until the 1880s; written descriptions must suffice. Henry Conklin hints at the work of barn raising in his book, *Through Poverty's Vale.*

> [In 1856] after the spring work was done, father and I hewed some timber and framed it and put up a barn for little Sammy, and he finished it off himself. When we got done for Sammy I went to work for Frank Flansburg building him a

FIGURE 5-20a. 1891 raising on the Alva Paddock farm, Kenosha, County, Wisconsin. Even though this timber-framed barn is larger than an English barn, the photograph is instructive of basic raising techniques and is one of the rare photographs that show the work actually being performed. State Historical Society of Wisconsin.

> barn also and we raised it on the fourth of July. The whole neighborhood was there from far and near, the girls included, and we had lots of fun and a splendid time.[9]

In his matter-of-fact style, Conklin never mentions the danger and difficulty of the work, but, lest one become carried away by the romance, a Mrs. Farnum of Pittstown tells of a sadder raising day: "Capt. Harmon, built a barn in 1802 or '3; at the raising, an adopted son of his, by the name of Butts, was killed outright, and Isaac Bishop was stunned, supposed to be dead. He recovered, but with the entire loss of the faculty of memory."[10]

The written descriptions of frame raisings from the early periods tend to be sketchy. In its spare way, Thayer Gauss's 1838 diary records the replacement of one of his barns which burned in East Bloomfield. Fortunately, J. Howard Pratt wrote in detail about a timber-frame barn raising around the turn of the twentieth century in his 1978 book, *Memories of Life on the Ridge*. Although the frame was sawn, the basic design and construction of this gable-roofed barn were typical of earlier English barns. Early in this century, John Henry Pratt, Howard Pratt's father, bought a farm without a barn and contracted for a new one with Jay Putnam, a barnwright and carpenter. Putnam did not need drawings; he knew the requirements and after some figuring produced a list of materials needed for a timber-framed barn without a basement. On a level piece of ground the foundation was laid, and Pratt describes what followed:

> The working platform is now clear of beams and ready for the frame of the first bent. For each bent three upright posts (8x8 inches) were used, set perpendicular to the sills (10x10 inches), and across the tops of these another beam was mortised. ..All the framing was done on the work platform in front of the barn and when the timbers were completely fashioned they were carried one by one and joined together at one end of the barn. The foot of each post was placed on the sill ready for the elevation of the other end.
>
> As the timbers of each bent were mortised and made ready they were placed on the preceding bent and united. When all were framed they appeared as if you had set up six dominoes in a close row and pushed them over. Number one lies partly on number two, number two lies partly on number three, etc. Now comes the "raising of the frame of the barn".
>
> Barn raising was always the work of many hands . . . the news spread throughout the neighborhood that at 1 o'clock the barn was to go up. ..Jay was the commander of the whole crew. He told each man what to do, when to lift and how much. First, three men were placed, each one with a crowbar, at the foot of the posts. The end of the bar was placed in the mortise cavity to prevent the tenon on the post slipping. Several men would lift on the top part of the three posts. They would raise it a foot then it would be blocked in its position. At each post a man always followed the elevating post with a block. Now Jay would shout, "Another foot, boys," and all would lift the beam one foot. Remember, they were bringing their top up; the bottom end remained stationary (see figure 5-20a & 5-20b).
>
> "Another foot, boys," would raise the posts again. When the top reached the height of six feet, pike poles were used. These were poles with an iron prong fastened in the end. Some were eight and some were twelve feet long. The iron point was jammed into the underside of the post being elevated, and the men pushed on the pole. "Another foot, boys!" and the beam rose the stated distance. Remember, someone always followed the post, as it rose, with a block so that the men were not holding the weight all the time.
>
> When the post reached an angle of 50 degrees, Jay shouted to his regular helpers, "Put on the stay boards, boys, and guy ropes!" Now the stay boards were nailed by one nail about the middle of the posts and were used to hold the posts upright when they became vertical. The guy ropes were used to keep the frame from falling over when the vertical point was reached. Now the elevating was slower and more cautious. Jay checked on the three men with the

FIGURE 5-20b. Detail of Paddock barn raising. This period photograph could hardly be better if the raisers had been actors who were coached into an optimum position. As the second-to-last bent is raised, men at position #1 use steel bars to guide the tenon into the mortise—a slip here could be fatal. At this point more than thirty men push with their arms and pike poles to raise the bent. At position #2 a man stands ready to guide a tenoned girt into the mortised bent post. When one or more treenails are driven home the bent will become stable and no longer a threat to those who muscled it into position. Two men have climbed the bent to the right and are ready to help with moving the second girt. Photograph from the State Historical Society of Wisconsin.

FIGURE 5-21. Raising the plate onto a timber frame in the East Bloomfield area at the turn of the century. If all posts and beams, tenons and mortises have been measured and prepared properly, the tenons will slip into the mortised plates without much persuading. Collection of the East Bloomfield Historical Society.

crowbars. Were they holding the foot of the posts in place? . . . "Keep those guy ropes tight and don't let the posts fall over!" Jay shouted. The tenons on the foot of the posts were entering the mortises on the sills guided by the iron bars. "Six inches more and she will be up," . . . He grabbed his level, and placed it against the corner post. "Just an inch more. Now another inch. There, that's perfect! Nail that stay board, Jim. Put three more nails into each end." Then he ran to the other corner post and checked to see if it was perpendicular to the sill. Another shove or two made it perfect, and the stay on the post was nailed securely. The center post was checked and securely stayed. The first and most difficult frame was erected.

The second frame went up much faster as the men knew what to do and when to do it. As it neared the vertical position a girt was placed in a mortise on the first corner post and the opposite end held up to enter the mortise of the second raising post. It went into place as Jay checked to see if the second frame was perpendicular. Then his helpers nailed stay boards on, and that frame was erected. Each frame went up a little faster, the last two hardly stopped from the time the first move was made until they were in their final positions.

A rest was now declared. The men sat around and quenched their thirst or smoked a cigar. Jay and his two helpers were busy, however. The plates were moved to the side of the barn ready to be placed on top of the posts. They were not so heavy, and by the aid of ropes and the pike poles they were raised to their position on top of the posts. When these were placed in position, the mortises of the plates dropping onto the tenons, the frame was locked together. The wooden pins were driven into position and the frame was complete (see figure 5-21.)

With the frame of the barn completed, all large heavy beams were in their permanent positions. The heavy work was over. Jay and his two assistants could finish the remainder of the work. It was only a little past five o'clock . . .[11]

Figure 5-22 is a model of an English barn, showing the three typical bays; this barn has no swing beam. As in most barns, floor joists under the driveway provide support for the threshing floor.

After the raising, some thought it best to allow the frame to season in order to eliminate movement before the barn was boarded and the roof rafters erected. Rafters were cut from saplings with the up-side hewn flat to take the roof boards. As with log barns, no ridge pole was used in most cases; instead the rafters were half-notched and then bored to take a treenail at the peak. As soon as possible the barn was roofed with wooden shingles. Early photographs and drawings indicate almost no overhang at the eaves in English barns compared to barns of the 1870s.

Most of the early barns in the Genesee Country were not painted. With the abundance of virgin timber in the area wood was not as expensive as paint, and for many the result was just as pleasing. William P. Wadsworth, a descendant of the pioneer Wadsworths, recalled that his father was fond of repeating this opinion. Certainly, a poor paint job can cause the wood to deteriorate faster than if it is left to form a natural patina.

FIGURE 5-22. A model frame of an English barn clearly showing the roof rafters supported by the purlins and the floor joists under the driveway. This model was constructed around the turn of the century by James Sweetheimer of Warsaw, New York builder and building mover. Now owned by his son, Ernest.

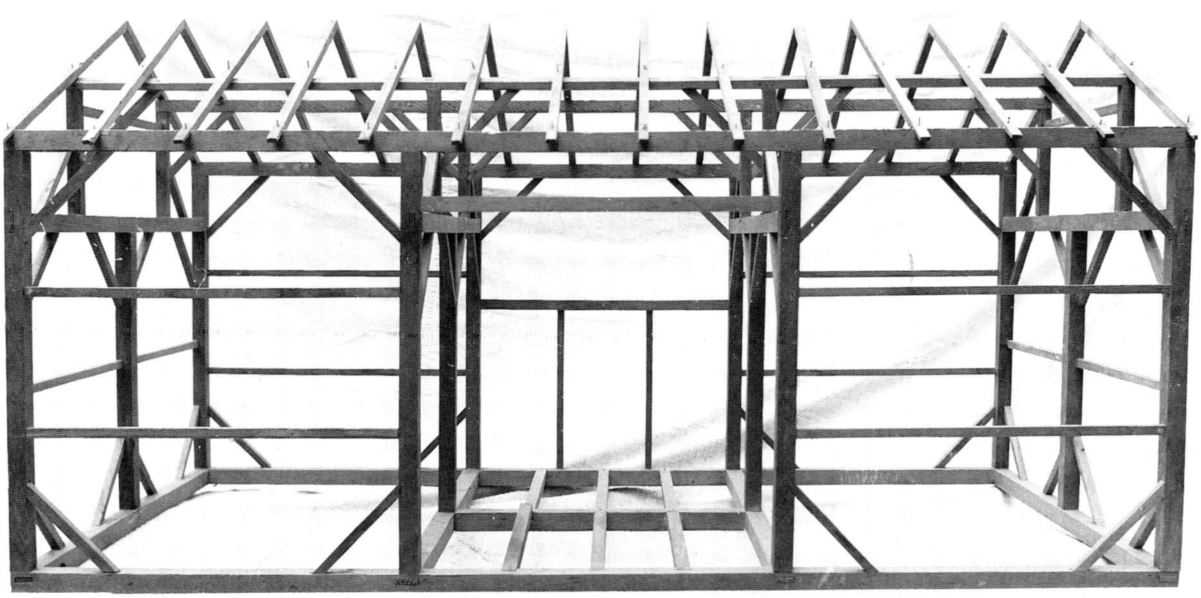

Some Typical English Barns

A barn that once stood on Lima Road in Geneseo was typical of the earliest small timber-framed barns in the area (see figure 5-23). Missing pieces of siding reveal the hewn frame beneath. According to the usual practice, this barn was roofed with hand-riven wooden shingles and—unlike present-day barns—sheathed with horizontal, random-width boards. The random-width barn boards are the result of sawing up a log so as to produce the least wastage. It is very

FIGURE 5-23. A small, early, timber-framed barn about 20 by 30 feet (now razed), once located on the Lima Road, Geneseo, New York. The hand riven shingles, shown here, were used on many barns and other farm buildings in the Genesee Country in the nineteenth century.

likely that the sheathing boards came from one of the early saw mills in the area, possibly from the mill on the Conesus Lake outlet near Ashantee, which was the closest saw mill. The pitch of the roof is fairly steep compared to later English barns.

Hand-split shingles gave a strong texture to the roofs of farm buildings; as wooden shingles are replaced by more modern materials the landscape is significantly altered (see figure 5-24).

The Nations Farm English barn, shown in the aerial photograph earlier in the chapter (figure 5-2), measures 43½ by 30½ feet with the sills resting on a dry foundation of rather large stones (see figures 5-25 and 5-26). The vertical sheathing is of random-width boards. Barns at this time were not often tight-boarded because it was clear to most that ventilation was good for both hay curing and the health of animals. The northern doors are hung on a "modern" track, while the barnyard doors are hinged in the original manner. The original wooden-shingled roof has been replaced with a metal one. Because there is little evidence of change in the interior, particularly in the frame, this English barn is of considerable interest (see figure 5-27 double page). If figure

FIGURE 5-24. Twin barns on the Fowlerville-Ashantee Road, Livingston County, reveal the significant change in appearance that occurs when wooden shingles are replaced with asphalt shingles.

FIGURE 5-25. Nations Farm complex. The English barn is the second large barn to the left of the farmhouse. Nations Road, Geneseo, Livingston County, New York.

5-27 is held so that a 90 degree angle is formed between the pages, the perspective is similar to what a person standing in the driveway might see.

As indicated, American versions of the English barn most often included a very impressive "swing" beam, big enough to support both the roof and the mow floors without the need for an intermediate post. A post would have limited the lateral free space at ground level needed for threshing with animals, or flooring. Flooring of grains in a small barn was facilitated by extending the center threshing floor through an end bay of the barn. In addition, the strength of the swing beam was sufficient to allow a temporary scaffold or loft to be built over the drive-through during the harvest season for storage of unthreshed grain or hay when the mows were full. The oak swing beam in the Nye/Hopkins barn in Pittsford measures 11 by 16 inches, the largest located in this study. The swing beam in the Nations Farm barn is 9 by 12 inches and the post that ties the swing beam to the upper beam is 9 by 10 inches (see figure 5-28). The latter post is mortised into the two beams and pinned with faceted treenails. The view from the haymow across the threshing floor shows the remains of the blind mortises into which the floor joists over the mow were placed (see figure 5-29). The barn, in the usual style of the English barn, was built without a ridge pole. Halfway between the ridge and top beam or plate are the purlins, parallel to the long axis of the barn. The purlins support the roof rafters and the purlins themselves are supported by the purlin posts.

In figure 5-30, showing the swing beam, the method of "letting" the beam into the post can be seen. Boxing or shouldering, also called a hitch, provided a narrow "shelf" to support the end of the beam, in effect taking the strain off the tenon. The degree of shouldering depended on the desire of the individual hewer or barnwright; some shoulders in other barns are several inches wider than this very narrow example. This "letting in" or boxing can more easily be seen in the diagonal brace above. The faceted shape of the treenails is visible above the end of the brace, as well as the hewer's scribe marks used in locating the augered holes. An old method of marking the various hewn members with incised numbers or symbols to permit proper assembly can be seen in figure 5-31.

The English barn on the Wade Farm on Little Road in the town of Avon is approximately the same size as the Nations Farm English barn but with several interesting differences (see figure 5-32). The barn's sills are set upon poles driven into the ground rather than on a stone foundation. (These wooden poles have since been replaced with concrete posts). Pole foundations are most resistant to heaving from the alternate freezing and thawing of the soil. In many instances the poles were charred in a fire before use to make them resistant to rot. In many English barns the original hinged doors have been replaced by newer track doors, such as those seen earlier on the Nations Farm, but the Wade barn retains one of the original hinged doors.

Both the Nations and the Wade English barns were constructed with a swing beam on one side of the driveway, but this construction was by no means universal. For example, the Merrimac Farm in Groveland has an early 28 by 48 foot English barn (see figure 5-33) with central posts on both sides of the driveway instead of the more usual swing beam. Interestingly, the post on the west side of the driveway was moved sometime in the past (see figure 5-34), considerably changing the interior space. This barn is asymmetrical, with an added bay on the western side (figure 5-33, left). In the addition, which can be located

FIGURE 5-26. Nations Farm English barn.

FIGURE 5-27. Interior of the Nations English barn. The threshing wall is to the right of the drive-through, separating the mow from the threshing floor, the swing beam to the left. The beams that parallel the drive-through support the smaller movable joists used as scaffolding to support sheaves of unthreshed grain.

To see this view from the perspective of a person standing in the driveway, hold the book so that the pages form a 90° angle.

FIGURE 5-28. Nations Farm English barn swing beam: the larger of these two beams is the swing beam, the size of which (9 by 12 inches) is designed to eliminate the need for a post in the center of the barn.

FIGURE 5-29. View across the threshing floor in the Nations Farm English barn. The swing beam is clearly shown, as are the purlins and their posts.

FIGURE 5-30. A detail of figure 5-29. Letting the swing beam into the post takes the strain off the tenon.

FIGURE 5-31. A hitch or shoulder (top left below the treenail) in the Gauss barn, slightly deeper than in the Nations barn. The system of marks used to match braces to posts and beams can be seen in the driveway bent of the 1794 Benjamin Gauss barn in Holcomb, Ontario County. Benjamin Gauss was the father of the diarist Thayer Gauss.

FIGURE 5-32. The English barn on the Wade Farm is sheathed with random-width boards and retains a hinged door. Little Road, Avon, Livingston County.

FIGURE 5-33. Merrimac Farm English barn, Groveland, Livingston County, New York.

FIGURE 5-34. Merrimac Farm English barn interior view across the threshing floor. Careful inspection of the near beam reveals the original position of the center post. Posts in this barn are typically 8 by 10 or 8 by 9 inches.

FIGURE 5-35. Merrimac Farm English barn interior showing sawn rafters in the "new" portion compared to the original rafters. Tenoned joining of the purlins and butt joining of the top plate can also be seen.

FIGURE 5-36. 1879 lithograph of the Pratt barn, Prattsburgh, New York. W.W. Clayton, History of Steuben County, 1879.

by a slight variation in the roofline, the roof rafters are sawn in contrast to the hewn original rafters (see figure 5-35). In addition, the purlin plates were tenoned into the existing mortised purlins and the girts themselves were butt-joined. Wrought iron straps were used to anchor the butt joints, a device commonly seen in the Genesee Country.

Throughout the Genesee Country, the wood-rich nature of the pioneer landscape is displayed in the types of wood used in early barns. The entire frame of Joel Pratt's barn in Prattsburgh is made of black walnut timbers. Pratt's 36 by 60 foot, four-bay English side-hill bank barn, built about 1817, is still in use on the south edge of the village (see figure 5-36; for a discussion of the bank barn see Chapter VI). The framing timbers typically measure 10 by 12 inches in cross section. The bent beams are unseamed 36-footers, all black walnut (see figure 5-37). Figure 5-38 shows a change in the framing that was made sometime in the second half of the nineteenth century to accommodate a hay track. (again, see Chapter VI) The 60-foot length of the Pratt barn required spliced longitudinal beams using scarf joints (see figure 5-39).

FIGURE 5-37. Interior of the walnut-framed Pratt barn, looking east across the central driveway toward the mow. The beams are mortised to take temporary scaffolds; a portion of such scaffolding can be seen to the extreme right.

FIGURE 5-38. At some point in the second half of the nineteenth century a hay track was added, requiring the removal of the collar beams, which tied in just below the purlin plates. The top "rung" on the ladder is all that is left of the collar beam. A diagonal brace of new lumber was later added to stabilize the structure.

FIGURE 5-39. The purlin beam in the Pratt barn is spliced with a scarf joint just to the left of one of the purlin posts, immediately above a knee brace. The sketch illustrates the nature of the scarf joint. Careful observation of the lower right side of the photograph reveals the stub of the collar beam that was removed to accommodate the hay track.

The Genesee Country Museum includes a three-bay barn in its collection of nineteenth century structures. Figure 5-40 shows this barn being rebuilt. Unfortunately, repairs to the timbers did not permit a raising of the frame in the nineteenth-century manner described by Howard Pratt.

Perhaps the simplest way of expanding the basic three-bay barn was by adding a shed (or two or three). These sheds lean on the original frame and might be found on each of three sides. The so-called "saltbox" profile is formed by the addition of a shed across the rear of a gable-roofed barn. An excellent example of this common addition is located about one mile south of Naples (see figure 5-41).

One additional small English barn with an interesting early history is the so-called "Mormon Barn," constructed in 1827 along the Canandaigua-to-Geneseo Stage Coach Route close to Lakeville and originally owned by David Wattles, postmaster (see figure 5-42). As this was the intersection of the stage routes from Canandaigua to Geneseo (east-west) and Dansville to Rochester (north-south), a number of buildings were necessary to handle the travelers' needs. In this period it was a fairly common practice for travelers with minimum funds to stay overnight in a barn, even without the owner's permission. Many farmers, including the owner of the Mormon Barn, accepted paying customers. Among the early travelers seeking lodging in the Wattles

FIGURE 5-40. A 26½ by 40 foot three-bay barn undergoing reconstruction at the Genesee Country Museum, Mumford. Originally located on Boughton Hill Road in Victor, New York, the barn was constructed in the 1820s and is of the type that lacks a swing beam.

FIGURE 5-41. A saltbox profile is created by the addition of a shed to the north side of this three-bay barn. Naples-Prattsburgh Road, Rte. 53. Additions such as these could be built by the owner, since the framing often was not as heavy as that of the original barn.

barn were Joseph Smith, prophet of the Church of the Latter Day Saints, his brother, Samuel Smith, and other Mormons travelling west through Livonia and Lakeville. They slept in the barn loft and held religious services on the barn floor. The name Mormon Barn seems appropriate, reminding us of a time when religious fervor swept the country, especially central and western New York State.

English barns continued to be built until well past the mid-nineteenth century. Eventually, hewn timbers gave way to sawn timbers in many new barns; as sawn timbers became more available, it was not uncommon to see both hewn and sawn timbers in the same barn. The large timbers seem to have been hewn on the spot, partly because of the difficulty and expense of moving logs to and from the mill. Eventually gable-roofed barns increased in size and incorporated other significant variations, making the term "English" obsolete.

FIGURE 5-42. The Mormon barn, located one mile west of Lakeville, Town of Geneseo, Livingston County, New York.

DIARY

Sheffield Winslow Peabody Springwater, Livingston County 1849–1869

FIGURE SP-1. Portrait of Sheffield Peabody made after he finished his diaries, pulled up stakes and left for the Oklahoma Territory. Collection of Betty Peabody Knoblock

S. W. PEABODY owned a farm near East Springwater (now Tabor Corners), Livingston County, New York. From the many entries in these diary volumes are some which provide an insight into Peabody's construction of two barns on his property.

Springwater, New York

1855

May 14:	*Foster and Hunt came on to build a barn . . .*
May 15:	*We drawed out timber A.M. Went over to the steam sawmill. Got some brace timber and studding.*
May 16:	*Rainy day. The boys worked in the woodhouse and the carpenters in the log barn . . .*
May 17:	*We drawed timber. Got one load of rafters from Abbey's mill.*
May 18:	*I drawed lumber and timbers. Ezra Coats scored timber for us.*

May 19:	*The carpenters have been here all the week.*
May 25:	*I went to Dansville . . . I got some nails and door hinges.*
May 26:	*I plowed and scraped for a barn foundation . . .*
May 28:	*We commenced to lay the wall under the barn.*
May 30:	*We layed wall . . .*
June 1:	*I worked on the wall. Brown layed wall.*
June 4:	*We worked on the wall.*
June 9:	*I raised my barn on the upper corner. We had plenty of help.*
June 11:	*Foster and Hunt covering barn . . . Rouse and father drawed lumber.*
June 12:	*Foster and Hunt and boys are to work on the barn.*
June 21:	*. . . Lorenzo Monk helped shingle on barn.*
July 14:	*. . . finished shingling the barn.*
July 21:	*We worked around the new barn.*
August 8:	*We fixed the bridge to the new barn. We drawed in ten loads of hay into the new barn.*

1861

July 17:	*Mr. Terry drawed timber for a barn out of the woods . . .*
July 25:	*We drawed 7 loads of hay into the log barn.*
July 26:	*We drawed 11 loads of hay into the log barn.*
July 27:	*. . . Filled the log barn.*
August 5:	*Mr. Terry and I finished drawing out timber for my barn.*
August 6:	*I plowed and scraped for a foundation for a barn.*
August 7:	*I drawed two loads of timber . . .*
August 8:	*I helped M. Crandall lay wall. Mr. Foster and two hands worked at framing the basement.*
August 9:	*. . . I raised the basement story.*
August 10:	*I pryed my foundation of a basement story of a barn, A.M. and raisd a barn, . . .*
August 14:	*We put up perlin plates on my barn.*
August 15:	*I helped put on the rafters.*
August 23:	*Jim Cole plowed and scraped dirt up to the barn doors.*
August 29:	*We drawed in some hay off the 13-acre lot into the new barn east bay.*

Diaries from collection of Betty Peabody Knoblock

SIDE-HILL BARN.

[M]ESSRS. EDITORS:—I send you a design for a side-hill barn, with an elevation, [if] you think of sufficient merit, you can give to your readers.

ELEVATION.—SIDE-HILL BARN.

The barn is designed to be built on [th]e side of a hill which will allow of [ex]cavation sufficient to form the base[m]ent floor, which contains five stalls [for] horses and six for cattle, with a [fee]ding passage between, into which is [thr]own the hay and oats through a [tra]p door in the floor above. From [thi]s passage the mangers on either [sid]e may be filled very handily and [wit]h much less trouble and less risk [of] being kicked, than when a person [ha]s to come up behind the animals to [ge]t at their heads. A may be used as [a c]arriage or waggon shed, or to store [aw]ay straw for the use of the cattle or [bed]ding for horses. R is a cellar for

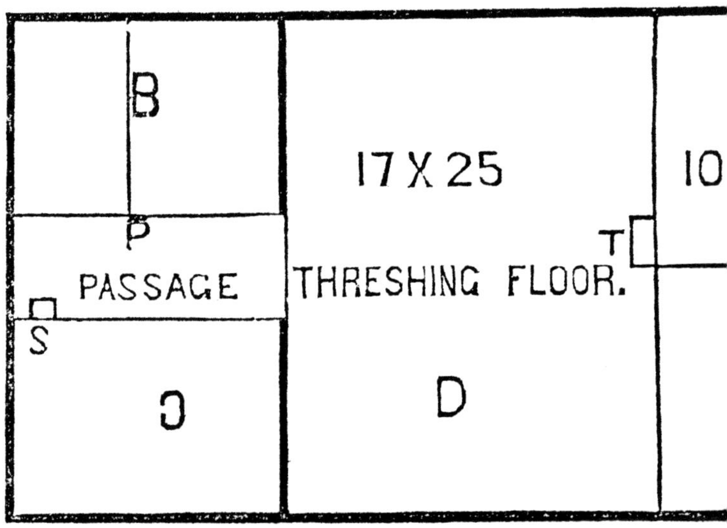

MAIN FLOOR.

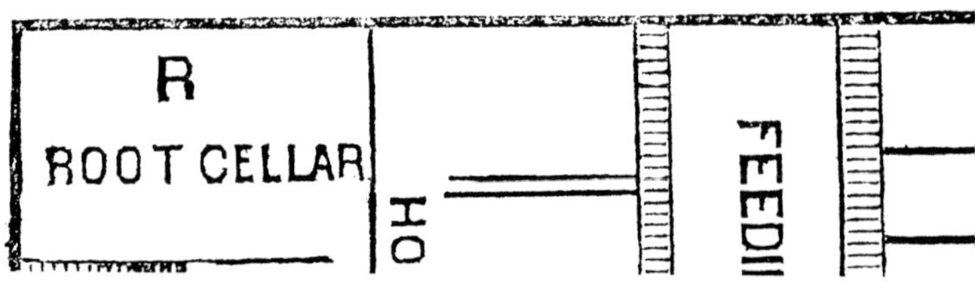

CHAPTER VI

Bank Barns and Other Enlarged Gable-Roofed Barns

Many factors contributed to the need for new barn designs; chief among them was the increasing mechanization of farm work, particularly the handling and storage of crops. Also important was improved transportation, which permitted greater access to distant markets and thus stimulated increased production. From the middle 1830s on all these pressures increased until new barn designs evolved.

Side-Hill Barns

Most of the earliest English barns were constructed on a low foundation with the sills just barely off the ground. In newer barns foundations increased in height until finally the extra height produced a basement level below the barn floor. Eventually barns were constructed against existing hillsides, permitting farm wagons easy access to both lower and upper levels. These side-hill, bank, or basement barns represent a logical evolution of the English barn in the Genesee Country; with their advent the term "English" began to disappear.

Figure 6-1 shows William P. Wadsworth, Master of Fox Hounds, leading the Genesee Valley Hunt along Jaycox Creek just north of Geneseo, with the hounds working just below the Culvert Farm. The gambrel-roofed barn dominates the photograph, but on the left is a gable-roofed side-hill barn larger than those seen in early chapters. Figure 6-2 represents a typical side-hill construction at the natural crest of a north-south striking hill near Lakeville. Wagons can be drawn up to the second level, and of course the basement level is ideally suited for cattle and horse stalls. Straw or feed can be pitched down to them without the need for climbing up to a loft.

Figure 6-3, a closeup photograph of the Culvert Farm, shows the masonry end-walls, with the sills supported on posts. This barn is an excellent example of the side-hill barn recommended in the 1840 *Cultivator's Almanac and Cabinet of Agriculture*:

> All barns should be set, if possible, over good cellars, for the purpose of letting the cattle have

Opposite: In 1852 The Genesee Farmer published a typical side-hill barn design.

FIGURE 6-1. A Thanksgiving Day hunt on the Culvert Farm, Nations Road, Geneseo, Livingston County, New York. The barn on the left is a 46-foot gable-roofed sidehill barn.

FIGURE 6-2. A side-hill barn one mile west of Lakeville, New York.

FIGURE 6-3. A side-hill or bank English barn on the Culvert Farm, Geneseo, Livingston County, New York.

shelter under them, or of letting the hay fall down below the floor for the greater ease of unloading. A good cellar under a forty foot barn is easily provided by a farmer, when he is building, if he has a good team and rocks in plenty.

Side hills, or sloping grounds, should be chosen for the purpose, so that water may not stand in the cellar, and that the loaded hay carts may be easily driven in above. A large bay, twenty feet square, and eight feet deep, below the floor, will hold an immense quantity of hay— one foot in height holding one ton of hay of the solid kind—and the carter is enabled to throw off his loads without any assistance, at any busy time in the day when help is sometimes worth half a dollar an hour.[1]

Some barns originally constructed on a low foundation were later elevated by careful jacking of the entire structure. A basement could then be created by building a new foundation under the raised structure. In fact, one F.W.L. of Monroe County wrote in the June, 1866 *American Agriculturist,*

> We see in all parts of the country the old-fashioned 30 x 40 foot barn, built years ago, with a hay mow at one end, stables and granary with loft over them at the other, and barn floor between. Such a barn is almost no barn at all, and having such a one in tolerable repair, I set myself to improve it without going into great expense. With a couple of Jack screws we raised it up eight feet from the ground, underpinned it with a good wall, and added 20 feet to the length at the end used for stabling, making the basement of the building 30 x 60 feet.

Figure 6-4, taken from an 1852 issue of *The Genesee Farmer,* shows a barn design very similar to the Culvert barn, except that the latter has equal-sized mows to the right and left of the drive-through. The designer of the side-hill barn estimated the cost at \$300–\$400.[2]

Figure 6-5 shows the sawn timber frame of a three-bay side-hill barn in Henrietta, looking much as it must have following its raising. Even though there is not much head room in the basement, it is clearly possible to enter the two levels from the ground. At the time of the photograph the frame was nearly intact except for the absence of the beam connecting the purlin posts on the near bent (the original configuration can be seen in the far bent).

While the exposed frame in Henrietta seems to have undergone little change, the Culvert barn in Geneseo has undergone many changes in its long

THE GENESEE FARMER.

SIDE-HILL BARN.

MESSRS. EDITORS:—I send you a design for a side-hill barn, with an elevation, which, if you think of sufficient merit, you can give to your readers.

ELEVATION.—SIDE-HILL BARN.

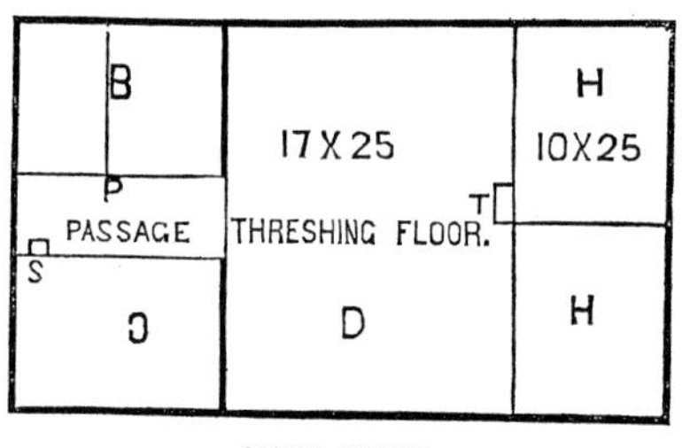

MAIN FLOOR.

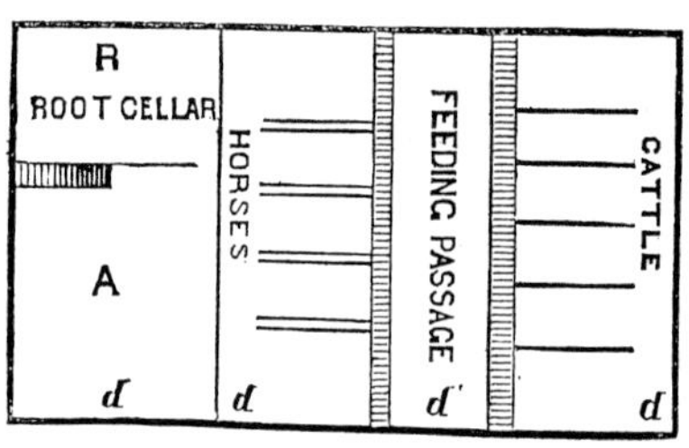

BASEMENT FLOOR.

The barn is designed to be built on the side of a hill which will allow of excavation sufficient to form the basement floor, which contains five stalls for horses and six for cattle, with a feeding passage between, into which is thrown the hay and oats through a trap door in the floor above. From this passage the mangers on either side may be filled very handily and with much less trouble and less risk of being kicked, than when a person has to come up behind the animals to get at their heads. A may be used as a carriage or waggon shed, or to store away straw for the use of the cattle or bedding for horses. R is a cellar for roots, &c.

On the main floor, H is a place for hay and oats. T, trap-door communicating with the feeding passage below. D, threshing floor, 17 by 25. C and B, cribs for corn and other grain. P, passage. S, steps leading into the basement.

The cost of this barn would probably be between three and four hundred dollars, and perhaps is as good and convenient as can be built for that sum. At least the plan is worthy of consideration. J. W. G.—*Hillsboro, Ohio.*

FIGURE 6-4. Side-hill barn design. The Genesee Farmer, January, 1852.

FIGURE 6-5. A side-hill barn frame at the northwest corner of Middle and Erie Station Roads in Henrietta, New York. The frame was razed in September, 1985.

life; indeed, the framing reflects much of the history of the barn (see figure 6-6). The person who hewed the main beams was rather careful, as indicated by the scoring marks and the squareness of the timbers. The roof rafters were hewn on one side and the bark left on the other, a common early practice. The roof boards, some of which have been replaced, bear the marks of a circular saw. The collar beams were removed long ago to allow the use of an iron hay track. This change made hay storage easier but created a structural problem. Loose hay compacts so tightly that it exerts tremendous outward pressure on the barn frame and walls, which can cause great damage, especially if the frame is no longer intact. To restore the integrity of the frame, diagonal braces were added; the new braces are easily identifiable because they are smooth-sawn and full of nails. The collar beams themselves were eventually replaced with iron tie rods which can be seen as a dark line paralleling the horizontal plate beams in figure 6-6.

FIGURE 6-6. Culvert Farm interior looking from one mow across the drive-through to the opposite mow. The hay track can still be seen along the ridge of the roof.

The Hay Track Changes Barn Design

Without a doubt one of the most significant improvements of the mid-nineteenth century, both for existing and for newly built barns, was the hay track, also known as a horse fork or horse track. Although the hay track was not patented until the final years of the Civil War, it and unpatented predecessors affected many standing timber-framed barns. In 1851, *The Genesee Farmer* carried a description of "The Improved Hay-Fork" (see figure 6-7), an unpatented device for easing the work of raising loose hay to the mow. A later development included a track along which a car could run. The horse fork allowed the hay load to be placed almost anywhere along the longitudinal axis of the barn.

FIGURE 6-7. *Horse Fork. The Genesee Farmer, July, 1851.*

Most systems employed a rigid track, which required careful maneuvering of the load on its way to the mow in order to clear the upper transverse beams of the frame. Often entire loads were prematurely dropped due to interference from the beams, which were therefore often cut, making use of the carrier easier but upsetting the structural integrity of the barn. Figure 6-8 shows mortised purlin posts from which the cross beams have been carefully removed. The frame of this bank barn, located eight miles south of Naples, is remarkably sound considering the alteration to the frame. The illustration, taken from Asher & Adams' 1876 *Pictorial Album of American Industry*, shows a Nellis' Patent Cable System for transporting hay, a method of hay handling not unlike the hay track but designed to avoid cutting the barn frame (see figure 6-9). The figure shows the hay being neatly transported past transverse beams, a special feature of the Nellis Cable System.

In 1864, A. J. Nellis began improvements on the Harpoon Hay Fork, invented and patented by Edward L. Walker. Nellis was eventually awarded many patents and purchased many others, at one time enjoying a virtual monopoly on the production of Harpoon Hay Forks. In 1892, Benjamin Butterworth showed various kinds of hay forks, including the "Tilting Horse Fork," in *The Growth of Industrial Art*. Figure 6-10 shows a Tilting Horse Fork from 1870 and figure 6-11, a Harpoon Horse Fork from 1884.

FIGURE 6-8. Altered frame in a bank barn to accommodate the operation of a horse fork. Route 53 south of Naples, Steuben County, New York. Razed in 1986.

FIGURE 6-9. Nellis' patent method of conveying hay, straw, etc. Asher & Adams, Pictorial Album of American Industry, 1876.

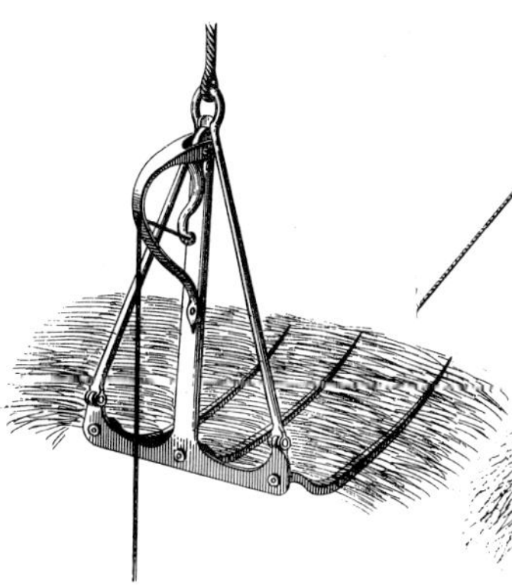

FIGURE 6-10.
Tilting Horse Fork

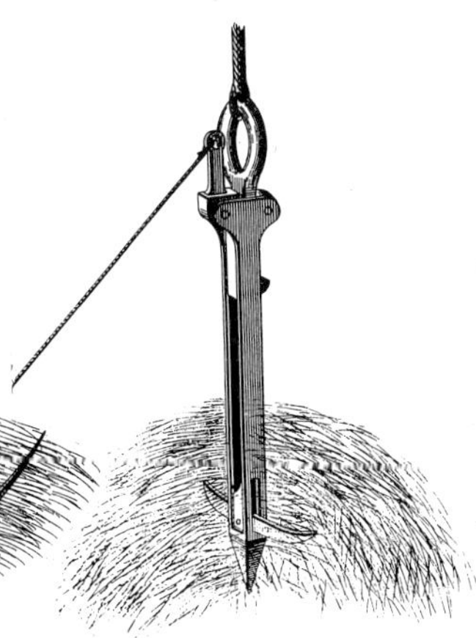

FIGURE 6-11.
Harpoon Horse Fork

An article titled "Hay Harvesting Machinery," published in the July, 1877 issue of the *American Agriculturist,* emphasizes the importance of the hay track or carrier:

> The labor of harvesting grain and hay has been more facilitated by machinery than any other farm work. But few young farmers can remember the slow and tiresome sickle, or the laborious cradle, with which grain was formerly reaped, or the scythe and rake with which hay was slowly gathered. The splendid harvesting machinery now in use in the field, is a lasting honor to American inventors and mechanics, who have originated and constructed so great a variety of it, and a remarkable help to the farmers whose labors are so greatly eased by it. But while

the labor, in the field has been simplified, the means for securing hay in the barn have been to a great extent neglected. The majority of farmers who have reapers, mowers, and rakes, still neglect the horsefork by which the time and labor of unloading are lessened more than three-fourths. This smaller service of unloading is of equal importance with the greater one of harvesting, for a whole crop may be endangered through delay in getting it under cover. A very excellent hay carrier is . . . the Anti-Friction Hay Carrier (see figure 6-12). It is made to run upon a wooden track which is fastened to each pair of rafters. The carrier and fork can be used to stack hay in the field as well as to put it away in the mow. A hay or straw fork, the Noyes Grapple Fork, which we regret we have not space to illustrate, and which may be used to raise hay, straw, chaff, clover seed, manure, or any other small stuff is also made by the same manufacturers, and is a very effective and desirable implement to use with this carrier.[3]

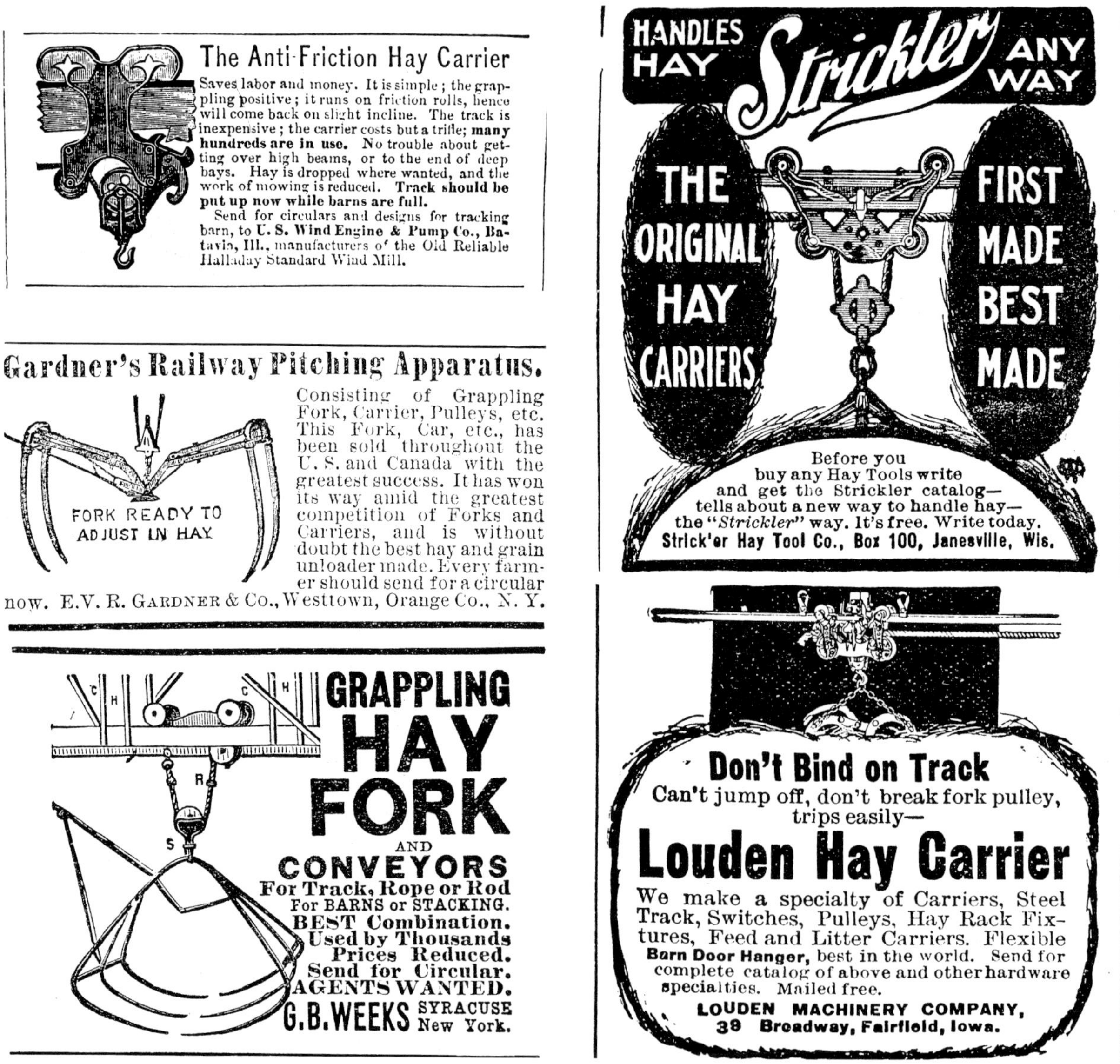

FIGURE 6-12. The Anti-Friction Hay Carrier and other cars as depicted in period advertisements from various agricultural periodicals and newspapers.

Eventually many companies were involved in the manufacture of hay tracks and their accessories. Gardner Weeks of Syracuse produced what he called his Grappling Hay Forks and Railway Hay Conveyers. In an advertising pamphlet around 1878 (see figure 6-13), the engravings depict haying in full swing, with the Weeks hay handling devices being used in the field as well as in the barn. Weeks claims that the entire cost of the poles, ropes, attaching grapple, stacking irons and fork is less than forty dollars (see figure 6-12 for a close-up of the grapple).

The operations depicted at the barn are further illustrated in figure 6-14; the first picture shows the wagon being unloaded in the driveway and the second, the wagon drawn up under the large door of the barn's gable end. The pamphlet claims:

> The track is free from the swaying or sagging of a rope or rod. It is firm, and the fork full is kept above obstructions. Rails of wood are more durable than a rope, and less costly than a rope or rod. The car is more easily shifted from one direction to another, and from one building to another. It is more convenient in use, as no extra operating cord is needed [an important factor]. The construction of the Car is such as to admit of elevating at any point over mow or way, or on outside of building.[4]

FIGURE 6-13. Engraving from a pamphlet advertising the Weeks hay products. Applications in both field and barn use are illustrated.

Rails of wood are more durable than a rope, and less costly than a rope or rod. The car is more easily shifted from one direction to another, and from one building to another. It is more convenient in use, as no extra operating cord is needed. It is more economical of time, as it is positive in its action, and causes no delay. No Pitching Apparatus is complete without embracing the carrying principle. Elevating alone is not sufficient.

FIGURE 6-14. Further illustrations and information concerning Gardner Weeks' Grappling Hay Forks and Railway Hay Conveyers.

Weeks discusses the construction and alteration of barns to accommodate the hay track:

> The success of the Railway Pitching Apparatus [hay track] and its great economy suggest the propriety of adapting and constructing barns for its use. A majority of barns will admit it with little or no change, requiring only the removal of purlin beams, which are usually not essential to the structure.
>
> Many are now building such barns, so as to take all hay, grain and litter into the loft from the outside by use of the Railway. It is desirable that there be a clear space eight feet in width and ten feet downward from the peak, and doors which shall give an opening in one of the gable ends, eight feet wide and eight feet under the track. To make sure of the latter there should be no cross beam nearer than ten feet from the peak.
>
> Above are outlined some of the more common modes of arranging (ca. 1878) the inside timbers of barns. In filling buildings [designed] in this manner, cross beams are scarcely an objection if they are ten feet below the peak of the roof.
>
> Driveways lengthwise in the peak of barns are highly objectionable, and for several reasons: They are usually very costly; it is hard getting the load in and tedious getting the team out, if it is backed out; it is difficult in such case to employ other than hand pitching, and the material is apt to be left in bad shape for mowing.

Indeed, driveways may be dispensed with in most barns, if only used for the purpose of filling.[5]

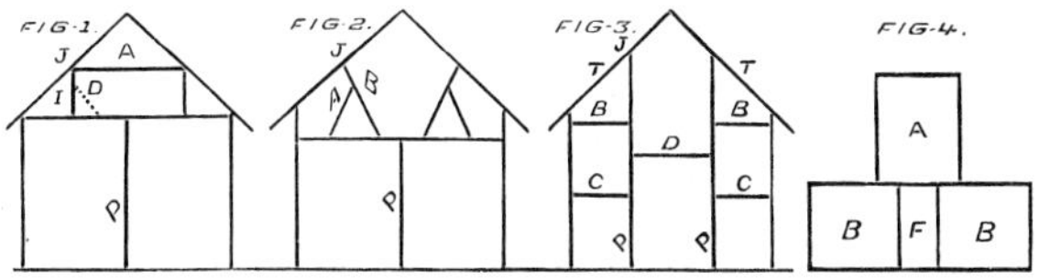

FIGURE 6-15. *Some of the more common modes of arranging the inside timbers of barns according to Weeks, circa 1878. In filling barns designed in this manner cross beams [collar beams] are not objectionable if located ten feet below the peak of the roof.*

Figure 6-16, one half of a stereo view, shows a horse fork in use. With an eye toward economy both the hay fork operation and a bull are shown in this custom-made stereograph. The team provides the power for operating the hay fork. With the most primitive horseforks the load was only lifted to the peak and had to be forked into the mow by hand. The cable system and later track systems, together with grapples and slings, allowed movement of the load right to the mow, and ultimately did lead to a change in the design of barn frames, as Weeks suggested.

FIGURE 6-16. *Horse fork operation, probably in the Moravia area northeast of Ithaca. Stereo card, Tuthill & Teed, Moravia, Cayuga County, New York.*

Robert F. Halsted wrote in a June, 1981 issue of *The Livonia Gazette* this revealing description of his experience with the hay track on the family farm:

> Hung just below the ridge beam of the roof by steel eye bolts was a rectangular "track" made of 4x4 maple that ran the full length of the barn. Suspended on the track by means of flat-surfaced steel wheels running along the top surfaces of the track was a "car"—a steel device with latches and pulleys built into it. It was this truck or traveling "car" that enabled hay or unthreshed grain to be unloaded from a wagon or truck in large quantities, relatively fast, and with minimal human effort. For years all the hay, etc. was handled with pitchforks here, even though the track and car were in place. I think the expense of the long lengths of rope plus the cost of the automatic fork, grapples, or slings kept the device from being used sooner; but, the necessary materials were finally obtained in the late 1930s and this certainly made the job easier.
>
> The "car" had an automatic latch that would lock it to the track when it was pulled to the center of the barn directly over the load of hay. Suspended from the car by long ropes and pulleys was the grapple, which consisted of four long gently curved hooks attached to one another at their tops. These hooks would be forced down deep into the hay load and locked into position. Then, a tractor or team of horses was hitched to the main hoisting rope, which was run through pulleys to a final one attached three feet off the floor. The rope was then extended out through the doorway. As the tractor or team was driven away from the barn, the grapples would lift perhaps half the load of hay at once and it would ascend straight upward toward the "car," which was still latched in place. When the load reached the "car," it would hit an automatic latch release, unlock the "car," and "car," load, and all would go whizzing along the track toward the end of the barn over the mow. When it reached the point where the load should be dropped, a special trip rope was pulled releasing the grapple. The load dropped into the mow, at

which point someone signaled the tractor to stop pulling and back up. The mechanism could then all be pulled back to the starting point and reset. I have used a lot of words to describe this unloading method, but after spending years grunting as I loaded many tons of hay into the mow—forkful-by-forkful—the grapple was certainly a wonderful thing to me.[6]

The Reaper Leads to Larger Barns

The two traditional staples of agriculture have been hay and grains, especially wheat, and both are well-suited to the Genesee Country. In fact, one of the major influences on barn design was the need for proper storage space for loose hay and both threshed and unthreshed grain, plus space for the work of threshing. Major changes in reaping and threshing during the nineteenth century changed the configuration of new barns, as improved mowing machinery permitted larger

FIGURE 6-17. Cutting grain with a hand sickle. In the distance stands the barn where the bundles will be stored and threshed, while to the right stands a mill where the grain will be ground. 1855 Lehigh County Agricultural Society certificate. Allentown, Pennsylvania.

quantities of hay and grain to be cut all at once, creating a need for more storage space.

Before examining the changes that reaping brought about in barn design, it is proper to discuss the evolution from hand to machine reaping. One of the oldest tools used for reaping is the hand sickle. In figure 6-17, workers are shown cutting grain with sickles while others are bundling and stacking the wheat in shocks. Great care had to be exercised with a sickle or the grain would be knocked from the head and lost in the field.

Figure 6-18 shows a long-handled scythe used in the mid-nineteenth century for haying. An improved scythe, known as the cradle or cradle scythe, was also widely used from the beginning of the nineteenth century for cutting grain and laying it in windrows (long rows). A series of lightweight wooden "fingers" paralleling the blade "cradled" the cut plants, allowing them to be gently lowered to the ground for bundling (see figure 6-19).

Many tried to perfect a mechanical reaper to replace the sickle and cradle. For twenty-one years Robert McCormick, father of Cyrus, tried to construct and market a reaper. In 1831, at the point of Robert's greatest despair and much to his displeasure, Cyrus began his own efforts to construct a mechanical reaper.[7] In 1834 Cyrus was granted a patent on his "Virginia" reaper with its serrated cutting bar, but farmers were not easily convinced of the value of a reaper and demonstrations were necessary. Usually conducted at county and state fairs and agricultural implement shows, public demonstrations are still considered essential to the sale of new farm machinery. A

FIGURE 6-18. Cutting grain with a scythe.

public competition between McCormick's machine and a reaper invented by Obed Hussey was held at Richmond, Virginia in 1843.[8] McCormick's reaper was declared the winner of the competition (Hussey's machine had problems cutting wet grain) and furthermore was much less expensive; McCormick sold twenty-nine reapers in 1843 for $100 apiece.

Historian Neil McNall cites a letter to William W. Wadsworth around 1845: "Not seeing you today I write this line to say that we got the reaper started formally to cut on the flats of Messrs. Mills at Mt. Morris and cut 12 acres in 6 hours in the neatest possible manner . . . Everyone pleased who saw it. Can you or your brother go over?"[9] Earlier, around 1835 or 1836, William W. Wadsworth designed a reaper of his own, but by all accounts it was not entirely successful.[10]

The first reapers required two people to operate them— one to drive and one to rake the cut grain from the platform— plus a number of people on the ground to bind the sheaves. Ketchum's Combined Harvester, shown in the 1858 *Genesee Farmer,* is a good example of a reaper that required an extra person to rake the platform clean (see figure 6-20).

Competition between Hussey's and McCormick's machines continued through midcentury. Joseph Hall, of Rochester, writes of his experience with both machines in the June, 1851 *Genesee Farmer:*

> I have been induced to take an agency for the sale of the Virginia Reaper, believing it to be the best machine for cutting wheat in use, and have determined to place one upon my own farm, after having purchased and used one of Hussey's Machines, and thrown it aside upon finding there was no economy in using it; it being necessary that the grain should be bound as fast as cut, requiring more help at one time than at others, to keep up with it; occasioning the stoppage of the machine at times, while at other times the men were not fully employed. Then again, the difficulty of cutting thin wheat, the wear upon horse flesh, and the impossibility of cutting at all

FIGURE 6-19. The cradle in use: the stalks cut with the blade (bottom of the cradle) are held by the "fingers" of the cradle and dumped gently to form a pile, which is then bound into a sheave. American Agriculturist, July, 1884.

FIGURE 6-20. Ketchum's Combined Harvester, The Genesee Farmer, 1858.

(as it has no reel)[11] when the wind is blowing with you, destroys, in my judgment, the utility of the machine.[12]

E. B. Holmes, congressional representative from Brockport, convinced Cyrus McCormick that the Globe Works of Seymour and Morgan could handle the manufacture of his reaper, and in 1846 Seymour and Morgan out-shopped 100 of these early reapers. As late as 1894 the Seymour & Morgan works were still advertised as: "the oldest Reaper Factory in the World."[13] Seymour and Morgan manufactured the McCormick machine through the 1848 season and in 1849 introduced their own machine, "The New Yorker," based on the McCormick reaper, with many modifications. McCormick felt there weren't enough modifications and fought an extended battle to force Seymour & Morgan to stop production. The introduction of the Seymour & Morgan "Triumph" reaper in 1873 ended the problem (see Chapter XIII for more on this subject).

McCormick increased production of his improved Patented Virginia reaper from 100 in 1846 to more than 1600 in 1849. By then McCormick had his own factory in Chicago. In an advertisement in the July, 1850, *Genesee Farmer,* McCormick states:

> One of the Machines, as lately improved, cut, in the last harvest, 300 acres, without a shilling's repair, and without sharpening or changing its sickle, and, without clogging in either green or wet grass, grain or weeds. Two of the Chicago Reapers [reference is to improvements made in Chicago to the Virginia Reaper] were sent to the Genesee Country for the last harvest, one of which was sold to Mr. Martin, of Victor, Ontario

> Co., who was so well pleased with it that he paid for it immediately after harvest, without waiting until December, the time upon which it was sold. The other was sold to Mr. Case, of Fowlerville, Livingston Co.—where several other Machines, manufactured at that place, without the latest improvements, were sold—and found so much superior to all other Machines, that this gave general satisfaction, to the purchasers of which he would refer. The Reaper will be delivered in Buffalo or Rochester—$30 payable on delivery, and $80 on the 1st of December thereafter, with interest; or for $105, cash, which is $10 less than the price at Chicago. It will be warranted to cut two acres of wheat or other small grain, in an hour, on tolerably smooth ground . . . and to save three-fourths of the grain lost by cradling . . . it will be warranted to be well made, of good material and durable. To be returned..if on trial by the purchaser, it fails to perform as warranted (see figure 6-21)[14].

After repeated claim and counterclaim, *The Genesee Farmer*'s January, 1853 editorial page remarked "we have always found that the farmer that used either Hussey's or McCormick's machine) pronounced it the *best machine*."[15]

Seymour and Morgan introduced the first "self-raking" New Yorker reapers in quantity for the 1854 season[16] (see figure 6-22). The rake mechanism periodically swept the quadrant platform clear of cut grain, eliminating one operator from the reaper.

As reapers became more complicated, it became essential that they be repairable without a lengthy downtime. Many early reapers had custom-fitted parts and required considerable time for repair. Walter Wood, of Hoosick Falls, in 1859 produced one of the first machines to offer interchangeable parts (see figure 6-23). Wood's own first design with custom fitted parts had been introduced in 1856. Wood's new machine could be repaired as soon as the replacement part was obtained.

Timeliness in reaping is essential to maintaining the quality of the grain (see figure 6-26): If it is cut too late, many of the kernels can be lost in the field. Improved mechanical reapers permitted larger quantities to be harvested at just the right moment. More acres could thus be planted with no loss of quality, creating a need for larger barns to store these bigger crops.

FIGURE 6-21. McCormick's Patent Virginia Reaping Machine. The Genesee Farmer, July, 1850.

FIGURE 6-22. A "New York" self-raking reaper manufactured by Seymour & Morgan, Brockport, New York. Circa 1854. The platform rake is shown just as it is about to sweep the platform clean.

FIGURE 6-23. The Wood Self-Rake Reaper, shown from the left. The machine could be repaired right on the farm. Circa 1860.

FIGURE 6-24. Self raking reaper in use on the Weidman farm, Springwater, New York. Note the basement barn. Smith, History of Livingston County, 1881.

FIGURE 6-25. Harvest scene showing reaping, raking, bundling and shocking. Farm Festivals, Carleton, 1881.

FIGURE 6-26. Hurrying to the barn before the storm. American Agriculturist, September, 1884.

Enlarging the English Barn Design

When a farmer needed a larger barn, it made sense for him to simply expand the basic design that had been dominant in the region—the English barn. The agricultural literature of the second half of the nineteenth century is filled with articles on how to enlarge a "common" or English barn (see figures 6-27 and 6-28). Larger barns were created primarily by increasing the width and height of the barn and by adding more bays to the length. A good example of an enlarged English barn design is the main hay barn of the Wadsworth Big Tree Farm, near the Genesee River in Geneseo (see figures 6-29 and 6-30). The 36 by 65 foot Big Tree barn was enlarged mainly by increasing its height and by adding a fourth bay. The bents were designed to provide an additional level and the second beam from the floor level is almost 17 feet above the barn floor. This added height brought problems: Pitching loose hay from wagons to the barn's third story required several men with pitchforks stationed at various levels to move the hay onto the platforms and into the mows. These difficulties in handling the hay made this barn a prime

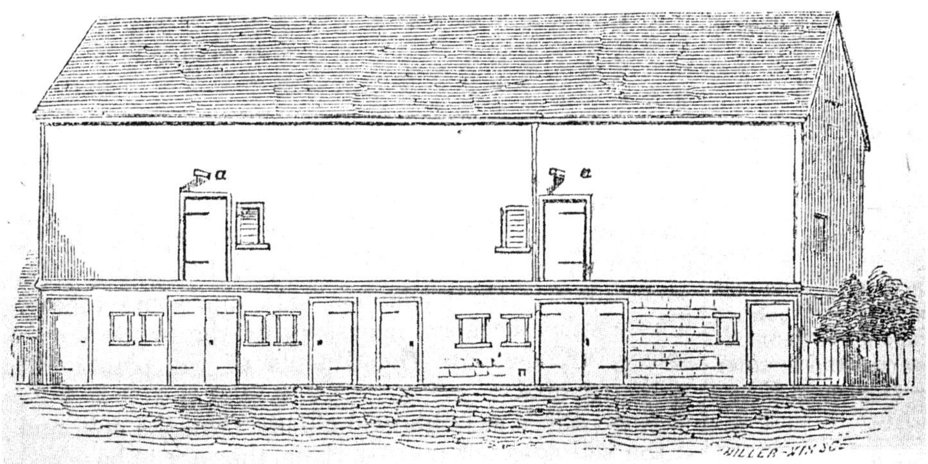

Fig. 1. ELEVATION OF A BASEMENT BARN.

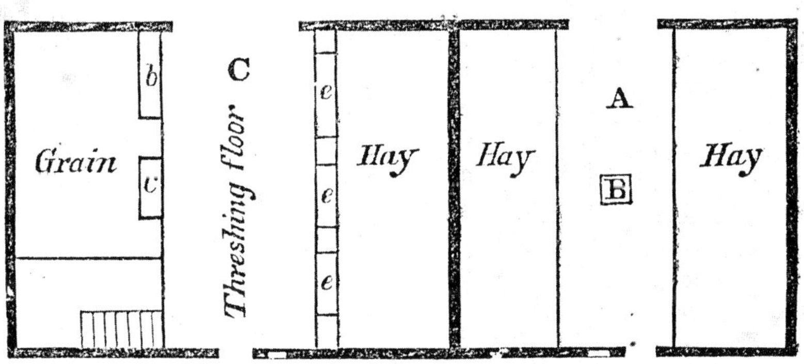

Fig. 2. PLAN OF THE MAIN FLOOR

FIGURE 6-27. Basement barn, The Genesee Farmer, February, 1851.

FIGURE 6-28. Enlarged basement barn in Henrietta, Monroe County, New York. Reeve Farm circa 1880. McIntosh, History of Monroe County, 1877.

FIGURE 6-29. Aerial view of the Wadsworth Big Tree Farm, on the Genesee River Flats, Geneseo, New York. The largest gable-roofed barn on the left is the enlarged barn.

FIGURE 6-30. Exterior of the Big Tree Farm main hay barn. Smaller barns combine with the main barn to make a U-shaped enclosure for the barn yard.

FIGURE 6-31. Interior of the Big Tree Farm main hay barn. The view is from the east end mow looking over the drive-through past the granary to the far wall. The hewn timbers in this barn exhibit some of the smoothest hewing one is likely to see.

FIGURE 6-32. Interior of the Big Tree Farm main hay barn. The view along the drive-through shows the added bracing, which compensates for the loss of the collar beams removed during the installation of a hay track.

FIGURE 6-33. Enlarged main barn on Pine Tavern Corner, Leicester, Livingston County, New York.

candidate for the hay track, which in its turn necessitated structural reinforcement (see figures 6-31 and 6-32).

The Big Tree barn is newer than those seen so far, judging by its size, its roof rafters, and—perhaps most important—its board and batten sheathing, which was popular from the 1840s through the 1870s. Details suggest that the barn was built in the mid-nineteenth century but no paper record seems to have survived.

While increasing the number of levels caused the most significant expansion in the size of English barns, adding additional bays was also a useful way of enlarging a barn. Careful examination of this barn on Pine Tavern Corner in Leicester shows a break in the roof where an additional bay was added to the original three-bay structure. The original barn had a centered cupola (see figure 6-33).

Another example of an added bay is the side-hill basement barn built by Robert Ennis in Alloway. Figure 6-34 shows the barn with its rather large cupolas forming a symmetrical structure. It should be noted that these large ventilating cupolas are characteristic of the Wayne County region but are not as common elsewhere in the Genesee Country. Following Ennis' son's takeover of the farm in the Civil War years, the 3-bay barn underwent some significant changes. The location of the many doors clearly indicates the addition of another bay on the north side of the barn (often as the number of bays increased the number of doors increased also). A smallish gambrel-roofed addition was made at right angles behind this new bay, but more important to this chapter is the alteration in the frame to accommodate a hay track (see figure 6-35). The original vertical purlin posts were replaced with posts placed at right angles to the rafters, allowing the load to be moved more reliably to the appropriate mow. This idea worked well enough so that inclined purlin posts were incorporated into new barn designs.

The large side-hill basement barn built by George Knickerbocker of Pittsford illustrates the stepwise expansion which often took place in enlarging a barn as needs increased (see figure 6-36). About 1835, Knickerbocker moved to Pittsford to take charge of the Erie Canal packet station at Cartersville. In 1842 he bought a 38-

FIGURE 6-34. Ennis barn exterior, Alloway, Wayne County, New York. Enlarged in the 1860s; the added bay is on the right.

acre farm with a small side-hill basement barn that he used for wintering canal horses and mules. This barn was expanded several times to provide additional space for driving horses from Rochester. About 1870, a barn was moved from across the road and added to the original section. (Barns were moved more frequently than might be supposed. Animals pulled the entire structure on rollers if conditions were right, or else the original structure was taken apart and re-erected on the new site.) The original center section of the long barn is mostly sawn while the northern section, being part of the moved barn, is mostly hewn. Soon after, two additional buildings were added at a right angle to the long barn. Around 1895 the final major forty-foot addition was made to the south end of the barn, bringing its length to 162 feet. At that time the cupolas were added.[17]

At the turn of the century as many as forty-five horses were kept at the Knickerbocker barn, while hay, straw, and oats were stored above the basement stalls, producing a very efficient arrangement. In fact, boarding horses was the main

FIGURE 6-35. Ennis barn interior, showing the inclined purlin posts.

FIGURE 6-36. The Knickerbocker barn. Pittsford, Monroe County, New York. The center section of the main barn is the original portion, the moved barn is on the right.

Knickerbocker family business until the proliferation of automobiles around World War I converted the business to general farming and dairying.

These are just a few samples of ways in which the basic timber-framed English barn could be altered and enlarged. Later, significantly new designs would be introduced and while these new designs would become dominant, the basic English barn would still be built by some. Perhaps the most outstanding characteristic of timber-framed barns is their continuous usefulness over a century and a half and their adaptability to needed changes.

DIARY

Some Winter Notes by L.E. (1859–1860)

The initials L.E. are written in the back of the diary, the only clues to the name of the diarist. L.E. lived near Geneseo and relates much of school and farm life in the area:

1859

Dec. 13	*"We were this morning..to the setting up of the horsepower & threshing machine. Threshed an hour and one half before dinner at the oats. . . . resuming this same work after dinner continued til 3 O'clock P.M. Early the eve' some of us started for Geneseo where I had a draft cashed by F. Omstead."*
Dec. 15	*"The oats on the barn [fl]oor threshed out—the chores done at the mill. —— & I commenced fanning it—being quite late when we commenced we did not finish."*

1860

Jan. 2	*"The cold weather continues The thermometer ranging from 2-6 above to 2-9 below zero. Cattle not housed are much pinched with the cold and make way with quantities of fodder. . . . commenced last ———— cutting feed for stock after which process the entire part of stalks and straw are eaten."*
Jan. 5 (Sunday)	*"This being the last day that I am expected to be at home for sometime it seems good that the rye rakings should be threshed accordingly the fore noon was thus employed."*
Apr. 10	*"..I have had plenty of manure to draw . . . had a fine shower this forenoon and commenced threshing the Hungarian Grass. This on the scaffolds. yet a No. of rats had found their way there and two of them were killed.*

Diary from the Genesee Country Museum Library

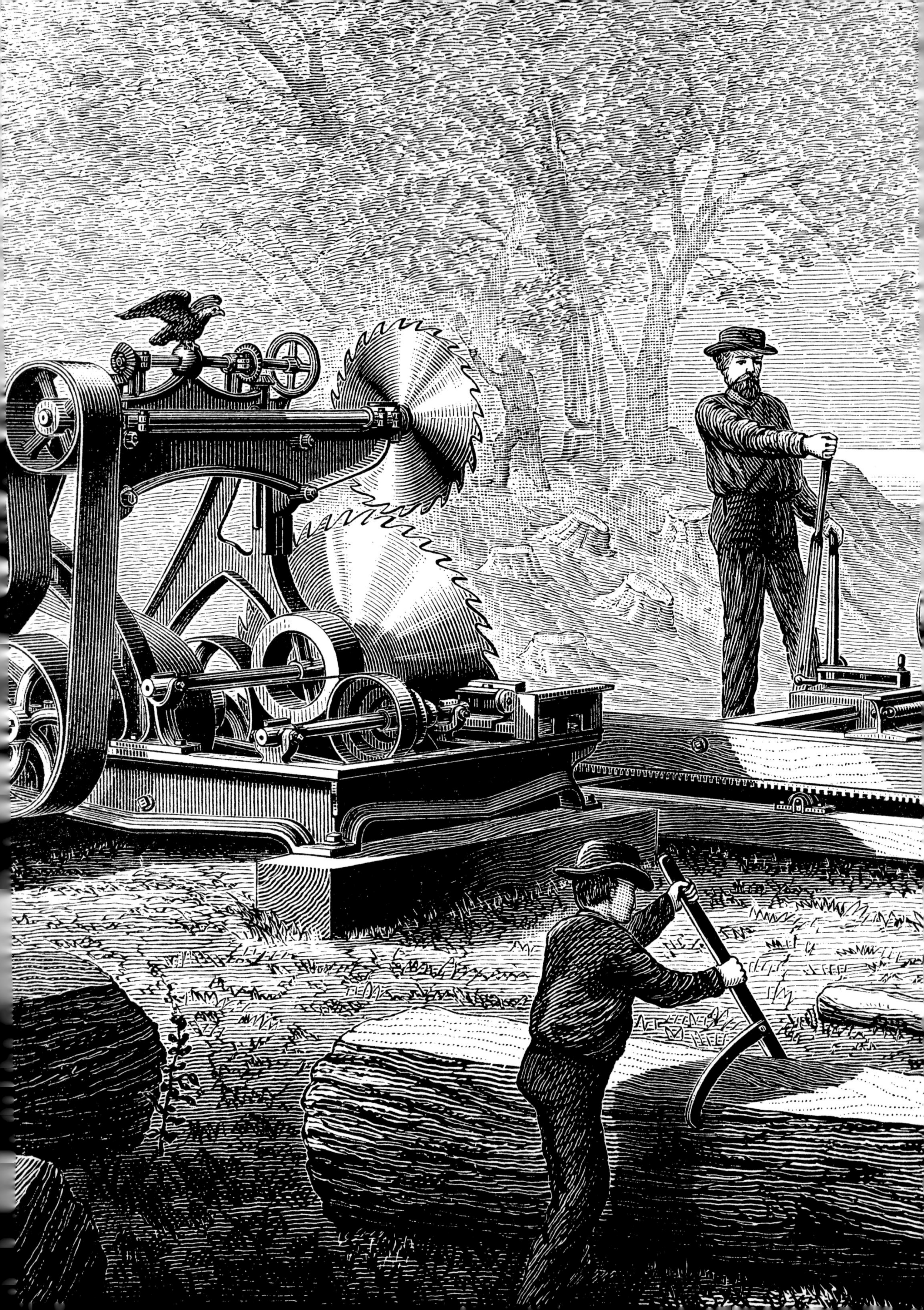

CHAPTER VII

Mills and the Sawn Timber-Framed Barn

THE NINETEENTH CENTURY saw a tremendous increase in the number of mills in the Genesee Country, which profoundly affected both farming practices and barn construction. Greater numbers of grist mills meant that expanded grain production could be accommodated locally, further increasing the need for more and larger barns to store the unthreshed grain. The proliferation of the sawmills led to a growing use of sawn timbers in barn framing toward mid-century.

The milling of lumber and grain by water power was so important to the changes found in barns built in the late 1830s and thereafter that it is necessary to look at the evolution of mills before continuing with the examination of barns themselves.

Land speculators knew the value of mills to prospective settlers, mills to grind a grist or saw a log into planks. While planks could be split out by hand, or corn and wheat ground in a stump mortar, these tasks were extremely time-consuming and not likely to lead to the kind of economic development the land speculators had in mind. Speculators would glean little profit from farmers unable to pay mortgages or rents.

A letter dated December 4, 1801 from Mr. Ellicott to Mr. Busti speaks of the settlers' need for mills:

> The saw-mill I have been erecting at Batavia, which has cost a great deal of labor, not being a natural seat, but a place where a convenience of this kind is absolutely necessary, will, the millwright informs me, be in motion by the 10th . . . at which period we expect to begin to make ourselves and the settlers comfortable floors, etc., which will be a great acquisition to our present situation.[1]

In 1789, Phelps and Gorham arranged with Ebenezer "Indian" Allan to build on a 100-acre tract of land at the Genesee Falls, at the future site of Rochester; the land would be Allan's if he built both a grist mill and a sawmill on the site. At about the same time, Captain John Ganson's mill was constructed just north of Avon. Turner describes the Ganson grist mill in his *Pioneer History of Phelps & Gorham's Purchase:* a "tub mill" on a small stream.[2]

> It was a small log building; no boards could be had; the curb was made of hewed plank; the spindle was made by straightening out a section of a cart tire; the stones were roughly carved out of native rock. It was a rude, primitive concern;

Opposite: A nineteenth century saw mill.

FIGURE 7-1. Sawmill and bridge at the Middle Falls of the Genesee River near Portage, Livingston County, New York. Stacks of seasoning lumber are visible behind the mill. All traces of the operations at these falls seem to have disappeared. Stereo half by L. W. Walker, Warsaw, New York, late 1850s.

> but it would mash the corn a little better than a wooden mortar and pestle; . . .[3]

Mills were essential if settlers were to be attracted to the area in large numbers.

The first decade of the new century was a period of intense mill building. All along the Genesee River, its feeder streams, and other creeks and rivers in the Genesee Country, sawmills and grist mills as well as fulling, carding and paper mills were erected. Some diaries indicate that so many mills were built they could not all compete successfully. One mill owner who was successful was William Bush, who reports that his boards were in such demand in 1806–7 that they brought $7.50 per 1000 board feet.[4] Elsewhere many milling operations were short-lived due to fire or uncertain water supply. One such mill, destroyed by fire on January 23, 1858, is shown in figure 7-1.

Typical of the medium-sized mills still standing in the Genesee Country is the sawmill in figure 7-2, built originally as a carding mill in the

FIGURE 7-2. Sawmill built by Herbert Wadsworth on Conesus Creek, Avon, New York, 1870s. The horizontal windows were added to this structure when it was converted to a private home in the twentieth century.

1840s. At one time there were three mills in close proximity on this site: a flour mill, a sawmill, and a woolen mill, all using water from the same dam. Further east along the Conesus outlet was a water-powered triphammer, built about 1800 and used for forging and manufacturing wrought iron.

Two major nineteenth-century historians of the Genesee Country, Orsamus Turner and Lockwood L. Doty, collected the reminiscences of surviving pioneers concerning early saw and grist mill operations. Some of those observations follow.

Jehiel Kelsey recollects:

> Judge Hosmer built a saw mill on the Conesus, as early as 1796, the first one in this region. The Wadsworths built one the same year on the same stream.[5]
>
> The prominent early merchant of Geneseo was the late Major Wm. H. Spencer. He was from East Haddam, Conn. Arriving upon the Genesee River in 1803, with his axe upon his shoulder, he was a Pioneer of "Fairfield" now Ogden; breaking into the wilderness on Rush creek, about a mile east of Spencer's Basin, he built a cabin, kept bachelor's hall, bought provisions of Mr. Shaeffer, carrying most of them in on his back; built a saw mill, and in a little over a year cleared fifty acres. Getting ready for his saw mill irons, he went to Connecticut, and brought them all the way from there with an ox-team.[6]
>
> Soon after settlement commenced, Mr. Williamson had erected a grist and saw mills, on the site afterwards occupied by Col. Rochester. David Scholl, who was Mr. Williamson's millwright at the Lyons mills, erected the mills. . . . The mill was burned down soon after 1800, after which before rebuilding, the neighborhood had to go to Bosley's mills at the foot of Hemlock Lake.[7]
>
> In 1813 the town [Avon] contained 5 sawmills, one grist-mill, 6 distilleries and one carding and cloth dressing establishment. There were 76 looms, in families, the yearly product of which was 21,325 yards of woolen, linen and cotton cloths.[8]

Grist Mills

Jared Boughton's* 1789 contact with the Ganson Mill at Avon and the Allan Mill on the mill tract are described by Turner in his *Pioneer History of Phelps & Gorham's Purchase*:

> Jared Boughton took the buckwheat and got it ground at Capt. Ganson's rude mill at Avon. His next milling expedition, (after wheat harvest,) was with a double ox team, to the Allan mill at Genesee Falls. Arriving within four miles of the river, (at Orange Stone's,) he came to the end of the road; any direct route to the River was through a dense forest, and low wet grounds; which obliged him to go around, and work his way over the range of hills east of Mount Hope. Arrived at the River, he belled his oxen and turned them into the woods, carrying his grain across and down the river to the mill.[9]

Both the isolation of Allan's mill and the importance of the facility to a man such as Boughton are apparent.

In 1803 John McKay, who had worked at Williamsburgh as a housewright, bought the mill that Williamson (of Williamsburgh) had con-

*Jared Boughton was the brother of Enos Boughton (both from Stockbridge, Massachusetts), who purchased the town of Victor, after both had explored the region in 1788, for twenty cents an acre from Phelps & Gorham.

FIGURE 7-3. Mills of all sorts line both sides of the Genesee River at Rochester, Monroe County, New York. The tail races blend with the blurred water tumbling over the Genesee Falls. Photography of the period could not stop the movement of the water. An anonymous stereo-half.

structed in 1799 for the Scotch settlers at Caledonia. The mill included 200 acres and was sold for $2,000. The mill stones, purchased in Albany, had cost $35 to transport to Caledonia. This was the first mill built west of the Allan mill; now marked by McKay Park, it also formed the nucleus around which the village of Caledonia grew.[10]

Clearly, grist mills were important not only for the local needs of individuals but also to "flour" grain so it could be shipped more readily to eastern markets. In the early years of settlement, when transportation was difficult, grain was often transformed into whiskey, which was easier to transport and apt to keep better. When local industries are listed, distilleries are as prominent as grist mills in the first third of the nineteenth century. The farmer, however, realized less from the sale of his grain as whiskey than from the same grain in the form of flour. At first farmers drove great distances to flour mills until demand increased their number. A partial list of mills includes the names of many of the pioneer families of the region: Allan, Ganson, McKay, Hosmer, Wadsworth, Ellicott, Williamson, and so on.

In 1800, Colonel Nathaniel Rochester, William Fitzhugh, and Charles Carroll traveled to inspect the Ebenezer Allan mill site in what is now Rochester. Recognizing the water-power potential of that area, they later bought the 100-Acre Tract for $17.50 per acre. In 1810, Rochester moved from Maryland to Dansville, where he owned almost all the land capable of water power and a number of saw and grist mills, as well as the first paper mill in the Genesee Country. During the four years Rochester lived in Dansville and later, in Bloomfield, he set about planning the layout and sale of lots in the 100-Acre Tract, the heart of the city that would bear his name.

Rochester grew from a few cabins in 1812 to a fast-growing boom town by 1827; in that year Captain Basil Hall, a retired British naval officer, commented that so furious was the work one might expect a standing tree to be cut down in the morning, stripped and sawed, all in time to become part of a building by nightfall.[11] In spite of the malarial "Genesee Fever" and the dense forests, people flocked to Rochester for two main reasons: It was an excellent milling site and the Erie Canal provided access to markets for the milled products (see figure 7-3).

Rochester was a natural mill seat, with a total fall of 260 feet within two miles, 100 feet more than the falls at Niagara. From seven mills with 24 pairs of stones in 1826, the milling facility grew to 22 mills with 100 pairs of stones by 1851, providing the enormous capacity of close to 500,000 barrels per year (see figure 7-4). The mills at Rochester were constructed mainly of stone, differing in that respect from the timber-framed mills found throughout the region. O'Reilly describes the operation of flouring along the Erie Canal "with machinery so adjusted that cargoes of a thousand bushels of wheat are elevated to a height of fifty feet and weighed in an hour and half. The boats, without changing position, in a similar brief period receive cargoes of flour."[12] Indeed, through the 1850s Rochester was the "Flour City," the premier flour-producing city in western New York.

Milling helped Rochester grow at an astounding rate: by 1830 it ranked twenty-first in population among U.S. cities, numbering 10,863 people. Rochester is often cited as the first boom town because its first lot had been sold only nineteen years earlier, in 1811. McNall identifies Buffalo as the next largest milling center, noting that while its water sources were not as good as those at Rochester, Buffalo became competitive during the age of steam.

Probably the largest mill site within the Gene-

FIGURE 7-4. Publisher C. W. Woodward of Rochester provided this stereo-half view of the flour mills along the west bank below the High Falls of Rochester. The concentration of these high-capacity flour mills at Rochester gave it the nickname "Flour City."

see Country that was not on the Genesee River itself was the Railroad Mills in Victor, Ontario County. This four-story mill was constructed in 1832 and lasted until it was destroyed by fire in 1883. McNall also mentions important milling sites at Batavia, LeRoy, Stafford (Genesee County); Mount Morris, Avon and North Dansville (Livingston County); Mendon, Perrington, Parma, Pittsford and Wheatland (Monroe County); and Cuba (Allegany County).[13]

Genesee Valley wheat and flour gained an excellent reputation during the first half of the nineteenth century. Wadsworth wheat, for example, was sold by exclusive contract in Boston, being preferred because of its high quality. At mid-century, local wheat production collapsed due to insect infestation and destructively wet weather (see Chapter X); the mills at Rochester were forced to obtain wheat from Canada and the Midwest in order to stay in business. With the collapse of wheat production, Genesee Country agriculture was characterized by experiments with alternative crops and farming, especially dairy farming.

Sawmills

The first sawmills were a blessing to those who split logs for planks or who used a whip or pit saw to produce sheathing boards. Two men on a pit saw considered 100 linear feet of planking a good day's work. The first water-powered saw mills were sash mills, commonly called "up and down" mills. These were essentially powered pit saws, with an advance mechanism to move the log along past the blade (see figure 7-5). Early blades worked at about eighty strokes per minute, and it was a joke among sawyers that they could start a log, go to lunch, and come back before the board was sawed. Although only about 500 linear feet per day could be cut with such a saw, *The Genesee Farmer* was still advertising sash saw mills for sale in 1850. Later improvements included "gang" sawing, where many blades were mounted side by side so an entire log could be sawn at one time.

FIGURE 7-5. An "up and down" saw was essentially a water-powered pit saw with an advance mechanism to move the log.

Logs were dumped into rivers at high water to move them to the mills. *The Livingston County Whig* carried a description of saw logs floating down the Genesee to Mount Morris in its October 19, 1847 edition:

> A Fine Sight.—Being at the River one day last week, we found our beautiful stream completely filled almost as far as the eye could reach, with as fine a lot of Saw Logs, as the most fanciful Carpenter would wish to set eyes on. We understand that they belong to R. R. Spellman, and were

rafted at the recent freshet from 20 to 30 miles up the River. What a bobbing and knocking about they must have made coming over the Falls at Portage. He was very successful—in a lot of some 4000, we learn that he lost very few.

Most of the water-powered sawmills remained the up-and-down variety, but when steam power was introduced in the late 1820s and early '30s, mills could make use of the faster sawing capability of the circular saw. The inventor of the first circular saw blade is not an uncontested certainty, but in 1813 the first United States patent for a circular saw blade was issued to Sister Tabitha Babbit, a Shaker. Although an advertisement of the period shows ox-powered saws, it was steam that allowed the full realization of the circular saw (see figure 7-6).

FIGURE 7-6. In this Emery & Sons Endless Railway Horse Power (here oxen-powered) both the reciprocating drag (left) and circular saws are being operated. One of the distinct advantages of the operation was that it could be taken to the log source, sometimes by the same horses or oxen that powered the mill.

FIGURE 7-7. Steam-powered portable sawmill. A team could skid the boiler and engine to a new location.

FIGURE 7-8. Green & McAuliffe steam-powered lumber mill at Fairport, Monroe County, New York. A saw log is being transported in the lower right hand corner and stacks of seasoning lumber can be seen everywhere. McIntosh, History of Monroe Country, 1877.

After the Civil War, most saw mills were steam-powered, their high-speed circular blades capable of producing thousands of board feet per day (see figure 7-7). In 1831 Frances Trollope hinted of the first steam-powered mills:

> Rochester is one of the most famous of the cities built on the Jack and Bean-Stalk principle. There are many splendid edifices in wood; and certainly more houses, warehouses, factories and steam-engines than were ever collected together in the same space of time; but I was told by a fellow-traveller that the stumps of the forest are still to be found firmly rooted in the cellars.[14]

A steam-powered lumber mill of considerable size was operated by Green and McAuliffe along the canal and railroad at Fairport (see figure 7-8). Compare it to the water-powered mill in figure 7-9. An even bigger sawmill complex, the Fox, Weston & Brownson mills southwest of Painted Post, covered about sixty acres; the gang mills (the mill location now is called Gang Mills) produced ten million feet of lumber, three million shingles and four million lath in 1868. By the end of the nineteenth century the faster, less wasteful bandsaw had replaced the circular saw in most large sawmills.

In the early days of sawmills, sawyers seldom ran a board through the mill an extra time just to produce a uniform board width. Even though standing timber was abundant, the sawing process itself was expensive and the random-width pieces could be used for many different purposes. For example, the first blade cuts through a log produce uneven "slabs," often used for firewood. The next material cut is the small-dimensioned framing stock known as "scantling," followed by wider boards and planks from the center of the log. As the sawing industry increased in size and capacity in the 1830s and 1840s, dimensioned lumber became commonly available. As a result random-width barn boards gave way to sheathing of more uniform width.

FIGURE 7-9. View of Stokoe & Wilson milling operation, Gorham, Ontario County, New York. Water power was essential to the operations at this mill site; when the water was low or frozen the milling slowed or even stopped. McIntosh, History of Ontario County, 1876.

Balloon Framing

In the mid 1830s a significant new kind of framing construction was introduced, called balloon framing by its detractors because it was supposedly no more substantial than a balloon. Because it was first introduced in Chicago, it is also known as the Chicago style of framing (see figure 7-10). Balloon framing or its variations is very common in modern framed construction, indicating that, in fact, the technique did work and did become accepted. Relatively light, small dimensioned lumber is used, such as 2x4s, 2x6s, and 2x8s. The vertical 2x4 inch studs are closely spaced (16 inches on center) to provide the same stability as heavy posts and beams. Of course, floor joists and rafters are necessarily larger, ranging from 2x6s to 2x14s.

Essential to balloon framing are machine-cut, machine-headed nails which became available in large quantities in the 1830s and largely replaced traditional wood joining. Balloon framing also required lumber of much more consistent dimensions compared to hewn framing. In timber-framed barns the hewn timbers were often not exactly the same in cross section; much depended on the uniformity of available logs.

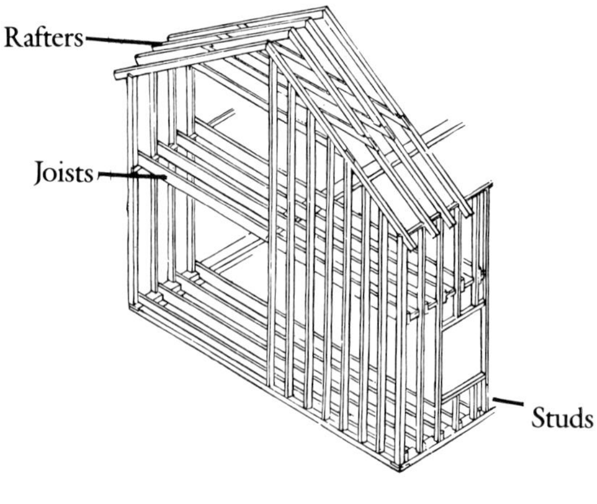

FIGURE 7-10. View of balloon frame (from Woodward). Major vertical members are 2x4s.

Often the vertical studs in early balloon framing ran the entire height of the building, as opposed to the single-story studs used in "deck" balloon framing today. In comparing barn roofs of various ages, perhaps the most impressive change from early practices is the switch from hewn-on-one side, bark-covered saplings to balloon-framed roof rafters with their close spacing.

Solon Robinson, in his 1866 *Facts for Farmers* . . . , devoted an entire section to encouraging farmers to build balloon framed barns, pointing out, "There is need of no stronger building than one made upon this plan."[15] Mustering support for his recommendation, he cites George Woodward's articles in *The Country Gentleman*. Both Woodward and Robinson believed that balloon framing might be hard to "sell" in areas that were wood-rich, but cite the thousands of successful structures built on the virtually treeless prairies. Woodward called timber-framing the "old fogy principle of framing" and claimed that with lumber becoming more expensive balloon framing was the only way to build. He further claimed, "Any intelligent man who can lay out a right angle and adjust a plumb line may do his own building, for it is without a mortice, a tenon, or brace, and a man and boy can do all the work." Writers such as Robinson and Woodward were very influential, and indeed the years following the Civil War saw widespread acceptance of balloon framing in the Genesee Country. Even the most reluctant might use the technique in roof rafter replacement.

The 1830s, a time of rapid technological advancement, saw many other innovations and inventions: Railroads were being constructed (Peter Cooper's *Tom Thumb* first ran in 1830); the Case Thresher Works were founded in 1831; McCormick's first reaper patent was issued in 1834; gas lighting was used commercially in 1839; Colt patented his revolver in 1835, and Hall and North began to mass-produce firearms. In 1837, John Deere made a steel-shared plow and Samuel F. B. Morse invented the telegraph. In 1839, Davenport invented the electric motor and photography was finally successful in Europe. With these changes, and others in the wind, it was bound to be a time of invention in agriculture and barn construction.

The timing of changes in technology provides markers that the historian can use to help unravel the mysteries of barn evolution. For instance, marks left by saw blades on planks, timbers or boards can give a rough clue to when a barn was built—as long as it is certain that the sawn lumber is part of the original construction (see figure 7-11).

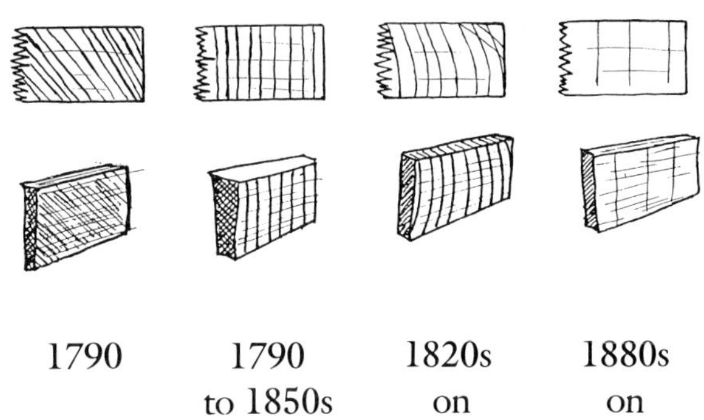

FIGURE 7-11. Saw marks indicate the method used and can help in dating lumber.

Metals and plastics are so much a part of twentieth century life that it is almost amazing to remember that the first barns, whether log or hewn-frame, were almost entirely constructed of wood, from sills to hand-riven shingles all the way down to the latches and hinges. Oak, spruce, ash, red elm, yellow and white pine, hemlock, black locust, maple, and even cherry and black walnut were used, each for its own characteristics. Of course, for some uses materials other than wood would have been extremely helpful: For example, wrought iron would have been a welcome replacement for wood in hinge pintles, where much wear can be expected. When a blacksmith was available, or his wares could be purchased, handmade nails, spikes and other forged hardware were used. However, nails and spikes represented a sizeable cash outlay, and in some instances, as when a new structure was built to replace an old one, the old nails and spikes were recovered by burning the structure to free the nails. One problem with the practice of burning a barn to recover the metal was that the nails became annealed (softened) in the process, making them difficult to work with in the new frame. In barn framing, the use of wood joinery lasted even into the twentieth century. These wooden joints were superior to metal fasteners over an extended period. Nevertheless, increasing industrialization led to the dominant use of metal fasteners and hardware, at first most clearly evident in the sawn timber-framed barns.

The three major nail types used in barns include the hand-wrought nail, the machine-cut nail and the "modern" wire nail (see figure 7-12). Blacksmith-wrought nails were available from the beginning of settlement and the first machine-cut nails were introduced just after the Revolutionary War (see figure 7-13). By the 1830s, machines were capable of heading and cutting the nail in a single operation, thus considerably increasing cut nail production and ready availability (see figure 7-14). Small quantities of cut nails are still manufactured today because some builders prefer them for their superior holding power. The wire nail was first manufactured in mid-century and was available in large quantities by the 1880s.

Nails, timbers and sheathing boards are essential, but barn flooring is equally important. An 1884 article in the *American Agriculturist* by L. D. Snook investigated this aspect of barn construction:

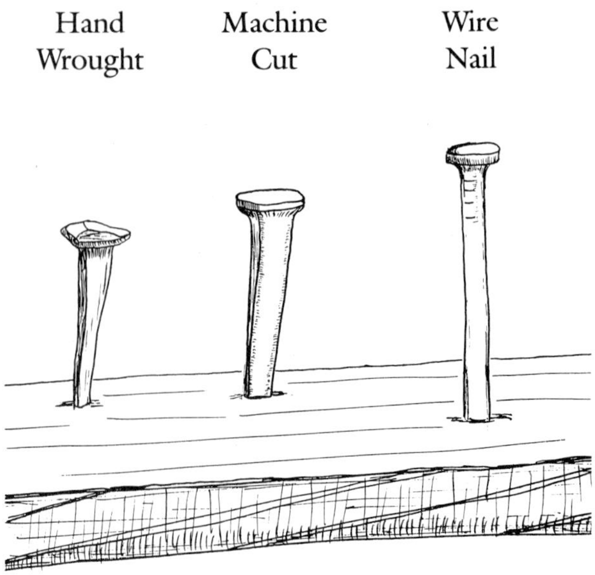

FIGURE 7-12. Major nail types used in barn construction.

> A barn floor needs to be strong to sustain heavy loads, with close fitting joints, to prevent grain, etc., dropping through. For the driveway, plank of pine, hemlock, or other soft wood is preferable for the teams, as oak or other wood gives a less secure foothold. They should always be laid crosswise, not only to give a better foothold for the team, but to distribute the weight of the load over a greater number of planks. The supporting joists should never be over two feet from center to center, better only sixteen or eighteen inches. They should be made as level and quite as carefully fitted as for house flooring, and not less than two inches thick; two and a half inches would be better, especially along driveways. The width depends first upon what is obtainable, and next upon the seasoning. If perfectly dry when laid, they may be of the full width even if two feet wide, but if not throughly seasoned eight inches wide will be sufficient. They should be cut a year in advance and so well seasoned as to shrink but little. The simplest way to make barn floor is to tongue and groove the plank . . . or . . . cut out the lower corner of each of the planks half an inch back and one inch deep, and as each plank is laid, put an inch strip in the groove, the next plank fully covering the strip. You thus obtain the benefits of a grooved floor, without danger of breaking away the tongue by unequal pressure from above. (In a similar method) . . . though much cheaper, one edge of the plank simply overlaps the one below, making a firm joint.
>
> Where the ends of a plank meet, a lock joint . . . is quickly made, which keeps them from

FIGURE 7-13. *Blacksmith at his forge.*

FIGURE 7-14. *Nailmaking machine. Machines eventually could cut the nail and head it as well, significantly increasing production.*

becoming unequally elevated. Many mechanics spike the floor down as fast as laid, but it is always best when practicable, to postpone nailing until just before fall rains, when the planks are dry and can be driven together by wooden wedges on one side. Flooring should be as free as possible from knots in the planks or other imperfections.[16]

This sort of sophisticated information on such items as flooring and framing was once readily available from barn builders and agricultural periodicals and was basic to the quality design of a barn.

Sawn-Frame Barns

In many of the first barns with sawn timbers, the framing and joining pattern remained the same as those seen in earlier hand-hewn frames, but in others the design of the frame itself is different. Thus, although the barn's exterior appearance may be similar to an early English three-bay barn or a side-hill barn, the interior may exhibit significant changes. Traditional post and beam bent construction begins to give way to truss framing (discussed in detail in Chapter XX). A good example is the Chanler barn, just off Jones Bridge Road in Geneseo, which has much the exterior appearance of a large, gable-ended English side-hill barn (see figure 7-15).

Once inside on the main barn floor, it becomes clear that this barn is different structurally from any examined thus far; in fact, it represents the next generation of barns. For one thing, the timbers and the rest of the lumber are entirely sawn. For another, the frame was expressly designed to

FIGURE 7-15. The Chanler basement barn, Jones Bridge Road, Geneseo, Livingston County, New York.

accommodate a hay track, using a series of vertical trusses and depending for strength and rigidity on its large diagonal braces, which produce a series of triangles (the most rigid forms in construction; see figure 7-16). The central purlin posts are tied to the exterior posts with beams and braces, providing an uninterrupted center aisle for access to the two mows. On the right, next to the threshing floor, is the smooth-boarded granary. The Chanler barn represents an 1870s design chiefly because of the character of the truss used in the frame.

The 28 by 100 foot barn on the Wadsworth Oxbow Farm north of Geneseo is a typical exam-

FIGURE 7-16. Interior of the Chanler basement barn. The view along the long axis between the sawn trusses emphasizes the open quality of this design. The resemblance to bridge trusses is strengthened if the photograph is turned on its side.

ple of a barn built to winter beef cattle[17] (see figure 7-17). Beef cattle became important to the Genesee Country with the coming of the railroads. At one time, the cattle were purchased in Buffalo, Chicago and Texas (yes, longhorns) and shipped by rail to the Genesee Valley to be housed in the Oxbow barn (see figure 7-18). Built on stone piers, this board-and-batten sided, gable-ended, two-story barn has served its purpose well in spite of the many changes that have taken place in farm technology over the years. The sawn vertical posts are tied together at the

FIGURE 7-17. Oxbow barn, South Avon Rd. Geneseo, Livingston County, New York.

FIGURE 7-18. Basement stalls line both sides of the aisle on the lower level in the Oxbow barn.

FIGURE 7-19. Second level of Oxbow barn.

first story level with substantial cross beams. Closely spaced joists support the floor and the rafters are balloon-framed. In contrast to the Chanler barn (figure 7-16), there are no interior purlin posts (see figure 7-19). The substantial diagonal braces were designed to prevent damage from the outward pressure of loose hay, as well as from raking (moving of the walls) in the wind. The cupola with its accompanying shaft, common to the Genesee Country, provided the ventilation essential for curing hay.

The Mulligan Farm on Barber Road in the town of Avon has a vast complex of barns. One dormered, cupola-topped, gambrel-roofed barn was built with a sawn frame in the late 1870s by Aaron Barber (see figure 7-20). The center aisle is

FIGURE 7-20. Mulligan Farm barn. Barber Road, Avon, Livingston County, New York.

considerably wider than that of the Chanler barn but otherwise there are striking similarities in the framing (see figure 7-21). It is apparent that the barn was originally built without a hay track, even though today there is one, complete with car. The purlin posts are mortised to take girts, but when the hay track was installed the girts were cut away to permit movement of the fork (see Chapter XI); subsequently, collar beams and tie rods were added to restore structural stability, followed by additional diagonal braces to increase rigidity. Although most cupolas in the area were functional, the one on the Barber barn was only decorative, having no access to the barn.

FIGURE 7-21, Opposite. Mulligan barn interior. The three collar beams identify the cupola position.

FIGURE 7-22. Preparing to raise the plate on the Herbst Farm on Middle Sodus Rd. near Lyons, New York. Photograph by Herman Wright.

From time to time photographs surface that illustrate important developments of the past. In this instance, August Belke built a three-bay barn close by the cobblestone farmhouse on the John Herbst Farm north of Lyons around 1910. Neighbor Herman Wright brought his glass plate Seneca camera to the raising, and two of Wright's photographs remain. The major change in this Belke design is found in the braces, which are much larger than barn braces of 100 years earlier and provide increased lateral stability. In figure 7-22, men have gathered on the planking at the second level, ready to place the plate onto the tenoned posts. Below, men and boys are carrying a timber into the barn. Figure 7-23 shows the plate after it has been raised into place.

As sawn timber became more available, sawn frames in barns became increasingly common. Sawn timbers might have cost more than hewn timbers, but in the post-Civil War period mechanization affected every facet of life and barn framing was no exception. Perhaps because of the cost, some barns were built with combinations of hewn and sawn timbers. Eventually hewn timbers were seldom used except when sound old timbers were being recycled, but as illustrated in Chapter V, the changeover was a long process. The majority of the wooden barns in the rest of this book will have sawn timber frames. In contrast to these wooden barns are the cobblestone barns of Chapter VIII, which are unique to the Genesee Country.

FIGURE 7-23. The plate has been successfully placed and now its opposite is to be installed. Photograph by Herman Wright.

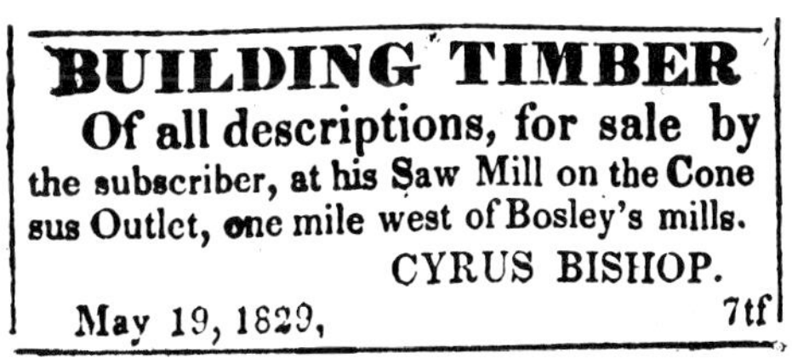

BUILDING TIMBER

Of all descriptions, for sale by the subscriber, at his Saw Mill on the Conesus Outlet, one mile west of Bosley's mills.

CYRUS BISHOP.

May 19, 1829, 7tf

CHAPTER VIII

Cobblestone Construction

MOST OF THE BARNS built during the nineteenth century in the Genesee Country were wooden except for their masonry foundations. The most notable exceptions are the cobblestone barns built prior to the Civil War. The inventive use of these round glacier-deposited stones produced barns, houses, and other structures with an appearance that is unique in the United States*

It is not known who invented this novel form of cobblestone construction, but with few exceptions the technique seems to be isolated in a radius of some 70 miles around Rochester. There are some regions in Europe and in the southern part of England where the setting of small stones reminds one of cobblestone construction, but the connection, if any, remains unclear.

Cobblestones are the result of repeated glaciation, the last some million years ago; they were formed as the glacier scooped up rocks and dragged, turned, and wore them into hand-sized rounded shapes. These polished or faceted striated rocks accumulated into drumlins, moraines and kames as the glacier melted and retreated. Glacial deposits can be beneficial, as in places where they have formed miles-long well-drained ridges especially suitable for roadbeds. (The Ridge Road, already discussed in Chapter IV, is built on one such deposit.) A negative impact on farming created by cobblestones is represented in figure 8-1, a tintype of a regional farm.[1] The cobbles could seriously damage a plow working through such a field. "Fetching the stones" was a job often given to children, but it seems no matter how often the surface cobbles are removed, the spring rains reveal more. Fences were often made from the cobbles, and at some point around 1825 they began to be used in foundations; eventually, entire structures were built of cobblestone.

Some kind of foundation for wooden buildings has always been necessary; a knowledgeable individual never built a log cabin or frame barn directly on the ground because the sills would rot out in short order from contact with the earth. Early structures used field stones laid up in a dry foundation, i.e. without mortar (see figure 5-19), producing a random appearance. As masons became available, a builder would likely contract for a mortared foundation of dressed field stone. Nineteenth century mortars in general tended to be much softer than those used today, restricting the number of courses that could be set in a day; a figure of three courses per day is often given. Each mason apparently had his own jealously guarded formula for mortar.

After the completion of the Erie Canal in 1825, many masons found themselves unemployed. Many had been brought from Europe and, rather than return home or find employment in other trades, a significant number turned to house building. Among the various kinds of

*the very few masonry barns of complete conventional masonry construction in the Genesee Country are not examined in this volume.

Opposite: Detail of the largest complex of cobblestone barns in New York State. Swift-Pulvino farm, Phelps, New York.

FIGURE 8-1. Tintype of a cobble-filled plowed field. Anonymous. Circa 1875.

houses built by these masons were those using the free material known as cobblestone, laid up in a unique way to form the distinctive characteristics of cobblestone construction.

Cobblestones allowed construction of a strong, visually interesting foundation or wall without the need for extensive preparation of the building material, such as dressing or sawing. The glacier had provided multiple copies of an interesting shape, contributing to an effect that probably could not be produced in any other traditional building materials.

FIGURE 8-2. Detail of a cobblestone wall with quoins. Quoins of locally quarried limestone are often 12 inches in height and are set in place first, ensuring a plumb corner. Because the quoins are fairly consistent in size, the coarseness of the cobbles can easily be measured in terms of the number of courses per quoin. Three courses of cobbles per quoin represents the coarsest example (the biggest cobbles) and eight or so per quoin represents the most delicate example.

Building With Cobblestones

Cobbles first had to be gathered together and then sorted into piles of similar size (and sometimes color), so that each course would be uniform. Sorting could be done by eye, but the results were more consistent with a "beetle," a board with variously-sized holes or wrought iron rings through which the stones were dropped.

Cobblestones were laid up with considerable exposure of the exterior stones; as much as half of each stone might project from the mortar, in contrast to rubble or rusticated masonry in which the stone was embedded in mortar except for its outer face (see figure 8-2). The rather extreme projection of these uniform exterior cobbles created a strong light-modulating effect: The appearance of a cobblestone wall changes dramatically as the light that falls upon it changes. Such would not have been the case had the masons not been sensitive to the visual possibilities of these multiple similar cobble forms.

Cobblestone barn walls (other cobblestone buildings too) were usually three courses thick. The inner course used traditional rubble masonry field stone construction. The center course was also rubble but of finer stones, perhaps a random placement of cobbles. Finally, the outer course might consist of four to six-inch cobblestones along with some longer cobbles. Figure 8-3 shows a cross section of a cobblestone barn foundation. Actual construction of cobblestone walls varied considerably, some consisting of a thin veneer of cobbles—at once the most fragile and the most decorative of the methods—while others were laid up as solid, thick walls, as much as twenty to twenty-four inches thick, with all the stones bonded together.

FIGURE 8-3. Detail of the cobblestone foundation of a razed barn. The three distinct courses (exterior, interior and rubble fill) are bonded together by mortar, forming a very durable foundation. East River Road, Rush, Monroe County, New York.

Each mason seems to have had his own way of tooling or finishing the mortar between the cobbles, usually in some decorative fashion. In fact, the work of one mason can sometimes be identified from building to building by the character of the tooling. Unfortunately the tooling does not identify a specific mason by name.

Cobblestone Foundations

Today one has to look carefully to find remnants of cobblestone construction in farm buildings. One can expect to find a foundation laid up with cobblestones if the building dates from the completion of the Erie Canal (1825) to the beginning of the Civil War (1861). In general, cobblestone construction ended with the Civil War. Two typical examples of cobblestone foundations include the gable barn along East River Road, south of Rochester (see figure 8-4), and the gambrel-roofed bank barn along the same road, just north of the old Lehigh Valley Railroad crossing (see figure 8-5). One shared characteristic of many barn foundations made of cobblestone is the elongation of the corner quoins. In figure 8-3, the corner of a now-demolished cobblestone barn foundation exhibits the elongated quoins and also reveals how the mortar held together the exterior course, interior course, and rubble fill.

FIG. 8-4. Cobblestone foundation of a gable-roofed barn. East River Rd., Rush, Monroe County, N.Y.

FIGURE 8-5. Cobblestone foundation of a gambrel-roofed barn. Elongated quoins such as these are common on barns. East River Road, Rush, Monroe County, New York.

FIGURE 8-6. Cobblestone house and barns, east of Phelps on Route 96, Ontario County, New York. Five cobblestone structures make up this complex. The larger barn to the rear of the property is 23½ by 70 feet, while the barn closer to the house is 29 by 38 feet. The four gable-ended barns were once separate from the house before the connecting addition shown here was added.

Cobblestone Barns

Although many houses were built of cobbles, relatively few barns used this construction. One notable exception is the Swift-Pulvino farm east of Phelps where all four connected barns as well as the house are of cobblestone construction. This cobblestone complex, built by Jonathan Swift in the early 1840s, is unique in New York State (see figure 8-6). Perhaps it is no wonder that the house and barns are constructed of cobblestone, since the farm's fields still abound in them. The walls are load-bearing, but it is of course necessary to frame the doors and the roof. The frames of the Swift-Pulvino farm barns are hand-hewn. In general the exposure of the cobblestones used on barns is not as great as the fancier exposure on houses. Figure 8-7 reveals a much-repointed east wall on the right, the entire face of the second barn, and the house in the background left. The connection of the four barns is shown in figure 8-8, where the view is of the west walls.

A solitary, large cobblestone barn stands

FIGURE 8-7. Detail and view of the west walls of the cobblestone barn complex east of Phelps, New York. The exposure of cobblestones on this and other barns is generally less than on houses.

FIGURE 8-8. The four barns are set at right angles to each other to provide for the connection of one to the next. The buttresses are a relatively recent addition to stabilize the walls.

FIGURE 8-9. Cobblestone Greek Revival barn, 27½ by 45 feet. Sodus Center Road northwest of Sodus Center, Wayne County, New York. In the cobblestone walls of the barn there are three courses per quoin.

FIGURE 8-10. Partial cobblestone barn on the Stonehedge Farm, East Williamson-Marion Road, Marion, New York. The cobblestone section very nicely represents Greek Revival detailing; the entablature is particularly striking.

northwest of Sodus Center, in the heart of apple growing country (see figure 8-9). While the foundation walls are of rubble, the rest of the barn is of traditional cobblestone construction with three courses of cobbles per quoin. This sidehill barn, banked on its south side, is gable-roofed with a center entrance. The ivy growth tends to obscure the Greek Revival detailing of the barn, which includes the entablature and the returns (Greek Revival barns are discussed in Chapter IX). This type of detailing can be expected in cobblestone structures, since their construction coincides with American interest in Greek Revival architecture. Excellent Greek Revival detailing can be seen on the cobblestone portion of the barn on the Stonehedge Farm in Marion (see figure 8-10).

Other cobblestone barns exist in the Genesee Country besides the examples shown; one of interest because of its alteration is the Maplewood Farms cobblestone barn, near Attica, which was converted to a gambrel roof sometime after its original gable roof construction. In addition a 30 by 45 foot cobblestone carriage house with Greek Revival detailing built around 1859 stands south of the New York Thruway in the Montezuma Wildlife Refuge. On a much smaller scale cobblestone outbuildings abound throughout the region. One exquisite example is a cobblestone smokehouse in West Bloomfield (see figure 8-11). As mentioned above, most often the cobbles used in farm buildings are coarser than those used for houses, with not as much attention paid to the exposure of the cobble or detailing of the mortar. However, this smoke house, with its fine overall detailing, is an exception. The white-painted classical Greek treatment of the eaves and gables emphasizes the interesting use of contrasting cobblestone shapes in the gable: The first course above the window is done in the so-called

FIGURE 8-11. Cobblestone smokehouse. West Bloomfield, Ontario County, New York. The walls contain four courses per quoin.

"herringbone" or elongated cobble and the gable is finished in stones smaller than those used in the walls.

Why did cobblestone construction disappear? Not for lack of material, since each spring rain brings a new "crop" of cobblestones to the fields. There is no doubt that the stones would continue to be used if the cost of laying up the cobbles were not so great. In this case, beauty and long-range durability have given way to economics.

DIARY

Philo Hampton's Book, Ossian, Livingston County, 1853–1877

PHILO HAMPTON, Ossian farmer, kept a diary which provides a brief look at wages, threshing techniques, livestock care and some very human feelings; anyone can understand his upset at not being invited to a dance or the need to rid himself of his hot winter long underwear. In these fragments of Hampton's words one feels the wintry blasts and penetrating cold of the Genesee Country:

Philo S. Hampton's Book Ossian, N.Y.

1853

August 1853 notation: John Stainbrook commenced work Aug. 1 for twenty dollars a month.

Aug worked	*twenty five days*	*$19.25*
Sept	*sixteen days*	*$12.32*
Oct	*nine days*	*$ 6.99*

April listing of purchases:	*Fixing buggy spring*	*50 cents*
	3 lbs. nails	*14 cents*
	lantern	*$1.00*

1875

Oct. 28: *Went from Bethany to Rochester Stayed all night in Rochester at the Clinton House on my way back stayed at Byron and went to Byron Center see if I could get a job of Engineering for the winter could not get it*

1876

Feb. 5: *States drew stumps today. Put up 20 bushels Oats at ther Cook Barn for John and 12 for States making 75 for John and 33 for States*

Feb. 6:	*Sunday the grainery spring a leak and about 10 best wheat ran out put it back in The cows got at it and eat much that it made them sick..*
Feb. 10:	*..The Thrashers came here today to thrash States Clover seed. There was a dance at Mr. McNinch's tonight I did not get any invitation King folks went and they got wet coming home . . . Foddered twice today.*
Feb. 11:	*It rained hard this morning..it rained so hard that they could not thrash. Mr. ———was here today Uncle John made a bargain with him to work land foddered at noon.*
Feb. 16:	*..roads drifted bad most impassable went horse back done nothing today but set in the house and done the chores.*
Feb. 17:	*Helped States thresh clover seed, Mr. Pl—cks machine . . .*
Feb. 26:	*States went to town horseback I cut down the hay mow at this barn so we could clover hay to feed and at the Cook barn also.*
Mar. 8:	*came home to day about noon talked of trading fiddle with W. Shutt..*
Mar. 11:	*Jim pulled stumps all day in the Blake-lot. I helped him..*
Mar. 21:	*done nothing but set in the house it is very bad weather snows and blows . . . went to town took a grist for John and two bushels wheat for himself . . . done the chores at these two barns*
Mar. 25:	*Our party was a success F——— and Carson played . . . making $3.00 that we payed the fiddlers*
Apr. 21:	*Got up and beat Jim 11 games to 3 [checkers] then came home about noon . . . began plowing in the field north of the house and in the calf pasture today*
Apr. 22:	*put up fence all the forenoon and salted the sheep all round and counted them.*
Apr. 27:	*Wm Wampole came about four O'clock and stayed all night..plowed the garden today.*
Apr. 30:	*It was very cold and snowing quite hard this morning. I went over to see..stopped C. Porters to inquire when he could saw Jim's lumber and to John Kriledlers [unclear] when I was coming back to get some seeds.*
May 4:	*Wm Wampole and Frank and I drew out manure from the Barn yard to the field where L——— is to put peas got out abut 28 loads*
May 5:	*done the chores in the forenoon today in the afternoon Wm. Wampole came and drew stone from the field north of the house to the stump fence worked all day.*
May 21:	*Sunday, took my underclothes off today for it very warm*
May 22:	*Cut some poles out in the woods to put on the fence then about 11 O'clock L and I went down to the Cook Barn and cleaned up 16 and one half bushels oats for seed run them through twice took out 4 bushels for feed . . .*

May 23: *It froze quite hard last night froze water on the North stoop and spoilt tomatoe plants that I just set out last night*

May 25: *Took the big two year old heffer down to Mr. Heyden and weighed her she weighed 1075 lbs. She is not as heavy as she was in April by 50 lbs . . . went fishing and caught three*

May 27: *In the forenoon I sowed clover seed on where L——— has oats in the field east of the house . . . dug up burdocks til noon then after went over to Mr. Ingals to help him put the roof on barn (got two letters from Gleanson Co.)*

Jun. 5: *Took 8 cows down to Fremonts to Bull the old cow Shut cow and Heyman cow the Heyman is all that took the bull*

[NOTE: June 15th begins picture selling said not very good, gave some away but on 19th orders more pictures.]

Oct. 26: *Went to town today in the afternoon to get some clothes to go to the Centennial with but Rice would not make them so I did not get any and gave up going bought an ax for $1.00*

Oct. 27: *Commenced cutting wood in the afternoon down in the back woods have got a job of cutting 50 cords for John at 45 cents per cord L——— went down on the flats after axles today.*

Nov. 16: *Luait [sp?] threshed his buckwheat today and Bill Wampoles machine he had 42 bushel Buckwheat and 39 Oats and Buckwheat I helped him three fourths day fodder[ed] his sheep in the morning*

Nov. 24: *Luait and I drew stumps in the afternoon Today I am 23 years old.*

1877

Jan. 31: *Peter Wampole threshed at the Cook Barn today and yesterday I took away grain from the Machine had 292 bushel oats..*

Mar. 17: *The Heyman Cow came in last night had a small white calf.*

Diary ends soon thereafter. . . . .

Philo Hampton, Diary, Unpublished, Ossian, N.Y., 1853, Genesee Country Museum, Mumford, N.Y.

CHAPTER IX

Greek, Gothic and Other Decorative Barns

FOR THE MOST PART, barns are rather practical in their designs, without much adornment. But in the nineteenth century, as the number of skilled craftsmen grew, transportation improved and communication with the East increased, some successful farmers became style-conscious; their new barns began to reflect the current architectural fashions in domestic houses and public buildings. For example, in western New York, Greek Revival architecture flourished from the 1820s until about the time of the Civil War. Here the Greek temple form and individual temple details were reproduced perhaps more often than anywhere else.

The Balcom farm house in Steuben County, circa 1825, is a Greek Revival structure with extraordinarily good proportions (see figure 9-1). The Balcom house dramatically exhibits the major characteristics of the prototype Greek temple, with its pedimented portico supported by Ionic columns (see figure 9-2). The accompanying small barn also exhibits Greek detailing in the orientation of its gable end to imitate a pediment, the low pitch of its roof, and its wide, tripartite entablature. Like many other Greek Revival barns, this one is clapboarded, in contrast to the vertical boarding in other barns.

In 1852 Aaron Barber (whose sawn frame 1870s gambrel-roofed barn was discussed in Chapter VII) built a Greek Revival house with a cobblestone foundation for himself in the Town of Avon. Three Greek Revival barns were built thereafter and appear on the Livingston County Map of 1858 (see figure 9-3). The Aaron Barber/Edward Mulligan barns all have hand-hewn post and beam construction. The cupolas were added later, apparently to conform to those on the newer gambrel barns. The house has prominent fluted Doric columns on the portico, capped by a good entablature located just under the eaves, an essential element that is shared by the barns (see figure 9-4). Like the Balcom barn, these are oriented so their gable ends face the road, the gables imitating Greek pediments. The entablature on the gable end is incomplete, leaving the corner vestiges of the entablature now known as returns. Greek detailing on the barns is rather spare, but distinctive enough to integrate the style of house and barns.

A large barn owned by Alfred Senft and located near Peoria has the major Greek Revival characteristics (see figure 9-5). Again, according to common practice the gable end faces the road. The pitch of the roof also replicates that of a Greek temple, and in this case the entablature is of good proportions, in imitation of the prototype. While the exterior is decorative, the timber frame is fairly conventional, with the mows

Opposite: Detail of the Coe Gothic barn, South Livonia, New York.

FIGURE 9-1. Balcom Greek Revival farm house, circa 1825, between Cooper's Plains and Campbell, Route 17, Steuben County, New York.

FIGURE 9-2. Typical Greek Temple: The Theseion, circa 428 B.C., Athens, Greece. Major characteristics of a Doric temple. Anonymous, nineteenth century stereocard.

FIGURE 9-3. The three Barber/Mulligan Greek Revival barns, Barber Road, Avon, Livingston County, New York. The near barn (20 by 23 feet) is the corn crib, the center (23 by 23 feet) a carriage house, and the rear barn (30 by 40 feet) a stable. The corn crib is still in use as such.

FIGURE 9-4. Barber/Mulligan Greek Revival house. The most obvious similarity between barns and house can be seen in the detailing of the entablatures.

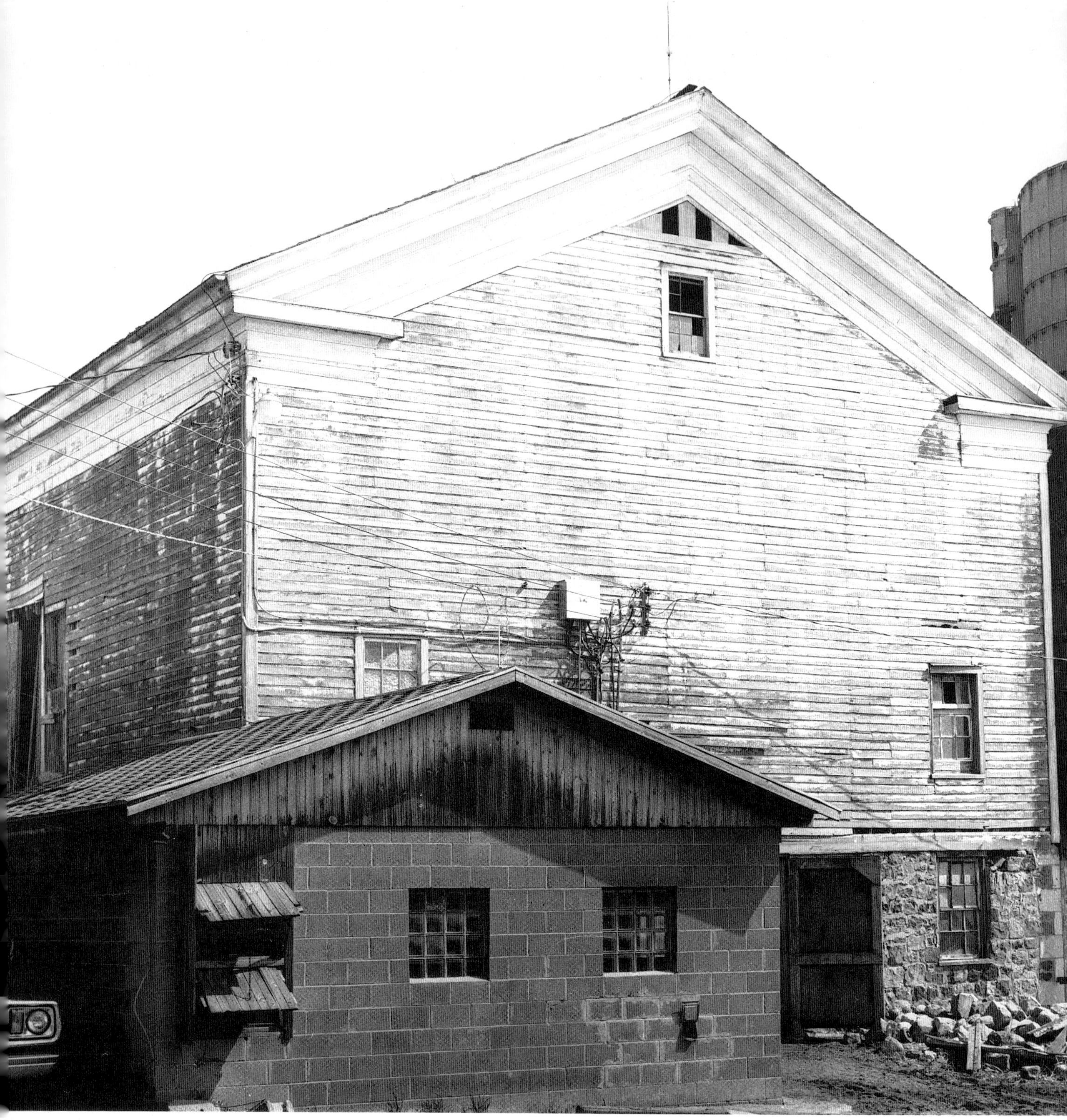

FIGURE 9-5. Senft Greek Revival barn, Simmons & Kendall Road near Peoria, Wyoming County, New York.

flanking the drive-through. Very similar in its proportions to the Senft barn is the Greek Revival barn located near Bath (see figure 9-6).

Figure 9-7 shows a lithograph of the Colonel Sardis Simmons Greek Revival barn in Richmond. Since the orientation of the house and barn is the same, it is easy to compare their Greek details. Further, notice that the barns in the foreground without Greek characteristics have quite a different feeling to them.

FIGURE 9-6. Greek Revival barn, Mitchelville Rd., Hickory Hill Campground northeast of Bath, Steuben County, New York. The major difference between this barn and the Senft barn is the contrast in siding, with horizontal boarding on the Senft barn and board-and-batten on the Mitchelville Road barn. The Mitchelville gable end does not face the road and is only lightly detailed; the entablature is the most dominant Greek Revival feature on this barn.

FIGURE 9-7. Lithograph from McIntosh's History of Ontario County, 1876, permitting close comparison of the sidehill Greek barn with the Greek farmhouse. Sardis Simmons Farm, Town of Richmond, Ontario County, New York.

Gothic Revival Barns

In the 1820s people in Europe, England, and the United States became interested in things Gothic, such as kings and queens, knight's tales, castles and cathedrals. Early in the century the poems and novels of Sir Walter Scott captured the imagination of people everywhere with lines such as these from *Ivanhoe:*

> The champions thus encountering each other with the utmost fury, and with alternate success, the tide of battle seemed to flow now toward the southern, now toward the northern extremity of the lists, as the one or the other party prevailed. Meantime the clangs of the blows, and the shouts of the combatants, mixed fearfully with the sounds of trumpets . . . The splendid armour of the combatants was now defaced with dust and blood, and gave way at every stroke of the sword and battleaxe.

Later, Rowena speaks:

> "I bestow on thee this chaplet, Sir Knight, as the mead of valour assigned to this day's victor" . . . The knight stooped his head, and kissed the hand of the lovely Sovereign by whom his valour had been rewarded; and sinking yet farther forward, lay prostrate at her feet.[1]

The people, places, and events chronicled in *Ivanhoe, Kenilworth, Rob Roy, The Monastery,* and other Scott novels caused many to sigh, wishing that they had been born in that age of chivalry when feuding, religious upheaval, clashes between lords and landholders, and courtly love were the order of the day. Since it is impossible for mortals to go back in time, it was thought a reasonable compromise to surround oneself with the trappings of this earlier period; in extreme cases this meant Gothic architecture. The castles and cathedrals of the twelfth to the fifteenth and sixteenth centuries served as prototypes for the Gothic buildings of the nineteenth, although in most cases details from different structures, perhaps centuries apart, were combined in the nineteenth century designs.

The Cathedral of Notre Dame in Paris, begun in 1160, exhibits some of the details borrowed for nineteenth-century designs (see figure 9-8). Perhaps the most conspicuous element of this style is the pointed arch, consistent with the steeply pitched roof and the upward-pointing fleche, finials, towers, and buttresses—all combine to produce the vertical impression that describes the Gothic style. Gothic architecture was the preferred revival style of early Victorians, one of the most influential structures being the new Palace of Westminster (Houses of Parliament), designed in 1835 by Charles Barry and A. W. N. Pugin to replace the old Houses of Parliament which had burned in 1834 (see figure 9-9). The illustration details the exterior stonework, showing the emphasis on the vertical with skyward-directed finials and the magnificent Victoria Tower, complete with cresting and encrusted with crockets.

It was not until the end of the Civil War that this country's first school of architecture was established at the Massachusetts Institute of Technology. Before that time, prospective builders could only refer to the many "builders books" that provided ideas and designs of all sorts. Samples of such books include A. J. Downing's *The Architecture of Country Houses,* which first appeared in 1850, and John Sloan's two-volume offering, *The Model Architect,* the first volume of which appeared in 1852. Downing's work appears to have been the more popular: By 1866 nine editions had been printed. Downing complained bitterly and often about the falseness of imitating the stone of prototype structures using wood or other substitute materials, but he did

FIGURE 9-8. Notre Dame Cathedral, Paris, begun 1160. The vertical dominates this structure from the pointed arches to the cathedral's spires, finials and pinnacles.

nevertheless provide sufficient examples to satisfy those wishing to build in the Gothic style. Interest in the Gothic existed for about forty years, with recommendations to "follow the Gothic" persisting until the 1880s. There is absolutely no evidence that a Gothic barn was better in any way; it simply appealed to the aesthetic side of the owner. Today many of the original details have been removed from Gothic buildings due to their delicacy and the high cost of upkeep; in some cases, entire structures have vanished.

An illustration in the June, 1878 issue of the *American Agriculturist* showing the dairy barn of J.E.S. Gordon is similar to the barns and stables of Westerly, near Piffard (see figures 9-10 and 9-11). These stables are examples of less decorated Gothic structures.

FIGURE 9-9. Victoria Tower, Westminster Palace, Houses of Parliament, London, constructed 1840–1860.

FIGURE 9-10. Simple Gothic detailing of a dairy barn from the June, 1878 issue of the American Agriculturist. This illustration was much reproduced during the next 25 years in all manner of books and periodicals.

In 1851 Major William Spenser finished the house known as Westerly on a farm of some 6000 acres. It was here in 1876 that Charles Wadsworth, a subsequent owner, is reported to have raised some of the first Shorthorn Herefords. Wadsworth referred to his entire operation as the Genesee Valley Stock Farm. Interestingly, a diamond-incised window pane at Westerly records the accidental death, from a chimney fire in the house, of ex-president John Tyler's daughter

FIGURE 9-11. Westerly Stables, Westerly, near Piffard, Livingston County, New York.

Julia. Before her death from smoke inhalation, Julia gave birth to a daughter in the house.

Clearly the buildings in the woodcut and photograph are similar, but Westerly was not modeled after the woodcut view. Rather, both were modeled after Gothic prototypes. The steep pitch of the roof and the vertical lines of the board-and-batten siding move the eye upward in true Gothic style. With the advent of battens the barn became significantly more weather-tight. Battens were produced in a variety of shapes to cover the joint between two adjacent boards. The simple batten form was the result of ripping one board into several narrow ones and nailing them over the board joints (see figure 9-12). Battens were frequently painted a contrasting color, further emphasizing the vertical lines. Red boards with white battens were referred to as "peppermint" barns in the Attica/Java area.

Vertical stripes were painted on barns during the Civil War, in at least one instance for patriotic purposes. Samuel Valentine had just finished his barns in Fishers, Ontario County, as the war broke out in 1861. In a patriotic gesture Valentine proceeded to paint the vertical siding of his barns with five inch wide alternating red, white, and blue stripes. These unique barns came to be called "Union Barns";[2] Rallies were held there and many were recruited into the Union Army on the spot. These rallies are recreated every Fourth of July at Valentown Hall Museum. Although the barns are now destroyed, a door displaying the faded paint is a part of the Museum's collection.

The oblique view of the west side of the barn at Westerly in figure 9-13 is a good example of optical movement created by the steep roof pitch and the lines of the battens. Cupolas continue the upward movement of the eye. After the Civil War, cupolas became much more prominent, frequently being added to existing barns as well as to new designs. The round-arched louvers and the extent of the overhang of the cupola roof are reminiscent of the Italian Villa style in domestic ar-

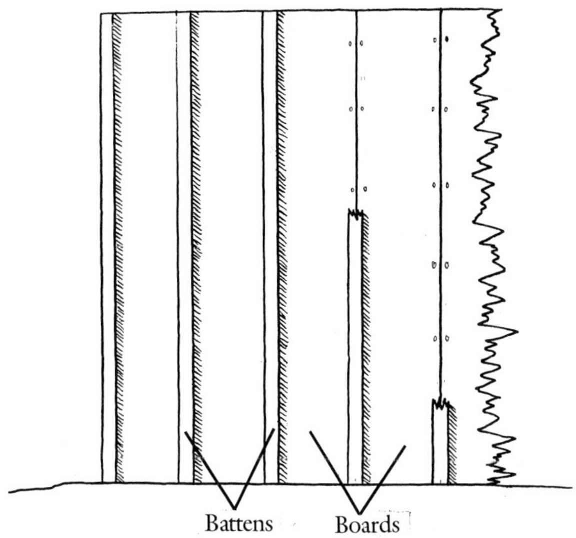

FIGURE 9-12. Board-and-Batten siding.

chitecture, which was popular in the post-Civil War years. (It would have been difficult to adapt all Italian Villa features to barn construction, partly because its decorative nature ran counter to the practical needs of a barn.) The one practical aspect of the louvered cupola is the fact that it helped ventilate a barn that had been made more airtight with battens. The elegance of the Westerly stable exterior does not prepare one for the hand-hewn frame clearly seen in the rafters and ridge pole of the haymow in figure 9-14. The nineteenth century box stalls were preserved throughout the long tenure of Porter Chandler but were removed in 1980 (see figure 9-15).

Although most Gothic barns were built as stables, the builders did have other concerns besides decoration, as shown in this 1866 discussion in the *Farmers' Barn-Book*, by Clater, Youatt, Skinner and Mills, which reveals period thinking about stable design and grooming:

> Before we proceed to the management of the horse in the stable, it will be as well to say something about the stable itself. This should have been constructed so as to contribute to the health and comfort of the horse; and be well-aired, dry, well drained, light, and sufficiently

FIGURE 9-13. Board-and-batten on Westerly stables, with the steep pitch of the roof leading the eye upward in a typical Gothic fashion.

FIGURE 9-14. Hewn rafters and ridge pole in Westerly stables.

FIGURE 9-15. Westerly box stalls.

spacious to allow the necessary grooming operations to be duly performed. It should also be sheltered from the coldest winds, and be easy of access.

The stable must be so paved that all the wet shall drain into the gutter, and out of that by another drain to carry every thing clean away.

Windows are very often too much neglected, or too small, or ill placed. Light thrown down from the roof, by means of skylights, which partially open or revolve, and can be readily opened or closed by means of a cord are excellent, as according good ventilation. In all stables, high windows answer well, when of sufficient size, and so placed that the light shall not fall directly upon and affect the horses' eyes. To obviate this as far as possible, they must be high in the wall, and in sufficent number to give a good light.

It is a mistaken opinion, that but little light is requisite in a stable. No horse was ever known to thrive in a dark stable; but many a good horse has had his sight seriously affected by this absurd and mischievous practice. It may answer a dealer's purpose to keep his horse from the light till brought out to be sold. When brought out from a dark stable into the light of day, a horse very naturally stares about him; he looks high, carries his head high, and appears as if he had a good deal of action and animation. Dark stables may thus suit particular purposes; but they invariably injure the horse's eyes; to say nothing of the necessity of light in the several grooming operations, . . .

The roof of a stable usually forms the floor of the hay-loft: in this case the ceiling must be at least ten to twelve feet from the ground, and higher, if more than four horses are kept in the stable. Where there is no loft above, the height should be greater; because in summer, the sun makes the tiles or slates hot, and the stable becomes like an oven; while, in winter, the cold, and sometimes snow lying on the roof, converts the stable into something like an ice-house. Such extremes must be avoided: they are prolifically productive of disease . . .[3]

The article helped influence stable design in the period.

The degree of matching exhibited by Gothic houses and their barns varies considerably from situation to situation. One such matching barn, reminiscent of a design from Downing or Sloan, is located in North Bergen. The relatively small board-and-batten, cupolaed barn was intended to suggest the fancier design of the house (see figure 9-16). The "correct" features of the board-and-batten, symmetrical farm house—pointed, arched windows, steep-pitched roofs, and delicate scrollwork—are also reflected in the stable, which has board-and-batten siding like the barn, but a lower roof-pitch and plainer appearance.

FIGURE 9-16. On the John Shields farm, Lake Road, North Bergen, Monroe County, New York: a finely detailed Gothic house whose elegance is imitated but not matched by the detailing of the barn.

FIGURE 9-17. The Coe House, circa 1850, South Livonia, Livingston County, New York.

On the east side of Livingston County was once the remarkable matching complex of buildings on the John Coe farm in South Livonia (see figure 9-17). The barn and original house were both primarily Gothic in their decoration, having been built about 1850. The house combined a basic Italian Villa design and rich Gothic detail, with its patterned slate roof, iron cresting and gingerbread trim. This house exists now only in photographs; it was destroyed by fire in March, 1927 (see figure 9-18). The barn still stands, however, displaying the splendor and elegance that Gothic structures can convey. The hood moulding of the louvered windows is arranged in imitation of the pointed arch characteristic of the Gothic. A textural richness is everywhere, from the slate patterns of the roof surfaces to the decorative bracing of the doors (see figure 9-19). Carpenters pierced the mouldings with their scroll saws to produce intricate designs, especially noticeable on the cupola and gables, where the pierced barge boards, or verge boards*, are built up with an overlay of lattice. These features are still remarkably sound in view of the 130 or so winters that have passed since their construction.

The barn was originally painted gray and

*barge/verge boards are vertical face boards that follow the roof edge of the gable—in Gothic architecture these are usually carved, sawn, or built-up.

FIGURE 9-18. Coe Gothic barn, Route 15, South Livonia, Livingston County, New York. The 40 by 60 foot barn has a frame of 6 by 8 inch timbers beneath its Gothic exterior.

FIGURE 9-19. Detail of Coe Gothic barn door.

FIGURE 9-20. Detail of Coe Gothic barn gable and cupola.

yellow but has been allowed to revert to a natural patina (see figure 9-20). Other Gothic structures were built in the Livonia/South Livonia area, indicating a strong regional interest in this style. Of these, the houses remain but the barns have succumbed to the ravages of time.

Most Gothic barns in the Genesee Country were built as stables in villages, towns and cities, generally on a smaller scale than the Coe barn. A suburban carriage house and stable, designed by A. J. Downing for M. Vassar of Poughkeepsie, appeared in Downing's 1842 edition of *Cottage Residences* (see figure 9-21). Downing included this design in articles written for several different publications and, such was its popularity, it was cited by many others even after Downing's death.

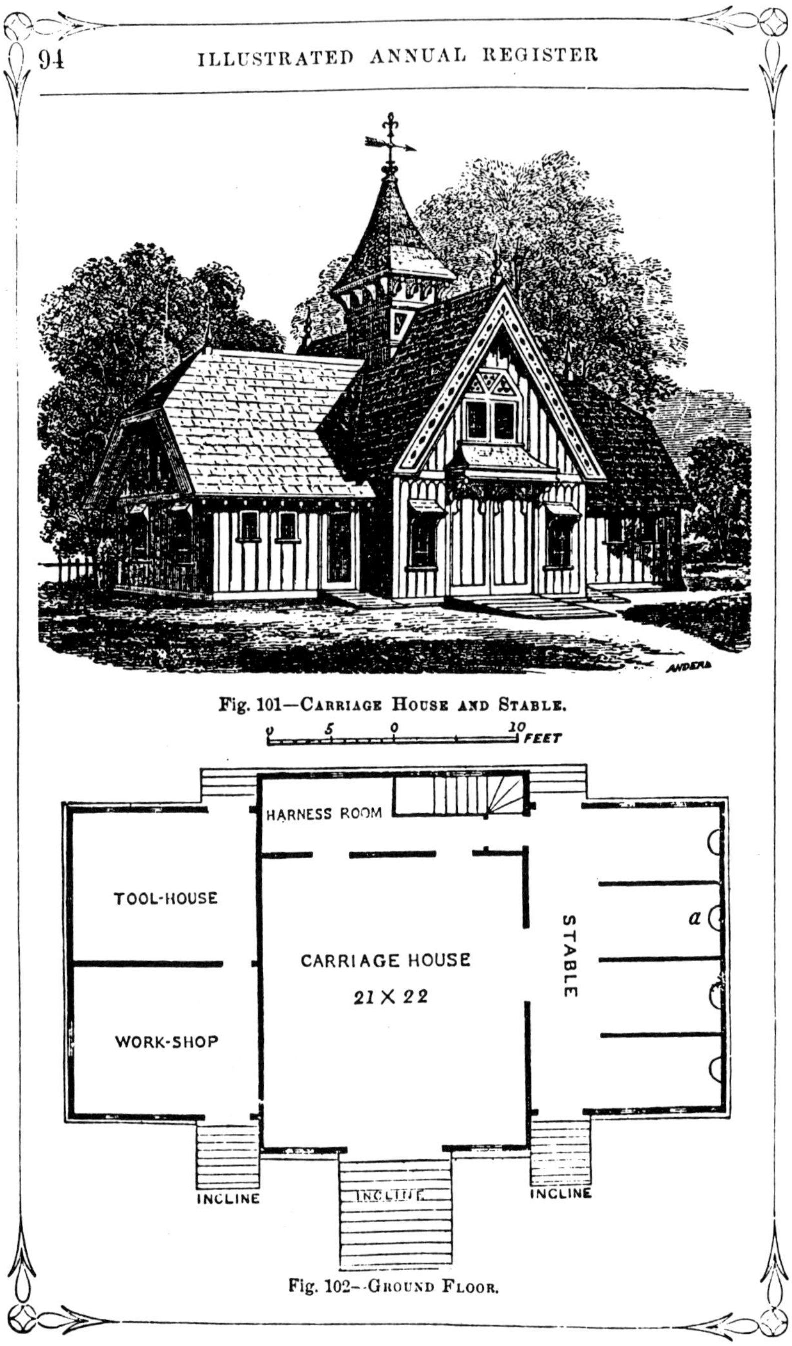

FIGURE 9-21. A. J. Downing's design for a carriage house and stable. This illustration appeared in the Horticulturist and The Genesee Farmer in 1853, Luther Tucker's Illustrated Annual Register of Rural Affairs for 1858, and Periam's Home and Farm Manual of 1884.

One can identify many similarities between the Downing design and the Coe barn, including the roof pitch, patterned slate roof, barge/verge board, and other details, although the Downing design is not suggested as a prototype for the Coe barn. Downing himself described the Vassar construction as a sidehill barn with the basement used for farm horses, cows and a root cellar. The timber frame is sheathed with matched pine boards and battened. Downing is apologetic about the cost, saying the Vassar structure cost $1,900 because of the high cost of lumber but could be built for $1,000 where lumber was not as expensive.[4]

Figure 9-22 shows a fine Gothic stable in Branchport. The steep pitch of the roof and the strong use of sawn Gothic detailing produce a fine overall textured effect. The gables are layered in decoration, from the pierced sawn vergeboards to the oval and sexfoil windows surrounded by pendanted cap-boards.

While board-and-batten was quite in keeping with the Gothic style, it was also used extensively on other barns built between the 1840s and the 1880s, such as the cow barn on the Wadsworth Homestead in Geneseo. This barn is more decorative than many of the other barns in this book, with the exception of "purer" Gothic barns (see

FIGURE 9-22. Pulver Gothic Stable, Branchport, Yates County, New York, built by Salomon Weaver.

FIGURE 9-23. Wadsworth Homestead cow barn, Geneseo, Livingston County, New York.

figure 9-23). Its roof is pitched relatively low because it was built after the Gothic fashion was almost past. Calvert Vaux, the nineteenth-century architect, wrote in 1864 in his book, *Villas and Cottages,* "From small Gothic barns good Lord deliver us!"[5] According to William P. Wadsworth, the barn on the Homestead Farm dates back to the early days of his father Major Austin Wadsworth's ownership, which began in 1874. The balloon-framed cow barn has clapboard siding at the first story level and board-and-batten siding at the second level; the resulting sharp visual change reminds one of the textural changes that occur in Italian Renaissance structures. Although fancy in appearance, details of the barn's double doors reveal a good combination of both decorative and functional purpose (see figure 9-24). The vertical and horizontal members are nicely beveled, allowing a play of light, while the boards of the door are set at a diagonal to provide both sheathing and stability. Figure 9-25 shows the interior of the cow barn as it appeared in 1982. The original metal patent stanchions are no longer present. The open door in the gable end was added later to accommodate conveyor loading of baled hay. Originally, dormers on the east side allowed hay to be pitched into the mow from a wagon drawn up alongside.

FIGURE 9-24. Wadsworth Homestead cow barn, detail of double doors.

FIGURE 9-25. Wadsworth Homestead cow barn.

The Italian Villa Style

The Italian Villa style in domestic American architecture first appeared in the mid-1820s, but seems to have been most prominent in the Genesee Country in the decade following the Civil War. As mentioned earlier in this chapter, the Italian Villa was a complicated form with heavy decoration. The most outstanding decorative element was the tower or roof-centered cupola, called a belvedere (see figure 9-26). Often there was a fondness for elaborate decorated brackets supporting the extended overhang of the eaves. Interesting geometrical play is found both vertically and horizontally with the spacing, detailing and numbering of windows and doors.

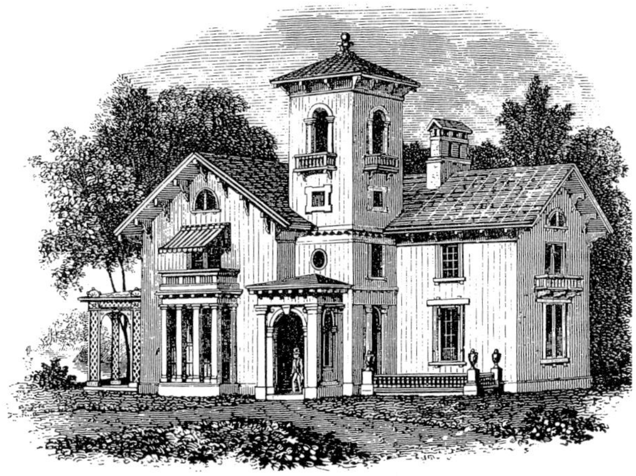

FIGURE 9-26. A. J. Downing villa in the Italian style. Cottage Residences, 1873,

The Italian Villa often was a bit eclectic and it is not uncommon to find both classical and Gothic detailing.

The barns on the L. M. Loss farm in Albion, shown in a lithograph from the 1879 *History of Orleans County,* reveal the fundamental visual connection between the bracketed Italian Villa farmhouse, with its central tower, and the louvered cupolas and bracketed eaves of the barn (see figure 9-27).

Similar connections between house and barn detailing can be seen on the 1876 Noah Cole/Gagnier Farm in Mendon. Figure 9-28 shows the east side and figure 9-29, the west side in a lithograph from McIntosh's 1877 *History of*

FIGURE 9-27. L. M. Loss Farm, Albion, Orleans County, New York. This marvelous complex also shows the formal garden associated with the farmhouse (a feature which Downing would have approved). Orchards blanket the upper left portion of the lithograph, while haying takes place in the field on the lower left and grain is being reaped and bundled in the lower right section. A New York Central and Hudson River Railroad passenger train hurries through the edge of the property. Lithograph from the Sanford History of Orleans County, 1879.

Monroe County. The brick Italian Villa has a belvedere rather than a tower; its round-headed windows are reflected in the rounded arched louvers of the barn cupolas. The barn has been reroofed and the brackets removed. This basement barn is also interesting because of the use of side lights next to the barn doors to provide natural light for the interior. A more common location for the lights would be in the transom position over the doorway.

McIntosh's 1876 *History of Ontario County* includes a fascinating lithograph of John Maltman's

FIGURE 9-28. Cole/Gagnier Farm, east side. Pittsford-Mendon Road, Mendon, Monroe County, New York.

FIGURE 9-29. Cole/Gagnier Farm, west side. McIntosh, History of Monroe County, 1877.

FIGURE 9-30. The barn and stable on the John Maltman Farm, Canandaigua, Ontario County, New York. McIntosh, History of Ontario County, 1876.

farm in Canandaigua. One can clearly see not only the layout of the entire farm but also the fancy stable and barn that are part of the complex (see figure 9-30). Both structures have intricate patterned slate roofs. The design of the stable, complete with horse weathervane, shares many basic features with the Cole/Gagnier Farm barn from the same period, including the arched, louvered openings in the cupola, which are doubled on the Maltman farm. However, the Maltman center gable does not form a projecting pavilion.

The practice of designing the barn to match the house was particularly common on larger farms or estates. After their marriage in 1834, James S. Wadsworth, son of the pioneer, traveled in Europe with his bride, Mary Craig Wharton. While in England the two became enamored with an English villa in Regent's Park, London and decided to construct a similar villa just north of Geneseo's Main Street. In 1835 Hartford House was constructed, named for Lord Hertford but spelled the way the name is pronounced in England (see figure 9-31).

The barns of Hartford House are an interesting complex of fairly small brick and timber structures stuccoed to match the main house (see figures 9-32 and 9-33). The roofing, also designed to correspond to the main house, is predominantly standing-seam metal, more commonly known as a "tin roof." This roofing technique became quite popular in the second half of the nineteenth century and has lived up to its original claims of durability (50–100 year warranties were not unusual). This use of many small structures is in sharp contrast to the more general practice of housing many activities in one large barn.

Although it may not seem so from the above examples, the practice of matching the architecture of outbuildings to that of the house was actually somewhat limited in the Genesee Country. Perhaps more typical was a combination of styles, as noted earlier among the various buildings at

FIGURE 9-31. Hartford House, 1835, Geneseo, Livingston County, New York.

Westerly, where a basically Federal house with Greek porticos is combined with Gothic barns. The major significance of the examples presented in this chapter is that they all exhibit exterior characteristics that were popular in house architecture when they were built, in strong contrast to the more utilitarian mode of most barn exteriors.

FIGURE 9-32. Barn complex at Hartford House. The various barns are, from left to right, the carriage house, a horse barn with blacksmith facilities, a separate horse stable, and a cattle barn.

FIGURE 9-33. The Genesee Valley Hunt begins at Hartford House. The field is moving from the house itself, shown on the left, to the barn complex on the right and thence to the hunting land beyond.

FIGURE 9-34. The Genesee Valley Hunt starts out at Nations Farm.

The Genesee Valley Hunt

When summer gives way to early fall, the mid portion of the Genesee Valley resounds to the sounds of the fox hunt, the notes from the hunt master's horn sounding above the baying of the hounds and the cacophony made by the hooves of a hundred horses or more. In figure 9-34, the Hunt gathers on the Nations Farm in Geneseo, wearing the blue and buff of the Colonial Army. For over 100 years the sport of fox hunting has helped preserve the character and lifestyle of the Valley. The preservation of farmland needed for the hunt has prevented the destruction of many a barn; in fact, some of the Valley's best-preserved older barns, with the fewest changes, are in fox hunt country.

The Genesee Valley Hunt was formed in 1876 by W. Austin Wadsworth and others of a like mind, making this one of the oldest active hunts in the country. People from all walks of life have ridden to the hounds: politicians, farmers, businessmen, educators and royalty. Theodore Roosevelt is one of many who have enjoyed the Genesee Valley Hunt because of the beautiful country and the stiff challenge that it offers.

In one major way the sport of fox hunting as practiced by the Genesee Valley Hunt is different from the same sport in England: The chase is emphasized rather than the kill. William P. Wadsworth, Master of Fox Hounds from 1932 to 1975, describes the hunt in his book, *Riding to Hounds in America*. Wadsworth's respect for the ability of the fox is evident, and he has always claimed that only a disease-weakened, tired or miscalculating fox can be caught. Although a good fox may lead a chase for as much as fifteen miles, it often escapes or goes to ground. In his chapter on the fox, Wadsworth notes,

> No man could list all [the fox's] strategems to throw hounds off his trail, but the most common are the sudden turn after a straight run, and running on a surface which he knows will not hold scent, such as a road or cart track, a railroad rail or fence rail, or even a fallen tree trunk, or a manured field. He loves to run through a flock of sheep or a herd of cows or pigs, and will often follow a deer for a while, hoping that hounds will stay on the deer line when he turns off it.[6]

The fox population varies according to disease and road kill, but it is a rare hunt when a fox is not sighted.

Hunting is not just a Saturday morning affair but requires considerable attention all year long, from the whelping of the pups in early spring to the end of hunting when the ground freezes in early winter. Because it is no longer economically practical for a single family to keep a pack of hounds, today the Genesee Valley Hunt is a subscription hunt. In addition to the considerable cooperative effort required to maintain a hunting pack, there is the constant need to maintain good relations with farmers so that their land may be hunted. Thus it is not surprising that a number of area farmers ride to the hunt.

At sunrise on the last Saturday in September the Genesee Valley Hunt resumes its tradition. The field meets at the Wadsworth Homestead at the south end of Geneseo and proceeds north through its streets to the hunting lands northwest of the village. The present Master of Fox Hounds, W. Austin Wadsworth, leads the hunting pack, the whippers-in and the field of well over 100 (see figure 9-35).

Hunts are mostly held in an area ranging from just south of Geneseo to the northernmost limit near Henrietta. It is within this region that many of the preserved barns included in this study have been located.

FIGURE 9-35. An opening hunt, Main Street, Geneseo, New York; almost 175 riders in the field trail the Master of Fox Hounds, W. Austin Wadsworth (third rider from the left.)

FIGURE 9-36. An early snowfall silhouettes William P. Wadsworth (on the right) and the hounds as they cross Wheeler Gully north of Geneseo on a Thanksgiving Hunt, 1972.

CHAPTER X

A Crop Sampler

Crop selection remains an ever-changing concern on a farm. From the pioneer period to the present, farmers have experimented with crops to discover those best suited to the climate and growing conditions of the Genesee Country. In addition, they have had to cope with problems in reaching markets and with changing demands for crops. As an outgrowth of changing crop patterns barn designs also change, as will be seen in this and subsequent chapters.

Corn, for Indian and settler alike, was one of the most reliable crops in the frontier days of the Genesee Country. In addition to using the ground kernels as meal for human consumption, farmers also used the stripped leaves and the stalks for cattle feed. More than 300 varieties had been planted in New York State by the 1880s, but corn began to decline as a major food crop at mid-century, when it was found that the Erie Canal could transport it more cheaply from the West. Later, however, a surge in the dairy industry led to a renewed interest in corn production in the 1870s, '80s and '90s. During this period corn was planted in even rows using seed drills; the resulting stalks were often dried in shocks by tieing together the cut plants, sometimes supported by several uncut plants, in a vertical shock. This allowed circulating air to dry the corn, which was then husked and stored in a corn crib. After shelling, it was the general practice to cook or steam the corn before it was fed to cattle and hogs. Corn prepared in this manner was considered a nearly-perfect feed for cattle.

Dairy farms today are often easily identified by their multiple large, cylindrical silos. However, silos were not always cylindrical and, in fact, were not even a part of the American farm until the last quarter of the nineteenth century. Prior to Genesee Country settlement by Easterners, the Iroquois had stored corn underground (see Chapter I). According to historian Ulysses P. Hedrick, Europeans were experimenting in the 1830s with the possibilities of fermenting green fodder in air-excluded chambers, but the idea was not introduced to American farmers until 1875. That year, a book about silos and ensilage by Dr. Manly Miles, of the Michigan Agricultural College, and the June issue of the *American Agriculturist* together stimulated interest in the French method of ensilage-making. The September, 1877 *American Agriculturist* presents a rather broad discussion of American practice with ensilage. An excerpt from the article explains the fermentation process:

> It has also been proved, that the fermentation which the corn undergoes, changes some of its cellulose in character, and renders it more digestible, as well as transforms some of its starch into sugar, and some of its sugar into alcohol. The preserved fodder posseses an agreeable vinous scent, which is not at all distasteful, but rather otherwise, to the cows to which it is fed.

Opposite: Detail of horizontal doors used to regulate curing on a Steuben County tobacco barn.

> When it is understood how easily this valuable fodder may be preserved, we shall look for the adoption of the method upon dairy and stock farms, where the food supply is a matter of serious consideration.[1]

Always controversial, the process had few early converts: An 1882 survey identified only ninety-two farms in the United States with silos.[2] Certainly one of the earliest silos in the Genesee region is the below-ground silo Herbert Wadsworth built at South Avon in 1884, at the farm now known as the "Silo Farm." In 1896 the D. M. Osborne company of Auburn illustrated in its catalog a square silo, built into the corner of the mow to make it easy to fill from an upper floor and to feed cattle on the lower floor (see figure 10-1). The Osborne publication also included five illustrated pages on how to construct a silo. Figure 10-2 shows a diagram of the end elevation of a square silo, showing a plan of the doors. Figure 10-3 shows a square silo on a farm located east of Jasper in Steuben County.

FIGURE 10-1. The D. M. Osborne Company catalog of 1896, touting Osborne machines for planting, cutting and binding corn. The catalog contains a full discussion of various procedures for making ensilage, including the construction of silos.

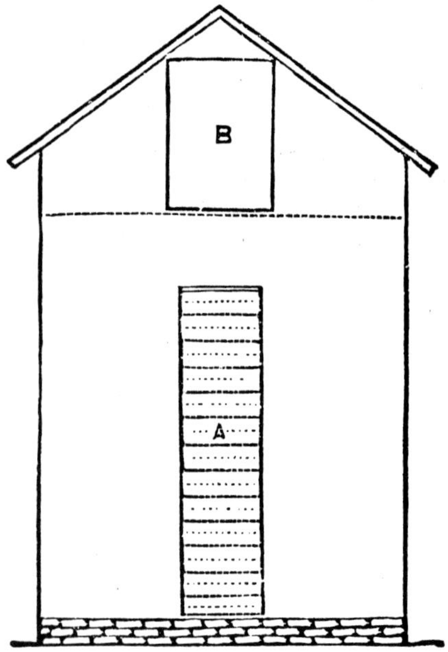

FIGURE 10-2. End elevation of a square silo from the Osborne catalog.

Resistance to the feeding of ensilage to dairy cattle is evident in a July 30, 1885 article in the *Livingston Republican*: "The editor of the Orange County, New York, *FARMER* does not believe, in spite of analysis and kindred tests, that a sour, fermented or putrefactive food can produce healthful milk, or be otherwise than injurious to the cow in the long run." This opinion persists in some to this day, but after initial uncertainty silos were built all over the United States and are now a common part of the landscape.

There were many ways of building silos, using many different materials, but two basic methods seemed to dominate early construction. In one method, 2x4s were laid flat (see figure 18-10). In the second method, a stave silo was made from thin vertical boards bound with round iron wire hoops to prevent spreading. Figure 10-4 shows the construction of a stave silo on the Albert Schleede Farm north of Lyons. Another example of a stave silo is from the George Wingate Farm in Livonia (see figure 10-5). These latter structures lack significant frames and eventually give way to the stresses created on the unfilled structure by

FIGURE 10-3. Square silo on a Steuben County farm east of Jasper on Rte. 417.

FIGURE 10-4. Building a stave silo on the Albert Schleede Farm, two miles north of Lyons on Maple Street Rd., 1920s. Photograph from the collection of Mrs. Ralph Schleede.

prevailing winds. Until the 1890s silos were generally built on masonry foundations, but as concrete became available as a building material in the 1890s, concrete silo foundations became common.

Figure 10-5, a 1905 photograph, shows silo loading on the Wingate Farm. The corn was cut when still green, pitched onto a wagon (usually four or five were needed) and hauled to the barn. The corn was then cut into pieces small enough to be blown up a pipe into the silo—approximately ¾ to 1¼ inch long (see figure 10-6). Horses, steam tractors and, later, gas tractors were used to power the chopper and blower.

To exclude air from the loaded silo, someone had to walk around and stomp down the chopped corn. Small boys often performed this task. With growing acceptance of corn ensilage, the number of acres planted to corn increased, since three to four times as much corn by weight can be grown per acre compared to hay, even when hay is harvested three or four times a year. A thorough examination of the evolution of silos could provide a subject large enough to produce a separate volume.

FIGURE 10-5. 1905 silo loading on the George Wingate Farm, Livonia, New York. The wagons draw the bundles (each with 8–12 stalks) to the cutter box, where the steam-powered blades chop the ensilage before it is blown into the silo. The blower delivery pipe can be seen on the silo to the left, and the steam traction engine to the extreme left. Photograph from the collection of Mr. and Mrs. Gordon Wingate.

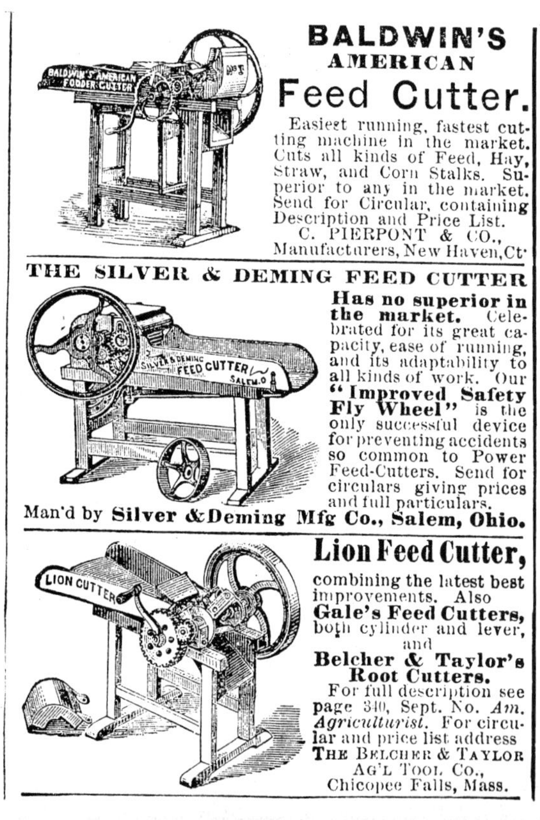

FIGURE 10-6. Feed cutter advertisements. 1877 American Agriculturist advertisements show box feed cutters of the sort used to chop the corn fodder.

FIGURE 10-7. Horizontally sheathed silo on Rte. 417 east of Greenwood, New York.

All that remains of a barn and silo complex east of Greenwood is part of the barn frame and the silo, which was originally constructed of 1x3s in the manner of a stave silo and subsequently sheathed in thin horizontal boards. An impressive number of nails were used to tie together the two layers of the silo (see figure 10-7). By the turn of the twentieth century many different manufacturers had entered into the silo construction business (see figure 10-8).

While the subject of silos could fill a book, their various roofs and caps could themselves be the subject of an essay. In addition to those already seen, the gambrel-roofed versions are of interest because they imitate the design of the barns they accompany (see figures 10-9 and 10-10).

Other methods of preserving corn were practiced: The Shakers were particularly noted for their dried shelled corn, and by the 1880s the canning industry was fairly well established, with canned corn among its principal products.

FIGURE 10-8. Patent silo advertisements from The Rural New Yorker, 1901.

FIGURE 10-9. Gambrel-roofed stave silo on route 417 near Woodhull, Steuben County, New York.

FIGURE 10-10. This square silo is capped with a gambrel roof similar to those on the main barns. Route 36, five miles south of Canisteo, Steuben County, New York.

(Canning and freezing continue to be important industries in the Genesee Country in the 1980s.) As the interest in growing corn grew, corn harvesters were developed to take advantage of the straight rows created by improved planting methods. Perfected corn harvesters came into use in the late 1880s and eventually were equipped with binding capabilities as well. With the old method of harvesting, illustrated in an Osborne Co. catalog (see figure 10-11), the most that might be expected was an acre a day. In 1896, Osborne estimated that its new binding corn harvester could handle seven to ten acres per day (see figure 10-12).

Farm folklore is filled with stories of husking bees held to ease the work of removing the husk from the dried corn. An 1858 patented corn husker, the Leavenworth, is shown from the pages of *The Genesee Farmer* in figure 10-13. The 1871 Philips Spiral Corn Husker was considerably advanced over the Leavenworth Husker. Philips claimed it could husk one bushel per minute (see figure 10-14). Machine husking was an enormous help, but shelling the dried corn remained a time-consuming job until the advent of

FIGURE 10-11. Nineteenth century illustration showing harvesting corn by hand.

shelling machines. The Genesee Corn Sheller was rather typical of such machines manufactured at the end of the century (see figure 10-15).

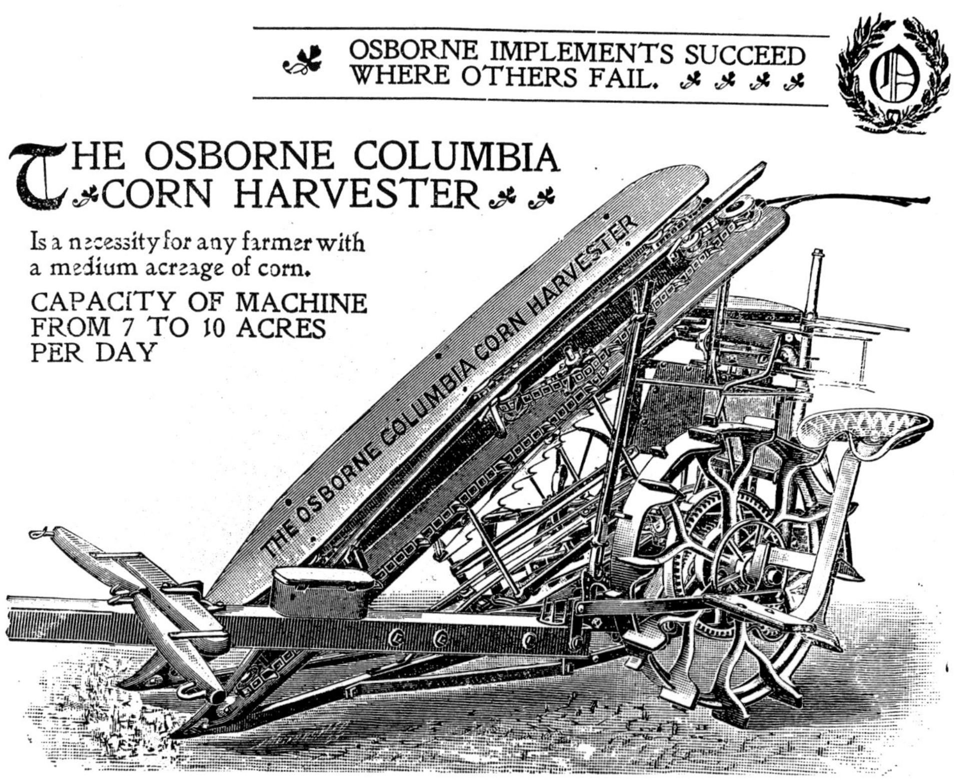

FIGURE 10-12. An Osborne Company advertisement, 1896.

FIGURE 10-13. Leavenworth's Corn Husker. Apparently the device was intended to be used in the field. The Genesee Farmer, 1858.

FIGURE 10-14. Philips Spiral Corn-Husker of 1871 represents an improvement over the Leavenworth. The new machine could handle ears of various sizes, remove the husks and silks, and leave the stalks in better condition for use as fodder. The American Rural Home, April, 1871.

THE GENESEE CORN SHELLER.

This machine has no superior. It is so adjusted by springs as to adapt itself to large or small ears of corn. Runs very easily. Built of best material. Gearing so constructed that you cannot get your hands injured. Boxes are solid, avoiding the loosening of bolts in frame, or allowing the gearing to get out of line, or cogs to slip. When preparing to shell, loosen springs by means of thumb nut as much as they will bear and shell clean.

No. 1 machine has single spout, and can be worked only by hand.

No. 1½ machine is same size and capacity as No. 1, with the addition of a screen and tray. Cleans for market.

No. 2 has castors so that when tilted back one person can move it to any part of the building. Has DOUBLE SPOUT, with shaker and tray, and cleans ready for market. Can be worked by hand or power. Shaker in front so that the delivery of corn and cobs are in sight of the operator.

GENESEE VALLEY MANUFACTURING CO.,

Mt. Morris, N. Y., U. S. A.

FIGURE 10-15. The Genesee Corn Sheller, manufactured in Mount Morris, Livingston County, New York, represented a tremendous labor-saving device. Advertisement, The Genesee Valley Manufacturing Co., 1898.

FIGURE 10-16. Comic post card touts the quality of corn in the region served by the New York Central Railroad. These post cards were altered to suit many areas in the U.S.

Granary of the Country

Wheat was grown almost from the beginning of settlement, but it was the period from the late 1820s to mid-century that earned the Genesee Valley the name "Granary of the Country." At this time Monroe and Livingston counties were consistently the largest wheat-producing counties in the United States (see figure 10-17). The reason for this productivity was given by Joseph Harris, editor of *The Genesee Farmer*, when he observed that the Genesee Valley was the best natural wheat-producing area in the world. The climate is right for wheat growing and annual flooding enriched the soil in the valley bottom (see figures 10-18 and 10-19).

FIGURE 10-17. Wheat in a stained glass church window. Carney Hollow Road, Livingston County, New York.

The great flood years in the nineteenth century are recorded as 1815, 1835, 1857, 1865 and 1896, years of tremendous destruction. In the 1865 flood the Genesee Valley and Erie Canals overflowed, highway bridges were carried away, the center of Rochester was under water, a major railroad bridge was washed out and the gas mains ruptured. Most years, however, the flooding was a benefit to the valley, renewing the land close to the river with new deposits of rich silt. The major problem to the farmer was the weed seeds in the deposits. Those who knew the valley built their houses and barns high enough off the ground so that the average flood would have little effect on them. This elevation is often not noticeable except during times of flood, since the valley is so wide that floodwaters normally were not deep. Indeed, often the only way to get around during a flood was on foot. Many stories exist of people walking from one side of the valley to the other, crossing the river channel on bridges submerged in the floodwaters. Today, with the Mount Morris Dam providing flood control, new structures are generally built without regard to flooding.

The following story of Horatio Jones gives some insight into early wheat planting (Jones's frame barn was celebrated earlier, in Chapter V). L. L. Doty relates:

> Leicester was the first of the towns west of the river in which a permanent settlement was made. Soon after the close of the Revolution, Horatio Jones, whose years of captivity here had made

FIGURE 10-18. Ripe wheat in the Genesee Country.

> him familiar with this region and therefore fully aware of its great excellence for agricultural purposes, prepared to settle on the flats . . . the south-east corner of this tract was at a point near the junction of the Canaseraga creek with the Genesee River . . . reaching the flats in July, they cut about nine acres of grass..and stacked the hay on the pioneer meadow. The same fall they plowed and sowed to wheat the ground they had previously mowed, and this nine acres is believed was the first crop of this grain planted west of the Genesee River. . . . As yet, however, there was no frame building of any kind in the town, but in the fall of 1796, Horatio Jones erected a frame barn, a little to the west of Jones's Bridge . . . and soon after he built a frame distillery at the Fort farm, as it is called, being the same farm that was occupied by Colonel Jones.[3]

Orsamus Turner, in his *Pioneer History of Phelps and Gorham's Purchase* of 1851, describes the efforts of two brothers-in-law of Horatio Jones:

> Enos and Jared Boughton..erected a log cabin, sowed a patch of buckwheat..and after sowing three acres of wheat, the whole party returned to Massachusetts [except for one caretaker].
>
> ..The stock of provisions [Jared and his wife] brought in, lasted with the help of the buckwheat that had been harvested the previous fall, until their wheat harvest.[4]

Jared Boughton's efforts to get his buckwheat to the mill at Avon were recorded in Chapter VII.

From these meager beginnings the Genesee Valley became by mid-century the wheat-producing center of the United States. Of course it wasn't just the quantity that mattered but the quality of the flour produced from this wheat. W. H. McIntosh, in his 1877 *History of Monroe County*, raved, "It is said of the Genesee wheat that it contains more saccharine matter than that from other localities, and will combine with less water in the composition of bread. The superiority of the flour is too well known to need remark."[5] Wadsworth flour, mentioned in Chapter VII, was sold under its own trademark, most of the flour being milled in Wadsworth mills, and

FIGURE 10-19. The Genesee River flats under water between Geneseo and Cuylerville, New York, in the aftermath of hurricane Agnes in 1972. Prior to the construction of the Mount Morris Dam the river flooded almost every year, renewing the soil.

sold at a premium price. About 1845 the major market for Wadsworth flour shifted from Boston to New York City.

O'Reilly's *Sketches of Rochester* was published in 1838 during the high point of wheat production. O'Reilly notes:

> Wheat is at present, and will probably long remain the great object of cultivation; and the quantities produced between the Ontario and the Falls of Nunda at Portageville, which may be considered the southern limit of the wheat country proper, would almost exceed belief; and in quality as well as quantity is generally considered much beyond that of any other section of the country. It was for many years supposed that the rich flats of the Genesee River were unsuited to the production of wheat, it being imagined that the growth would be so luxuriant as to produce lodging of the grain and mildew, and the consequent destruction of the crop. To a certain extent this was and still may be true on some of the more moist and recent alluvial sections; but the general introduction of the harder-stemmed varieties of wheat, in place of the former kinds of red wheat, such as the white flint in the place of the red chaff and bearded reds, has in a great measure obviated these difficulties, and the flats are now as celebrated for wheat as they formerly were for corn. Of this the following instances, and they might be multiplied to almost any extent, will be perfectly conclusive:—
>
> In 1835 Messrs. P. and G. Mills cut from twenty-seven acres on the Genesee Flats near Mount Morris, 1270 bushels of wheat, or forty-seven bushels to the acre. In 1834 the same gentlemen cut from eighty acres three thousand two hundred bushels of wheat, being forty bushels an acre. The most beautiful field of corn we ever saw was in the summer of 1833, on the farm of W. C. Dwight, Esq., on the flats a few miles above Geneseo. There was one hundred and seventy acres lying in one body, and from it he harvested twelve thousand eight hundred bushels of shelled corn. In 1834 the same gentleman had twenty acres of wheat, which averaged forty-eight bushels per acre, and two acres of the best of which produced fifty-two bushels per acre. The elevated country on the east and west of the river is scarcely inferior in the growth of wheat; the greatest amount we believe on record as the well-authenticated product of a single acre having been raised by Mr. Jirah Blackmore, of Wheatland, being sixty-four bushels per acre.[6]

It might serve to put the above production figures into better perspective to mention that in 1796 Thomas Jefferson, a very careful and enlightened farmer, figured eight bushels of wheat per acre to be a good harvest on his Virginia land.[7] By 1840, according to the sixth U.S. Census, the average wheat production in New York State was fourteen bushels per acre, the average in Monroe County closer to twenty bushels. Appropriately the town of Wheatland was consistently one of the largest wheat-producing areas of Monroe County. By 1850 both Monroe and Livingston counties were still producing in excess of one million bushels of wheat per year.[8]

This preeminence ended in 1855, when excessive rain caused the crop to rot in the fields. The

Hessian fly and the wheat midge ruined the next three years' crops and the year after that, in 1859, a killing June frost proved disastrous (see figure 10-20). One of the first published observations of the Hessian wheat fly was in western Vermont, reported in the 1820 *New England Farmer.* The fly moved across the Hudson and Mohawk valleys to the Genesee about 1855. *The Livingston Republican* of July 30, 1885 described the dual pests that affected wheat production:

> The Hessian fly lays its eggs in autumn, and again in spring. The midge..lays its eggs in the flower or ear. The eggs hatch, and the little violet-colored maggots live in the inside of the soft grain to devour it. A well-filled ear of wheat, near harvest time, hangs down; if you see one with its head erect, by examining the kernels you will probably find the midge. If about the time the wheat is shooting into ear, or later, you see the plants "crinkling down", or withering up, on examination you will find maggots at the joints. This is the Hessian fly, either in the larval stage or as the pupa. Early sowing, and early varieties, are the recognized preventers against serious injury from the midge; but late sowing is desirable when there is danger from the Hessian fly.[9]

Even after control of insect damage became possible through changing the wheat varieties planted, fertilizing, and carefully controlling planting time in relation to the life cycle of the insects, the era of wheat growing prominence in the Genesee Country was over, and that distinction went to the Middle West.

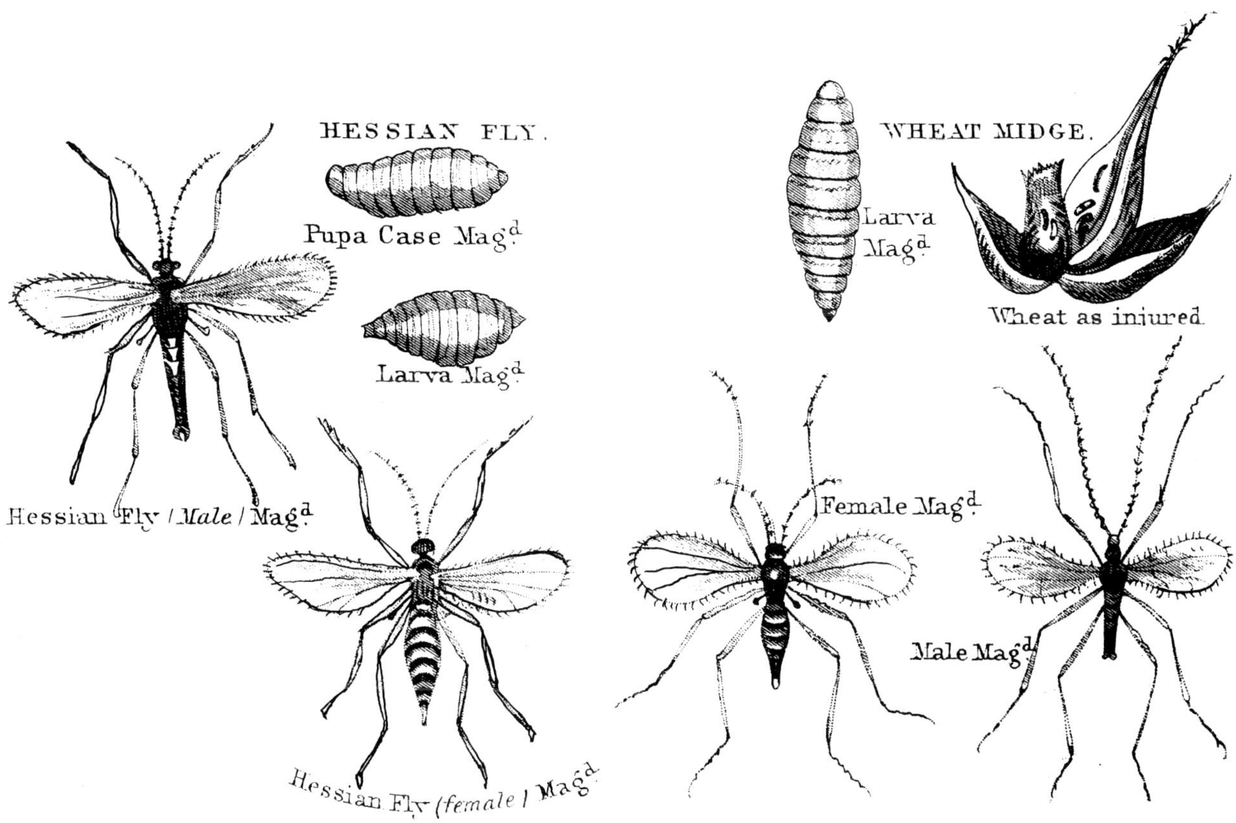

FIGURE 10-20. Drawings of both the wheat midge and the Hessian Fly, as illustrated in Solon Robinson's 1866 Facts for Farmers. *It took considerable time to understand the life cycles of these insects so that destruction of the wheat could be avoided.*

Where wheat continued to be grown in the valley, fertilizing and better soil preparation helped to protect it against insects as well as enhance productivity. Fertilizers included manure, guano, and plaster (gypsum). Plaster had to be brought into the area until extensive deposits were discovered near Scottsville and Mumford. The oldest practice of improving the land consisted of leaving it fallow: Following the wheat harvest the land was plowed two or three times, harrowed half a dozen times and left fallow for a season, which supposedly allowed it to recover its fertility. This system, which was expensive, was replaced by summer-fallowing: In the fall the fallow land was planted to wheat and in spring grass was sown on the wheat land. The winter wheat was harvested in early July and the grass allowed to grow, with the land remaining unplowed until the following spring. Summer-fallowing did little to improve the land. Later, experiments proved the superiority of crop rotation, in which clover and grasses were planted alternately with grain crops. Clover alternating with wheat at a three-to-one ratio was often mentioned as a good rotation to prevent land exhaustion. On the other hand, heavy fertilizing had the advantage of allowing wheat to be grown each year.

Beans

In the Genesee Country a wide range of beans, starting with white beans, replaced wheat as the staple crop and are still important in the 1980s. *The American Rural Home* discussed in 1871 the replacement of wheat with beans, emphasizing the compatibility of the two crops. Beans were harvested, cured in the field, bundled, stored in the barn, and threshed whenever equipment was available. The yield of beans per acre was roughly the same as the yield for wheat, although the market value of wheat was roughly twenty-five percent higher. Space in barns for unthreshed bean storage was a concern, especially in cases where a variety of crops were grown, all requiring considerable barn space before threshing. One especially interesting aspect of nineteenth century bean growing was that, while beans were used for cattle feed, most Americans did not find them palatable. For almost twenty years the bean industry lagged, until a taste for beans was acquired by the Civil War troops (beans were a part of their rations, like it or not). When the taste for beans spread following the war, New York State became the country's largest bean producer.

Clover, Hay and Oats

Clover was planted to feed livestock after the fanning mill made the production of quality clover seeds possible. Both clover and timothy hay became important crops by the mid-nineteenth century because of the increased use of transportation horses in non-farming areas (see figure 10-21). Haymarkets were a common part of large towns and cities, supplying large quantities of hay and oats for both commercial and private stables. Oats, of course, were a major crop in this region, as elsewhere. The market for feed used by transportation horses lasted for 100 years in the Genesee Country, until approximately 1920.

Hay was sometimes stored in stacks, both in the field and in the barnyard. The hay was bundled and carefully laid up to keep the stack

FIGURE 10-21. Red Clover.

FIGURE 10-22. Stacking hay on the Hubbard Farm, North Chili, Monroe County, New York. McIntosh, History of Monroe County, 1877.

upright and to protect it from the elements. Often the haystack was formed around a pole, but other wooden forms were also used. In some places roofs were built over the haystack for added protection. It is difficult to know if artistic license was employed in figure 10-22, but the haystack in the lithograph is as high as the barn roofs and still growing.

The Burning of Barns by Lightning and Combustible Hay

When a wooden barn burns a loss is felt; all know that the replacement barn, if any, will be different. Unfortunately, barn fires are not uncommon. Storing hay in the mow without properly drying it first can create sufficient heat to cause the barn to burn, and lightning is a constant threat (see figure 10-23).

The storage of green hay in a barn is always dangerous, because green, wet hay supports the growth of bacteria that generates large amounts of heat. Drying or curing the hay before it is stored is something each farmer has to understand, and nineteenth century periodicals frequently dealt with this important issue, since a fire due to improper hay storage could lead to the loss of the barn. York Historian Mary Root made an interesting list in her *History of York, N.Y.*, recording some of the various births and deaths of local barns beginning in 1844, a time when the English barn was still the dominant design. In these ways the barns were winnowed away:

July 13, 1844, Lightning struck and burned H. Long's barns [Gratwick Farm.]

October 11, 1844, H. Long had a big barn-raising.

May 18, 1845, fire started in hotel barn of J. McLean, it burned Hayward's barn.

June 29, 1850, Ray Hitt had barn-raising.

June 17, 1852, F. A. Gray had a barn-raising. [these barns burned in 1936.]

July 24, 1859, Large barns of Neil Stewart burned, with Alex Reid's two horses, buggy, etc.

May 30, 1872, The large barns of Orange Sackett burned. Small boys playing with matches started it.

August 15, 1877, As the threshing machine was being moved out, the barns of Patrick Nolan were burned, together with hay, oats, wheat and the threshing machine.

September 3, 1877, Fire destroyed the barn of John Gilmore. Also the same night, the barns of Alex. Hutton were burned and with it the threshing machine of Mr. Doremus.

October 11, 1877, Fire destroyed the barns of William Lowrey.

November 22, 1877, Alexander Hutton had a barn-raising.

February 27, 1878, Frank Clements' barns were burned.

July 18, 1878, Guthrie's barn burned during an electric storm.

September 3, 1878, Rev. Van Eaton's barn burned.

July 9, 1886, J. W. McArthur's barn burned during a storm.

August 26, 1888, The barn on the Stewart Carson farm burned.

FIGURE 10-23. Lightning strikes on the Genesee River Flats.

January 16, 1893, Frank Shannon's barn burned at Greigsville.
July 31, 1893, Severe electrical storm caused the burning of the barns of C.C. Wadd, shop and barn of Geo. Bauer at Linwood, and the barn of James Conway at Greigsville, which spread to the barn of William Ramsey.
May 5, 1898, Eli Sherwood's barns burned.[10]

See Figure 10-23.

One method farmers used to combat the build-up of heat was the inclusion of windows high in the gable ends. Further ventilation was provided with working cupolas, a design feature that became increasingly prominent following the Civil War. Other safety precautions included drying the hay thoroughly and even treating it, as the very first *Genesee Farmer* in 1831 mentioned, in an almost cryptic comment:

> Farmers should not forget that a little salt sprinkled upon their hay, as they pack it away in the stock or mow, not only enables them to put up their hay sooner with safety, but contributes to keep their stock in better health when fed with it.[11]

J. Cozzens, in an 1846 letter, published in a report made to the New York State Legislature the connection between barn fires, hay storage and lightning:

> . . . I believe it to be a fact, that more barns are burned by lightning just after they are filled with new hay, than after the hay has become seasoned, or when it has become purely dry, or when the barns are empty.
>
> Hay, when first stored, has always more or less moisture in it; it has also an essential oil; it is also always more or less hot from the field, and when put in the barn in large quantities, its juices and moisture are undergoing decomposition; and the compound is giving out gases which are forming new compounds; these ascend from the barn, and like all other exhalations reach some height in the air, so that when the clouds which pass over, are plus, and earth about the barns is minus, the fluid, *oleous* gas, and vapor, may act as a conductor, and attract the electric *fluid,* and carry it directly to the heated hay in the barn.
>
> I am pretty certain that if hay was salted when housed, it would be beneficial . . . [12]

The *American Agriculturist* of September, 1877, describes the curing of hay, as opposed to the drying of hay:

> Every year the problem comes up, how green will it do to get in hay? and every year our practice is to get in with less and less drying. Cut between 4 and 8 PM; the next day, if heavy, shaken up and turned before noon, and once after, then raked up at 3 PM, and got in the same afternoon, is pretty quick work, and to accomplish it, and have the hay cured enough to keep without salting, requires a hot, drying day. I have just had excellent luck, cutting late one day, raking into windrows at about noon, cocking up by three o'clock, and getting in the next forenoon, and using salt or not, according to the condition of the hay. This hay is cured and little dried, and there is very little work about the operation. Salt has a wonderful effect, I do not understand it. The hay "sweats" well, but if salted rarely molds, or gives off any but fragrant odors. It does not seem to me that the salt could come in contact with the hay so generally as to affect it all, and yet every lock seems to be preserved.
>
> I have never used salt upon corn-fodder, but I shall do so this year, confident that four or five quarts to the ton will aid greatly in curing it, and prevent the necessity of so much drying. The finer the salt, the further it will go in salting hay. A handful of salt will penetrate through three feet of hay in the mow, as I have proved by experiment, and I presume if the hay were very dry, it would rattle down a good ways farther. [13]

A possible explanation of the beneficial action of salt is that it inhibits the bacteria that decompose wet hay and thus produce heat.

Potatoes, Tobacco, Cabbage and Hops

Potatoes are important to western New York and have been since the development of blight-resistant varieties in the second half of the nineteenth century. One concern of potato growing is that stored potatoes, because of their density, have been known to damage the structure of barns built of light timbers for other purposes; floor joists are particularly vulnerable. Several Genesee Country hay barns converted to potato storage were later discovered to have cracked main floor beams due to the added weight. Figure 10-24 is an advertisement with a potato motif from a nineteenth century Harris Seed Catalogue.

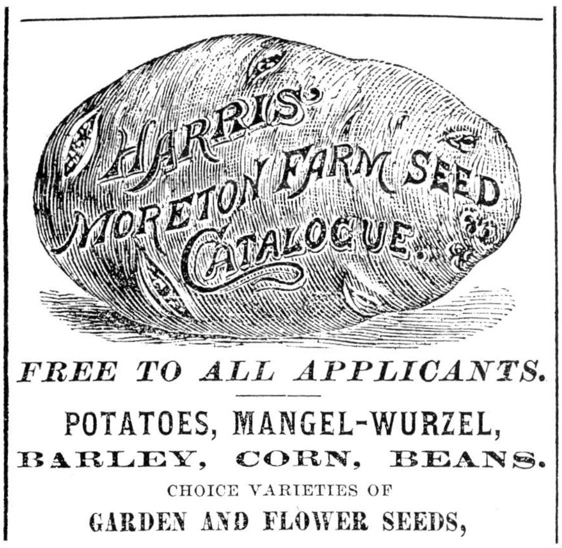

FIGURE 10-24. Potato from Harris Seed advertisement.

Tobacco was grown in small quantities from the very beginning of settlement on many Genesee Country farms for the farmer's personal supply of chewing tobacco. Gradually, larger quantities were grown commercially. Figure 10-25 shows chewing tobacco hanging in the Yowell barn, South Lima. Larger tobacco-growing operations existed on an east/west line just south of Lake Ontario and along a similar line from Chemung County on the east to Steuben County in the west. The tobacco barns, with their characteristic narrow vertical doors, have now disappeared in the northern Genesee Country but can still be seen in the southern portion of the region.

Tobacco growing in earnest in the Genesee Country seems to have started in the mid 1850s and was later encouraged by shortages created by the Civil War.[14] J. Howard Pratt mentions in *Memories of Life on the Ridge* that the sandy soil between Gains and Ridgeway was particularly well suited to tobacco and that production peaked at about half a million dollars worth during the Civil War. South of Keuka and Seneca Lakes, and east and west of Corning, tobacco was grown in quantity on the best river bottom land from the 1850s until the 1940s. Prior to the Civil War, the large market for Genesee Country tobacco was New York City.[15] McIntosh mentions that in the 1870s a pound of tobacco had about the same value as a bushel of wheat.[16] For some, tobacco was grown when a farmer retired from general farming; the practice was viewed as a kind of nineteenth century "social security."[17]

Solon Robinson's *Facts for Farmers* (1866) and Jonathan Periam's *Home and Farm Manual* (1884), both much used in the region, discussed the cultivation of tobacco and the all-important tobacco barn (see figure 10-27). Periam very carefully recommends a tobacco barn with four or five tiers on which to dry the stalks. The most obvious characteristic of these drying barns is their vertical doors, used to control ventilation

FIGURE 10-25. *Chewing Tobacco curing in Yowell barn, South Lima, New York.*

FIGURE 10-26. *A "Hand" of Tobacco.*

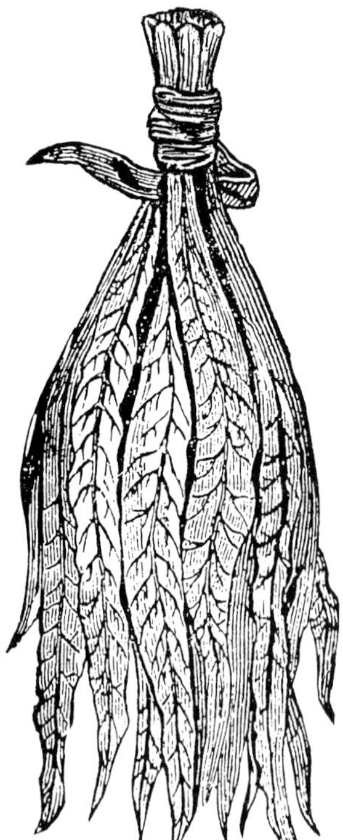

during drying. In damp weather the doors are closed and, in windy weather, adjusted to prevent the leaves from blowing about. The October, 1872 issue of the *American Agriculturist* also specifically mentions the ventilated cupolas essential to the curing of tobacco. Multiple cupolas are an additional exterior sign of the tobacco barn.

FIGURE 10-27. *Periam's tobacco house illustration, 1884. Of the tobacco barn illustrations used in this volume, Periam's most strongly reflects those once found in the Genesee Country, because of its conspicuous series of vertical doors; however, the expected ventilating cupola is absent.*

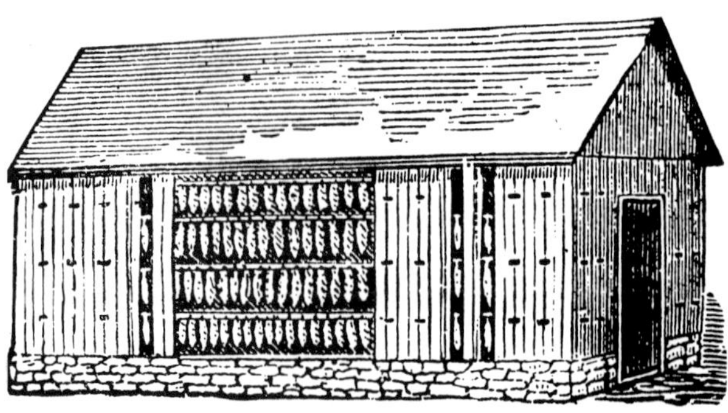

Planting and Harvesting Tobacco

The tiny tobacco seeds were planted early in a cold frame because the New York State growing season is too short for tobacco. Weeds are a significant problem in growing tobacco, so after the steam farm engine became practical the soil was steam-sterilized before planting to kill weed seeds. The rule of thumb for transplanting tobacco plants is to set them out when both days and nights are warm at about 6,000 plants to the acre.

Considerable cultivation is required as well as attention to devastating cut-worms and tobacco worms. The plant is pinched back when it blossoms and its suckers are removed to encourage the growth of true leaves. After topping, the plant matures in about three or four weeks. Periam writes, "Uniform size of leaves, and a stiffness of the leaf, making it liable to break by bending and handling, are the surest signs of maturity."[18] The stalk is cut with a heavy knife or hatchet and all the plants laid on the ground in the same direction, care being taken not to break off the leaves. When the tobacco has wilted it is drawn to the barn to be cured.

The tobacco was placed on poles three to four inches in diameter and laid horizontally across the long axis. Some were end-hewn on one side to prevent rolling, others were sawn. Men on the wagon below pitched the plants up to the twiner, who tied them to the poles by the stalk.[19] The poles, long enough to bridge the barn bents, were often placed across the barn axis. In some tobacco barns the poles were placed at right angles to each other on different levels. A 14-foot long pole would hold 50 to 60 plants. Figure 10-28 reveals the various hewn beams that supported the drying poles in the McCarthy tobacco barn east of Addison. Four men were needed to hang tobacco on the purlin, fewer on the lower levels. Some barns had charcoal or coke heaters to supplement natural curing. In December or January, depending on the weather, the tobacco was dry and could be removed from the poles. A damp day was selected, Joseph Kane comments, "because it was less likely for the tobacco to crumble" when handled.[20]

The cured tobacco was placed in a stripping room (see figure 10-32) with a kettle or pan of boiling water, the steam from which kept the tobacco leaves moist enough to prevent crumbling when they were removed from the stalk. Nineteenth-century agricultural periodicals are filled with articles describing the importance of proper curing and handling to the tobacco's value.

Next, the tobacco was sorted into three grades and bundled into a "hand" of twelve to sixteen leaves, then placed in boxes or cases ready for sale. The humidity had to be carefully regulated until the cases were removed by the buyer. Most buying was determined by inspection, some by contract. A buyer would first inspect a "hand" visually, spreading the leaves, then hold a lighted match under a leaf. If the flame stayed lit while moving along the leaf, the tobacco was ready.

Tobacco was used in many forms; it was chewed, especially by farmers when haying, threshing and so on because tobacco kept them from "drying out"; it was smoked in pipes and cigars; and it was milled to produce snuff. Cigars were often handed out to those who helped in a barn raising. Cigarettes became somewhat popular in the 1870s with the advent of cigarette-rolling machines; forty years later, World War I expanded their use enormously because they were

FIGURE 10-28. Interior of the McCarthy tobacco barn, east of Addison, Steuben County, New York. The many closely-spaced cross beams exist to handle the poles to which the tobacco is tied for curing.

given to the troops, often establishing a life-long habit.

Cigar making was probably the largest cottage industry in the United States at the end of the nineteenth century. Cigar production occurred both in the regions where tobacco was grown and in the cities and villages where immigrants tended to settle. The demand for cigars was high (see figure 10-29). Small cigar factories were established throughout the Genesee Country. In his book, *Corning and Vicinity*, Uri Mulford says that in 1866 there were twelve cigar factories in Corning with an annual production of over half a million cigars.

In the Genesee Country the outstanding tobacco works was the William S. Kimball & Co.'s *Peerless Tobacco Factory,* manufacturer of smoking and chewing tobaccos and cigarettes, located in the triangle formed by Court Street, Canal Street and the Erie Canal in Rochester. Kimball started during the Civil War with cigars and chewing tobacco, and until 1890 he was one of the top five producers in the United States. "Peerless" fine-cut chewing tobacco won first prizes in the Paris, Vienna, Melbourne, New Orleans and Atlanta Expositions.[21] At Kimball's peak more than 1,200 workers were employed in Rochester. Cigarette making commenced in 1876 with an annual production of fifteen million, and two years later Kimball received an award at the Paris Exposition for the finest cigarettes in the world. Cigarettes were first rolled by hand (by women because their wages were lower); by 1887 the use of machines boosted production and over 300 million were sold. Kimball's cigarettes were called "Satin, " "Athletic," "Four-in-Hand," "Cupid," "Superlative," and "Fragrant Vanity Fair"; his "Old Gold" brand lasted well into the twentieth century. In 1890 Kimball & Co., along with four other major tobacco companies, merged to form the American Tobacco Company.

FIGURE 10-29. Tintype of four men with their cigars.

FIGURE 10-30. Andrew Dzybon tobacco barn, Route 352, Town of Corning, Steuben County, New York. Wagons loaded with tobacco were drawn into the barn through the large doors on the gable end. The vertical drying doors on all four sides of the barn could be opened during the day and closed at night or when weather conditions warranted. Forged hook and eye latches secure the doors.

FIGURE 10-31. Tobacco barn with vertical doors, some open. This barn is located on the U.S. Department of Agriculture Big Flats Soil Conservation Service property on Rt. 352, Chemung County, New York. The end doors, through which the tobacco wagons entered the barn, can be seen.

FIGURE 10-32. 125 by 38 foot tobacco barn, Addison, Route 417, Steuben County, New York. Interestingly, this tobacco barn, typical of others with horizontal doors, is sided in board-and-batten, making the barn snug when the doors are closed and nicely open when the horizontal doors are raised. The small structure at right angles to the main barn is the stripping house.

The combination of vertical doors and cupolas on the gable-roofed tobacco barns in Steuben and Chemung Counties makes them easy to identify. The tobacco barn on the Andrew Dzybon Farm in the Town of Corning measures 55 by 150 feet, with 27 vertical doors to the long side. There were once five cupolas on the structure (see figure 10-30). Inside, the hand-hewn frame has room for five tiers of poles; on the eastern side of the structure (not shown) is the required stripping room.

On the U.S. Department of Agriculture Big Flats Soil Conservation Service property in Chemung County is a 25 by 110 foot tobacco barn, no longer used for that purpose, with 22 doors on each side; the doors are easily seen in figure 10-31 because of the light and because they are being opened in imitation of their use in previous years.

Somewhat unusual but by no means unique is the 125 by 38 foot tobacco barn in Steuben County, with its system of horizontal doors (see figure 10-32). The gable-roofed barn has been expanded, as can be seen in the roof line and the evidence of reroofing. The horizontal doors are hinged so they swing up when opened. This view plainly shows the stripping house (complete with chimney for the required stove) in which the cured tobacco was removed from the poles, steamed, stripped, sorted, bundled and cased.

Cabbage

For many it comes as a surprise to learn that as late as 1884 most cabbages were imported into the United States. Cabbage was primarily used to make sauerkraut. Joseph Harris writes in the July, 1884 *American Agriculturist* that if American farmers would only plant cabbages on a large scale they could easily be competitive with the imported varieties. Farmers in western New York accepted Harris's challenge and cabbage continues to be raised to this day. It is recorded that the 1897 cabbage crop brought $4.00 per ton and it took 3,000 railroad cars to move the cabbages from Rush and West Rush alone.[22]

Hops

Hops, once an important crop in New York State, were grown in the Genesee Valley as well as in the valleys to the east and required specialized hop-drying barns. At that time, small breweries were everywhere because beer could not be stored long and was widely consumed as one of the many substitutes for drinking water, which was much mistrusted. The October 22, 1831 *Genesee Farmer* says,

> There is a very great proportion of the best farming lands of the United States which are not supplied with pure and wholesome water . . . I do not know of any thing that I can so safely recommend for the use of laborers, as a common drink, which we think will serve so well to quench thirst, and be conducive to health, as *small beer*, made in the good old way with *malt, bran, hops* and *pumpkin*.[23]

Lydia Child, in *The Frugal Housewife* (1832) observes, "beer is a good family drink. A handful of hops, to a pailful of water, and a half-pint of molasses, makes good hop beer." Mrs. Child's *Table Beer* was stored in corked stone jugs to which a lump of white sugar had been added producing a drink which is "..brisk and pleasant, and continues good several months."[24]

The beer produced commercially was different but also used the dried female flowers of the hop plant. Barley malt was used to produce the mash from which the beer was made. Barley, still grown today, was a rather important crop in the Genesee Country. After the starches turned to sugar, the liquid was boiled in a copper vessel with the hops, which imparted a pleasant, bitter taste to the beer and also acted as a preservative. After the liquid cooled and settled, liquid yeast was added and the mixture allowed to ferment. The beer was then stored several weeks or months before marketing.

FIGURE 10-33. Hop picking often was a family affair. It required great care because leaves and stems mixed in with the hops reduced the value of the bale. Anonymous. Photograph about 1900.

FIGURE 10-34. 1870 Flannigan Farm hop barn. Once located on Stone Rd. in Greece, it has been removed to the Genesee Country Museum.

Hop growing seems to have become a large scale business when the English crop failed in 1822. Also, the Erie Canal opened up markets to the east and west of the Genesee Country. McIntosh describes hop raising in East Bloomfield:

> Hop-raising is the leading agricultural interest of the town. One of the first successful experimenters in this branch of farming was George Thompson. Charles Page purchased his farm and has carried on the business over twenty-five years. He has one of the largest and best worked hopperies in this locality.[25]

Various kinds of hop houses or barns had to be built close by the hop yards (see figures 10-33 and 10-34). The hop vines grew around poles which were removed when the hops were to be picked. Vines planted one year would produce the next and in years thereafter. The pickers, mostly women and children, could pick about forty bushels per day, earning about a dollar for the effort (see figure 10-33). The picking boxes had to be taken to the drying house immediately because the hops would spoil in the boxes in a few hours from overheating.

FIGURE 10-35. Hop yard with a drying barn in the background at the Genesee Country Museum. The hops are almost ripe.

Careful drying was critical if flavor was not to be lost. Figure 10-34 shows a typical 1870 hop house, from the Flannigan Farm in Greece (now located at the Genesee Country Museum). This hophouse is typical, with a two-story sawn timber frame, the ground floor divided into the drying kiln (here located at the rear) and the bale-storing room where the dried hops were stored after being pressed into 200 pound bales. The patent hop press (not shown) was located on the upper floor, here identified by the floor joists at the top of the photograph, and the pressed bale exited the bottom of the press onto the lower floor. The smoothest and best bales were pressed on rainy days when the high humidity made the hops pack better (see figure 10-35). On the drying floor immediately above the fireboxes (arched openings in the brick furnace wall), open slats supported loosely woven cloth, which allowed the dry heat to circulate. The hops had to be carefully turned to permit thorough drying, which took almost half a day. In some areas of New York the hop barns were quite distinctive due to the methods developed to provide heat to dry the hops; the building at the Genesee Country Museum is a simple, gabled structure.

Hedrick notes that New York State production of hops rose to more than 2 million pounds in 1848, peaking in 1879 at about 21,600,000 pounds.[26] The *American Agriculturist* of 1877 describes western New York and Wisconsin as the two great hop growing districts in the United States. A fungus in 1920 effectively ended hop-growing in New York State and almost all hops used in brewing in the United States now come from Oregon and Washington. The discontinu-

FIGURE 10-36. Hop barn interior. Flannigan/Genesee Country Museum. The brick kilns are fired at the rear of the ground floor. Heat rises to the second level where the hops have been spread out on the cloth-covered joists.

ance of hop growing ultimately led to the destruction of many a hop barn through disuse.

Space limitations allow just this hint of changing crop considerations, change which affected the designs of barns and other specialty farm structures. Horticulture is so important to the Genesee Country that Chapter XI will be devoted to an overview of apple, peach, and pear growing, seed production and nurseries.

CHAPTER XI

Horticulture

SPECIALTY BUILDINGS AND BARNS were a part of fruit growing in the Genesee Country almost from the beginning of settlement. The combination of excellent natural conditions and the heritage of the Iroquois orchards gave horticulture its early start. In their 1779 campaign Sullivan's troops had noticed the fine quality of the orchards planted by the Senecas. It had been part of the army's orders to destroy the fruit trees, laden with nearly ripe fruit, by girdling them. After the destruction of war came the replanting of orchards during settlement. Most early farmers planted their own small orchards; orchards on a commercial scale in the Genesee Country were a post-Erie Canal phenomenon. The combination of naturally fine growing conditions with reliable transportation to markets was required before commercial fruit growing could be undertaken.

The depth of concern for planting orchards was presented in Chapter III, in the 1832 ledger notation from James S. Wadsworth to his outdoor clerk, Gardner Scott, in which Scott was advised to promote the setting out of fruit trees on the tenant farms.

Lake Ontario has a moderating effect on the climate in its vicinity, which both helps produce superior fruit and protects it against late fall and early spring frosts. The climate also improves the keeping quality of the fruit, increasing its market value. Figure 11-1 shows the orchards associated with the Sumner Farm in Shelby. In 1838 O'Reilly identified the reasons that the Genesee Valley became an important fruit growing region; he also described the experiments with silk production in the same area:

> It must be evident, from the nature of the soil and peculiarities of climate in the lower Genesee Valley, that it is admirably adapted to the production and perfection of the various fruits and vegetables raised in our latitudes.
>
> It is found that the various kinds of hardy fruits, such as the apple, pear, plum, quince, cherry, &c., are of the best varieties and easily cultivated; and that many of the more delicate fruits, such as peaches, apricots, nectarines, grapes, &c., attain a size and richness of flavour rarely equalled in our northern latitudes. . . . the southern shore of the Ontario is emphatically a Fruit Country. A great variety of ornamental trees and shrubs, which are unable to withstand the early frosts and severe cold of the valleys of the Hudson and the Connecticut, succeed without trouble in the vicinity of Rochester and the Ontario. That the Valley of Genesee is adapted to the growth of silk would seem clear from the fact that the various kinds of foreign mulberries, such as the Chinese, Broussa, and Italian, withstand the usual cold of our winters without injury, but also that the wild mulberry is found on the Upper Genesee and many of its branches. [The mulberry leaves were fed to silkworms.[1]

Agricultural periodicals were full of ideas about how silk could be produced for profit, but experimentation with silk ultimately was not commercially successful and the fad faded.

Early settlers often brought seeds and stones to

Opposite: Wayne County apple orchard in late winter.

FIGURE 11-1. Orchards on the Sumner Farm, Shelby, Orleans County, New York. Sanford, Historical Album of Orleans County, 1879.

start apple and peach trees. It was, at best, a long-term and uncertain process from seed to quantities of fruit in the orchard. Producing trees from seeds meant varietal uncertainty, since every apple tree grown from seed is potentially a new variety. In contrast, the present practice of grafting ensures the reliable propagation of specific varieties, each one derived from a single parent plant. Earliest records indicate that one simply referred to the fruit as "apples" or "peaches" and not by the name of a specific variety. Not being content with chance, John Martin of Rush returned to Maryland in 1809 for family visits and acquired apple grafts, which he used to produce the first grafted orchard in Rush. However, in his 1857 book, *The Fruit Garden,* Patrick Barry warned that planters should not depend entirely upon grafted trees because in the very uncertainty of planting seeds a better variety, perfectly suited to the region, might be created.

A combination of both techniques was employed by more enlightened planters in developing orchards. The Chapin orchard in Bloomfield was a mixture of stock from seeds planted about 1800 and grafted stock. The Chapins developed three of the region's best known apples from these early seeds: Early Joe, Melon, and Northern Spy, which is still a popular apple. Chapin's production in 1875 was about 3,000 barrels.[2] Hedrick notes that one of the first large western New York commercial orchards was begun in 1833 by Lewis Allen on Grand Island near Buffalo, with more than three miles of peach and apple trees along the shore.

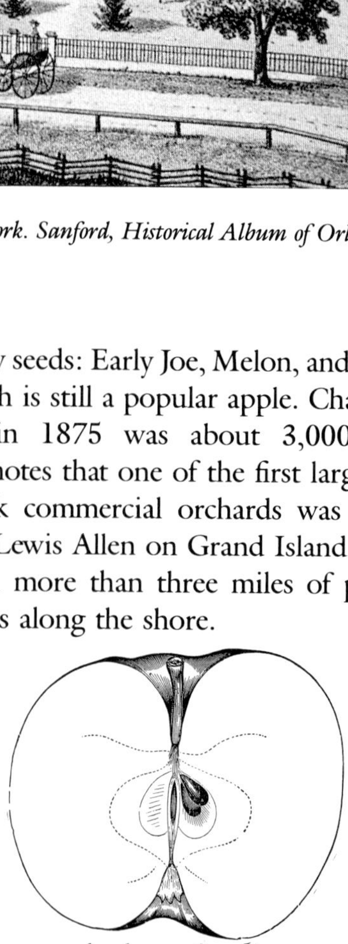

In the very early days of settlement, fruit, particularly apples and peaches, was generally preserved by drying. Cider and peach brandy were also commonly made (at least until temperance became a big issue) as a method of preservation. Later, improved packing and transportation allowed the distant sale of the best apples as fresh fruit. Hedrick mentions that P. C. Reynolds, originally from Duchess County, established two orchards in Ontario County where about ten percent of his trees were grafted. In the 1840s Reynolds sold apples from his grafted trees to a man in Palmyra via the Erie Canal; he received 75 cents per barrel and that price included the barrel.[3]

Because water was usually not drunk, most farms had a cider cask in addition to small beer. When fermented, this cider was known as hard cider and if properly bottled had the appearance of sparkling champagne after a year or so. Apple jack was a brandy distilled from hard cider. Figure 11-2 shows a patent cider mill of mid-century.

Apples could be dried for home use by paring, slicing and placing near a fireplace or stove, a delicious-smelling but slow process. Some farms built their own dry houses while others depended on commercial dry houses or evaporators, which supplied both American and European markets. Byron Halsted wrote in his 1904 book, *Barn Plans and Outbuildings,* "The evaporated apple in-

FIGURE 11-2. Hutchinson's Patent Cider Mill.

dustry centers in a few towns in Wayne County, N.Y., bordering the southern shore of Lake Ontario" (see figure 11-3). Actually, evaporators were also found to the west of Wayne County, while throughout Wayne, Monroe and Orleans Counties the farm dry house was a common sight. With the coming of mechanization, apples could be prepared on a production line: pared, cored, inspected, bleached and sliced. The apple slices were placed four to six inches deep in the dry room on a slatted floor over a coal fire (see figure 11-4). After several hours they were turned to complete the drying process, then bagged for shipment (see figure 11-5). A few dry houses in the region remain in operation to this day. Figure 11-6 shows a vignette of mixed farming in which fruit growing and dairying are important elements.

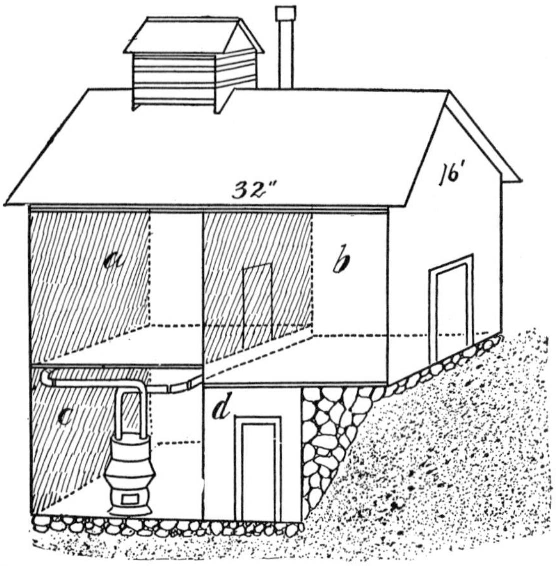

FIGURE 11-3. Halsted illustration of an apple evaporator. In the outside room (top right) the apples are received and processed, then dried in the inner room (top left) over a coal furnace. The Halsted evaporator was estimated to cost $350 in 1904.

FIGURE 11-4. 1871 Tippecanoe Apple Paring Machine.

Apples and peaches were also excellent for fattening hogs, which were not as fussy as humans abut the condition of the fruit. One of the major reasons that fresh fruit was at first not sold off the farm was the difficulty of keeping it sound and getting it to market. Later, in 1831, *The Genesee Farmer* commented on this problem:

> It is passing strange that those who have been at the trouble of raising fruit for sale should not recollect that, one bushel of sound fruit will fetch more than three after they have been bruised to pumice. We have seen in our market the week past peaches offered for sale, which, had they been brought packed in chaff or bran, would have sold readily at one dollar and fifty centers per bushel; but they were put in barrels standing on end and brought some fifteen miles over the rough road in a waggon, so that all those which were mellow, and which would have been desirable, had they been carefully packed, were rendered unfit for the table, and the owner was glad to dispose of them for fifty cents per bushel. Apples, pears, and plums are often brought to market in the same way.[4]

Some resisted packing fruit in sawdust or chaff because of the flavor they imparted. Two of the factors determining the selection of apple varieties were their ability to keep well in storage and to withstand shipping. Certainly something had to be done to insure that the apples produced would reach the market safely in some suitable condition, because the 1864 apple harvest in Monroe County alone yielded almost 500,000 bushels.

For almost 50 years pears received much attention, rivaling the general interest in apples. At mid-century, Hedrick notes that one grower realized $9,200 per acre, or $9.00 per barrel, for his pears. Pears were grown by the gentleman farmer and commercial specialist with equal diligence. Horticulture books and agricultural periodicals were filled with sections on pears. Other fruits were also grown throughout the region; plums, for example, captured the imagination of growers to such an extent that people became "plum tired."

The fruit canning and processing industry, which continues to this day, was once even more

FIGURE 11-5. Workers at the Samuel Burns apple dry house on the Alton-Sodus Point Road in Wayne County line up for the photographer. As was common in period photographs, items related to various facets of the operation are lined up in the front row including an apple paring machine. If the appearance of the boarding is any measure the dry house has recently been enlarged. Undated. Collection of the Wayne County Historical Society, Lyons, New York.

prominent in the Genesee Country. Processing houses were generally located along the main transportation arteries, which then included canals, roads and railroads. Nicholas Appert, a Frenchman, is credited with the invention of vacuum canning in 1795. As with many inventions it wasn't until later (1815 in America) that the process was used to preserve fruit and vegetables, mostly for home use. An 1832 Boston cookbook, *The Cook's Own Book,* contained recipes for canning pickles, candies, cheeses, syrups, butters, jams and jellies. Because glass containers at first were quite expensive they found limited use in both home and commercial canning. The less expensive Mason jar was patented in 1858.

By the end of the century, commercial canning of fruit was commonplace in the Genesee Country. In Geneseo the Jam Kitchen had its own buildings, special jars and labels, and a large staff. Of special interest was a jelly jar with slanted sides and a top wider than its bottom so that the entire contents of the jar could be easily removed. Most

FIGURE 11-6. Seth Jones Orchard in Kendall, Orleans County, New York. Illustration from Sanford's Historical Album of Orleans County, 1879.

of the fruit came from the Wadsworth fruit farms south of Geneseo. Today little remains of the orchards or the factory. During World War I such canning kitchens were in great demand to supply jams, jellies and crystallized fruit to troops serving in Europe, and in fact some operations existed solely for the duration of the war.

The New York State Horticultural Society, founded in 1818, spawned other such organizations, including the Domestic Horticultural Society centered in the Finger Lakes Region. There were few horticulture books before these societies were formed but the number quickly grew and in fact there are too many to be fully discussed here. William Prince's 1828 book on fruit was probably the first published in New York. John Thomas and Charles and Alexander J. Downing were the most prolific contributors to the various pomological (fruit growing) periodicals. A. J. Downing's 1845 work, *The Fruits and Fruit Trees of America,* was one of the most influential books on the subject for fifty years. Downing also published highly successful books on architecture (see Chapter IX).

Seed Production and Nurseries

Philo Hampton repeatedly mentions preparing seeds throughout his diary. Traditionally, each farmer set aside some of the current year's harvest as seed for planting the next spring; however, throughout the nineteenth century seeds became increasingly available from commercial sources. These seed companies offered improved varieties and promised better seeds from proven varieties. For Shakers in the Genesee Country, first at Sodus Bay and later at Groveland, seed production was a major part of the cash-producing ventures intended to sustain their community. (The Shakers are more fully discussed in Chapter XVII.) Sometime in the early 1830s the Shakers began to market their seeds, appointing a trustee from within the community to travel the outside world selling seeds, brooms and other Shaker products (see figure 11-7).[5]

In 1862 James Vick, a pioneer in the systematic growing of flower seeds, established a seed firm in Rochester. Vick was also involved in the publication of *The Genesee Farmer* after Luther Tucker gave it up in 1839 in order to publish *The Cultivator* in Albany. Vick's seed business was divided between the 25-acre seed production "Home Place" on East Avenue and a flower farm near Lake Ontario. The "Home Place" operation was served by several Greek Revival barns and many sheds. Vick's Flower Farm consisted of some 75 acres on the Charlotte branch of the New York Central Railroad, to facilitate transport. Among its barns was an imposing double cupola-capped, gable-roofed hay barn, needed in part to store feed for the horses used in the operation (see figure 11-8).

Vick's East Avenue and Floral Street seed warehouse was 60 by 160 feet and five stories high including the basement, in which bulbs were packed and stored. The first floor was used for the sale of seeds and plants, the second floor for the business offices (annual postage ran to $50,000), and the third floor for all orders written in German, a department requiring 50 workers (Vick spent considerable time searching Europe for superior seeds). The fourth floor contained the firm's printing facilities, with a capacity of some 10,000 catalogs per day. By the mid '80s

FIGURE 11-7. Samples of nineteenth century Shaker seed packets. State Museum of New York, Albany, New York.

FIGURE 11-8. Vick's Flower Farm. McIntosh, History of Monroe County, 1877.

Vick was considered the foremost flower seedsman in the United States. The firm was purchased from his sons by the Burpee Seed Company in the 1920s. Figure 11-9 shows one of Vick's advertisements and figure 11-10, an advertisement for another local seed company, Hiram Sibley & Co.

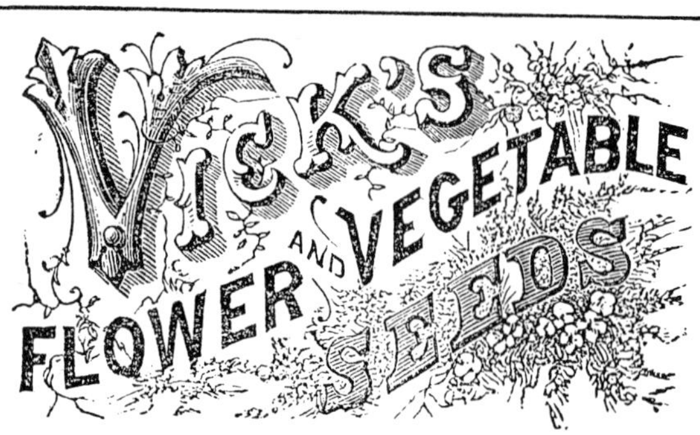

ARE PLANTED BY A MILLION PEOPLE IN AMERICA. SEE
Vick's Catalogue—300 Illustrations, only 2 cents.
Vick's Floral Guide, Quarterly, 25 cents a year.
Vick's Flower and Vegetable Garden, 50 cents; with elegant cloth covers, $1.00.
All my publications are printed in English and German.
Address, **JAMES VICK**, Rochester, N. Y.

FIGURE 11-9. Vick advertisement.

Joseph Harris, who founded the Harris Seed Company in 1879, had also been editor of *The Rural New Yorker* in 1855 and editor and publisher of *The Genesee Farmer* for eleven years before selling it to the *American Agriculturist* in 1865.[6] The Harris Seed Company grew rapidly, soon requiring a postal substation to handle the volume of incoming and outgoing mail. The Harris philosophy of seed production was based on careful selection and cultivation of quality varieties combined with buyer education, a contination of ideas expressed during his periodical publishing days. The Harris Seed Company remains an important seed producer today and continues in the belief that quality seeds plus sound planting information will lead to superior results.

Hedrick believes the first nursery in Rochester to have been that of Reynolds and Bateham in 1834, which was sold to Ellwanger and Barry in 1840. Barry was also an editor of *The Genesee Farmer*, from 1844 to 1852, and editor of the *Horticulturist* in 1853 and 1854. Ellwanger & Barry Nurseries, along with Alonzo Frost & Co., the Genesee Valley Nursery, Briggs Brothers,

FIGURE 11-10. Hiram Sibley & Co. advertisement.

HIRAM SIBLEY & CO.

Will mail FREE their Catalogue for 1882, containing a full descriptive Price-List of Flower, Field and Garden

SEEDS

Bulbs, Ornamental Grasses, and Immortelles, Gladiolus, Lilies, Roses, Plants, Garden Implements. Beautifully illustrated. Over 100 pages. Address

ROCHESTER, N.Y. & CHICAGO, ILL
179-183 East Main St. 200-206 Randolph St

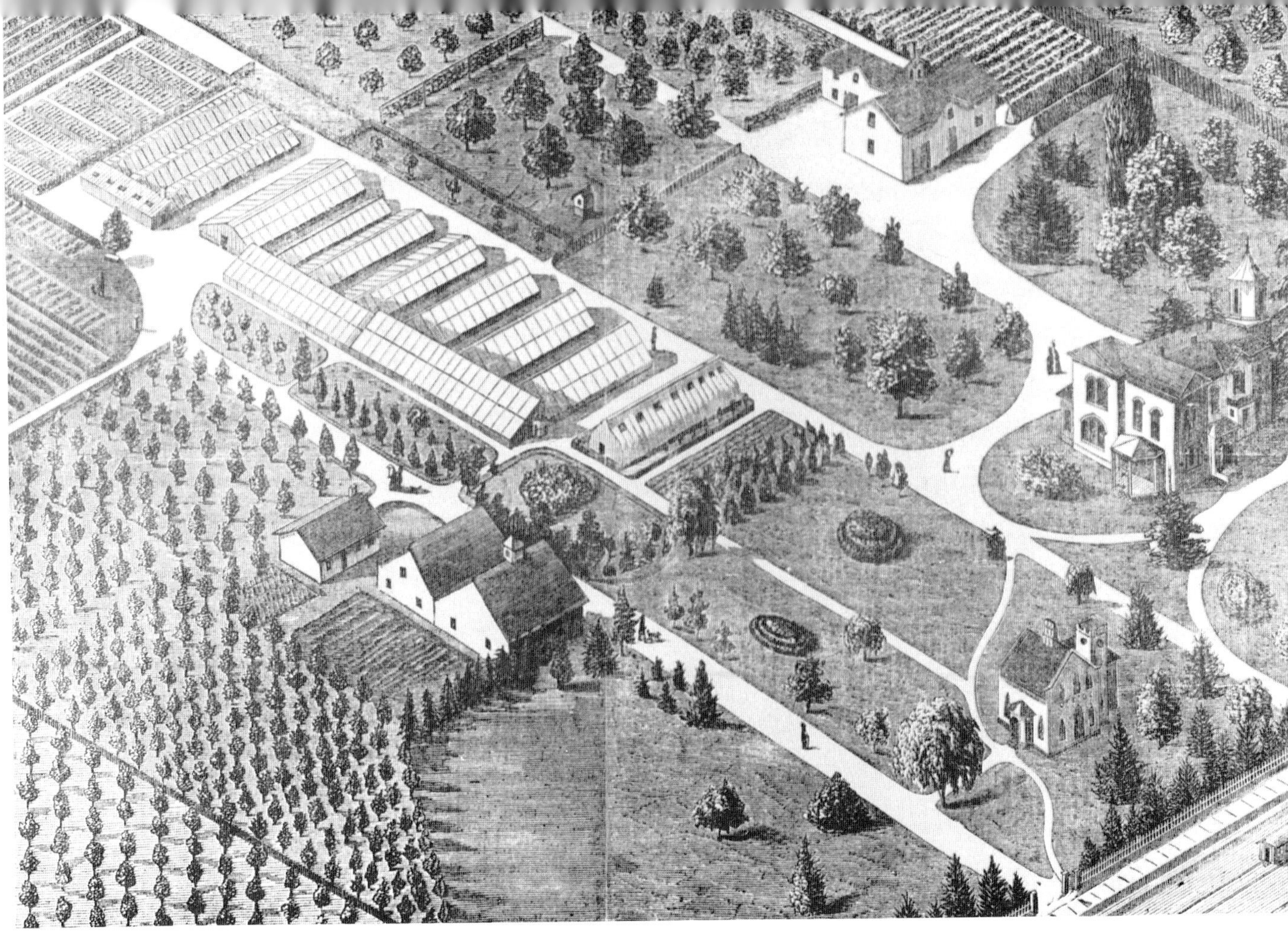

FIGURE 11-11. Ellwanger & Barry Nursery on Mt. Hope Avenue, Rochester, New York. Engraving from Moore's Rural New Yorker, October, 1870.

William Reid, Joseph Harris, L. P. Gunson, Hiram Sibley, James Vick and others combined to produce a volume of business so large that it was considered appropriate to change the nickname of Rochester from "Flour City" to "Flower City"; the latter name has remained. In the 1850s there were some 18 nurseries in Rochester and with the phenomenal growth of Ellwanger and Barry's nursery western New York became the nursery center of the United States.

The Ellwanger & Barry Nursery grew from seven to more than 650 acres, making it the largest nursery in the world at that time. Figure 11-10 is an 1870 view of Ellwanger & Barry's Mount Hope Nurseries in Rochester. The Gothic office, Italianate house, and carriage house to the rear can still be seen. 30 horses were used regularly and in peak seasons 400 to 500 people were employed. Ellwanger and Barry claimed "half a mile of greenhouses," some of which can be seen in Figure 11-11. Fire destroyed the original barns and sheds in 1859. In the same year the Ellwanger & Barry catalog listed 400 kinds of pears, 300 apples, and 70 cherries (see figure 11-12 & figure 11-13). The October 13, 1859 issue of the Rochester *Democrat and American* newspaper tells of a record single railroad shipment of nursery stock totalling thirteen boxcars.

FIGURE 11-12. Ellwanger & Barry Advertisement

In 1918 Ellwanger & Barry ceased operations, but several remarkable buildings still remain from the business. The Patrick Barry Italianate villa, completed in 1857 in limestone-trimmed brick, is after a design by Gervase Wheeler (see figure 11-14). Immediately to the rear of the villa is the carriage house, which retains many of the details of the main house (see figure 11-15). The windows are slightly arched, the roof bracketed, and the materials contrast according to the style. The villa's round arched windows are reflected nicely in the louvered cupola of the carriage house.

FIGURE 11-13. Ellwanger & Barry Advertisement

HORTICULTURAL.

PEAR TREES.

IN addition to the large stock of Trees of our own raising, we have just received a large quantity of first rate kinds selected in Europe the past winter. Among them are

White Doyenne or Virgalieu,
Gray Doyenne,
Bartlett,
Louise Bonne de Jersey,
Vicar of Winkfield,
Fondante d' Automn or Belle Lucrative,
Duchess d' Angoleme,
Beurre d' Arremberg,
Colmar d' Arremberg,

And many others of the highest reputation. The above can be furnished on quince stocks, of large size, strong, and well branched, suitable for pyramidal garden trees, that will bear the second year after planting.

We have also strong and beautiful standard trees on pear stocks, of such excellent sorts as

Beurre Bosc,
Beurre d' Arremberg,
Brown Beurre,
Gansell's Bergamot,
Chaumontel,
Winter Nellis,
Napoleon,
Van Mon's Leon le Clerc,
Ne Plus Meuris,
Beurre Rance,
St. Germain,
Dunmore,

And many others, besides a full supply of all other Garden and Orchard Fruits. Orders promptly attended to for less or more.
☞ Prices lower than usual.
[4-tf] ELI WANGER & BARRY.

The seed barn of E. S. Hayward, a little-known seed-producer of the time, is known through its description in the 1871 *American Rural Home* (see figure 11-16). The barn, now gone, was located on Webster Avenue off Goodman Street in the town of Brighton (now Rochester). The article clearly identifies the modifications required by Mr. Hayward for the production of seeds, with special emphasis given to handling and drying:

MR. HAYWARD'S SEED BARN

The accompanying illustration gives a view of the barn of Mr. E. S. Hayward, which, although used as a horse-barn, grain-barn, and carriage-house, was built with particular reference to his immense seed-growing operations, and is probably one of the most perfect barns for that business in the country. It is 50x70 feet on the ground, with 22 foot posts, and a gambrel roof, which gives a height of 38 feet from sill to gable. There are five floors all laid with matched pine, with the exception of the upper two, which are above the first plate, and laid with slats, and used entirely for drying seeds, which, when the weather is not favorable, have to be dried under cover. Their principal use is for drying his crop of sweet corn for seed, of which he has twenty-five acres this year. Each floor is reached by staircases, and an opening in the center from the lower floor to the gable, through which hay, grain and seeds are taken from the wagon by the horse-fork, and discharged on the different floors. Perfect ventilation is obtained by doors some eighteen inches or two feet wide, running the length of each floor, and hung with hinges,

FIGURE 11-14. Patrick Barry's Italian villa on Mt. Hope Avenue, Rochester, New York, completed 1857.

FIGURE 11-15. The symmetry of the carriage house behind the Barry villa contrasts interestingly with the asymmetry of the villa itself.

FIGURE 11-16. Engraving of Mr. Hayward's seed barn, Brighton, Rochester, Monroe County, New York. The American Rural Home, 1871.

> so as to be turned up when desirable. On the lower floor, which is wide and roomy, stands an Emery treadpower and threshing machine, which is used to thresh the grain and seeds, and also to drive a machine for cutting hay, straw and corn-stalks. A pump also on this floor, furnishes water for use in the barn. This barn is filled and emptied two or three times during the season. Sometimes all the floors are occupied in curing his onion crop. The frame is covered with matched pine, battened and painted.[7]

Of additional special note in the Genesee Country outside of Rochester is the Geneva Nursery of W. & T. Smith, which was begun in 1846 (see figure 11-17). Flowering plants, ornamental trees, shrubs, roses and bulbs were grown on 400 acres. The Smith operation and that of T. C. Maxwell were the leading Geneva nurseries, together making Geneva a nursery center that by the end of the century had surpassed Rochester, according to Hedrick.[8]

Farther south in the Genesee Valley, the Kelly Brothers Nursery was established in 1880 in Dansville and continues in the same family to the present day. Dansville was the other major nursery center, along with Geneva, outside Rochester and Monroe County. As with seed production, the nurseries were able to offer superior varieties at reasonable cost confounding the many skeptics who had speculated that such companies could never compete with the industrious farmer who saved seed and propagated his own plants and trees.

FIGURE 11-17. W. & T. Smith Geneva Nursery, Geneva, New York, contained some 400 acres. McIntosh, History of Ontario County, 1876.

77

CHAPTER XII

Railroads Change The Genesee Country

HARDLY HAD THE ERIE CANAL been completed and talk of the Genesee Valley Canal begun when excitement arose in the region for a new form of transportation: the railroad. Successes in England and Wales with steam locomotives kindled the imagination of many. In 1831 *The Genesee Farmer* wrote of this amazing invention:

> Rail Road, - Among modern improvements, none promise to have a widerspreading or more powerful influence, physical and moral, on our country, than the invention of rail roads.—On these, locomotive engines can be propelled at the amazing speed of from thirty to fifty miles per hour; a speed exceeding that of the fleetest race horse, and approaching nearer to that of birds than any thing which a century ago, we should have dreamed of.[1]

There were two major considerations in building early railroads: first, the all-weather transport (turnpikes and canals could not claim to be usable in severe winter weather or in times of extended rainfall) and second, the potential for economical, high-speed operation. For farmers the railroads would provide year-round, rapid access to new markets and thus permit the sale of new products.

Throughout the period of rail building people had mixed feelings about the advantages of the railroad. Although some were opposed to the basic idea of railroading, others who had been bypassed by the canals did not want to be left out again by the railroads. It was not uncommon for villages or towns to support the construction of part of a rail line just to make sure they had a "connection." Local governments had to obtain permission from the state legislature in order to subsidize railroad building, but such permission seems to have been readily obtained.[2]

Similarly, at the onset of railroad fever farmers were strong supporters of the railroads, seeing them as necessary to the continued good health of farming in the Genesee Country. If they failed to support the railroads, other geographic areas might take over. Crops might change but at least markets would be assured. In addition, manufacturers of agricultural equipment in the region supported railroads for marketing their products. In spite of the area's large agricultural community, local sales were known to be insufficient by themselves, and the manufacturers needed access to a larger national market.

Considerable capital was required to build these connecting rail lines and many early projects fell through, including the 1832 attempt to build a Dansville & Rochester Railroad. More

Opposite: Two highly polished ten wheelers pause for their portrait on the 234-foot high, 800-foot long Erie Railroad Bridge at Portage. In 1860, about the time this photograph was taken, this bridge was considered the largest wooden bridge in the world. Anonymous. The Clark Rice Collection.

successful was the New York & Erie Railroad (later to become the Erie), which was chartered in 1832 and the route first surveyed in 1834. The line, which was to run for almost 500 miles, began four miles outside of New York City. The diagonal route across the state from the Hudson River to Lake Erie was projected at a cost of about six million dollars for a single track line including locomotives and cars. A proposed rail feeder line that was actually constructed was the Rochester & Genesee Valley Railroad, incorporated in 1856 to build a line from Mount Morris to Avon. In 1859 the first train ran over this line and it eventually became part of the Erie Railroad.[3]

The public's view of railroading is formed by the passenger train, but with very few exceptions the transportation of freight is the railroad's lifeblood. The first short lines were essentially freight haulers connecting remote regions with terminals on the Erie Canal. For instance, the Tonawanda Railroad (also sometimes spelled Tonnewanta), chartered in 1832, was to connect Attica with Rochester and the Erie Canal via Batavia. This first line initially ran with horse power, but by 1837 a steam train was making the 32-mile run from Rochester to Batavia in two hours at a fare of $1.50 (see figure 12-1). Other very short lines were built to circumvent the Rochester falls on the Genesee by connecting the city with Lake Ontario ship landing warehouses below the falls. In 1838 O'Reilly compares travel times between Rochester and Buffalo as follows: "The railroad can be traversed in three, or at the most in four hours, while the stages consumed from fifteen to eighteen hours, and the canal-packets about twenty-four hours in passing between the two places" (see figure 12-2).[4]

FIGURE 12-1. 1838 steam locomotive built by Rogers Locomotive Works for the Tonawanda Railroad. The single pair of drivers plus the four wheel lead truck worked well on the early trackwork of the railroad. Two similar locomotives received in 1836 from Baldwin had preceded the "Batavia."

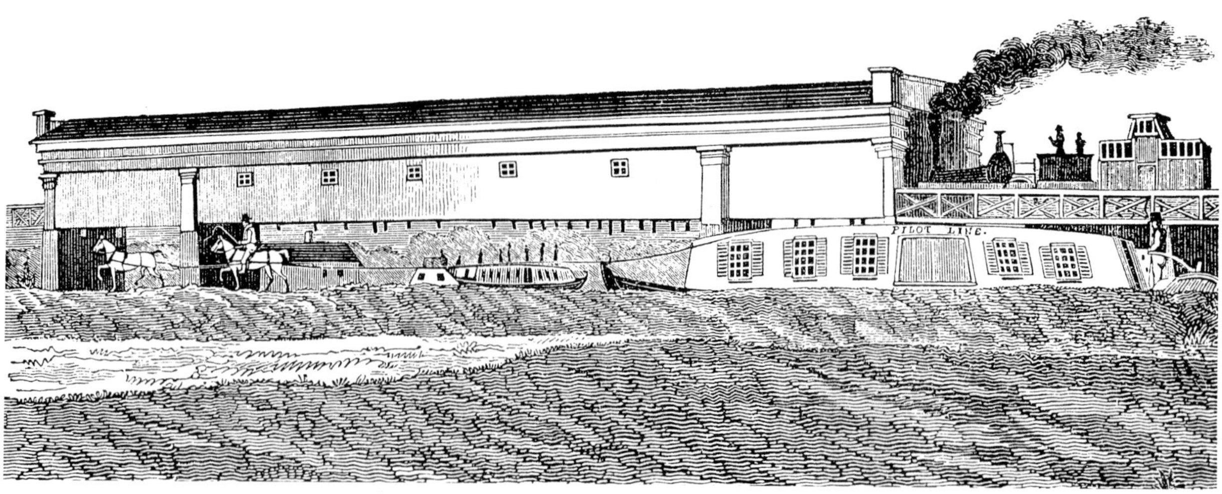

FIGURE 12-2. A light freight train crosses the classically detailed bridge over the Erie Canal on the Tonawanda Railroad at Rochester. The caboose-like car is a passenger-baggage car with a capacity of 24 people. O'Reilly, Sketches of Rochester, 1838.

As more and more rail lines were built, the connection with canals became less important than connection with through railroad lines to major cities and markets. Railroads connecting New York City and Buffalo were contemplated. A bit of "railroad fever" can be found in O'Reilly's project for Rochester-to-New York travel: " . . . we may anticipate that, in the course of three years, the journey between Rochester and New-York will be made by railroad and steamboat within twenty-four hours, or between sunrise on one day and the same period on the following day!"[5] Finally, by 1853, ten railroads were combined to form the New York Central Railroad connecting New York City with Buffalo via Albany, Syracuse and Rochester, but these consolidations did not occur overnight. Several lines had previously been combined in 1843 to permit travel by train between Albany and Buffalo with a running time of 25 hours. The one-way first class passenger fare between Albany and Buffalo was $10 (see figure 12-3).

The lengthy trip was affected by speed limitations over strap rails; these were wooden rails with iron straps fastened on as wearing surfaces. Strap rails were quite dangerous, especially if the strap loosened, bent up, and pierced a car. Such a loose strap, called a "snake head," derailed a locomotive on the Auburn & Rochester Railroad between Batavia and Rochester, according to the *Livingston Republican* of May 10, 1848. Other inconveniences were suffered at first in places such as Rochester, where the refusal of local merchants to allow a connection between rail lines necessitated the transfer of cargo by wagon and cart for distances as short as four city blocks.[6] By 1843 the legislature insisted on the connection of railroads at Rochester and the trip between Albany and Buffalo was thus reduced to 15 hours.

Before the individual railroads were connected and open to Buffalo, the New York lines east of Rochester were connected to Boston via the Western Railroad. In December, 1841, Boston city officials traveled by rail to Albany to celebrate the connection of the two cities. The Boston officials brought gifts to Albany and in turn received gifts including "a barrel of flour, the wheat for which had been threshed at Rochester on the previous Monday [they returned to Boston on Wednesday], while the barrel itself was made from wood which, on the threshing day, had been growing in the tree." The barrel interior had been charred to prevent wetting the flour, which was made into bread in Boston and eaten at a banquet there.[7] Industry and speed were on the minds of many.

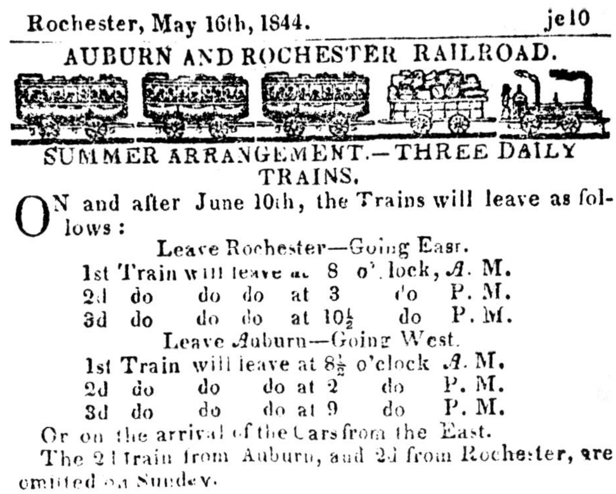
Rochester, May 16th, 1844. je10

AUBURN AND ROCHESTER RAILROAD.

SUMMER ARRANGEMENT.—THREE DAILY TRAINS.

ON and after June 10th, the Trains will leave as follows:

Leave Rochester—Going East.

1st Train will leave at 8 o'clock, A. M.
2d do do do at 3 do P. M.
3d do do do at 10½ do P. M.

Leave Auburn—Going West.

1st Train will leave at 8½ o'clock A. M.
2d do do do at 2 do P. M.
3d do do do at 9 do P. M.
Or on the arrival of the Cars from the East.

The 2d Train from Auburn, and 2d from Rochester, are omitted on Sunday.

FIGURE 12-3. Auburn and Rochester Railroad advertisement. Rochester, May, 1844.

Wood was a major material in the construction and operation of the railroads. An enormous number of square-sawn timbers were made into railroad ties, although the very first ties were often only cut logs. Wood was also used to fire most steam locomotives for almost the first 30 years of railroad operation, even though the first locomotives had burned regionally scarce and expensive bituminous coal. Relatively inexpensive wood was used until the price of coal became competitive, about the time of the Civil War. Farmers often had contracts to supply the railroads with cord wood for the locomotives. The Rochester, Lockport & Niagara Falls Railroad, a relatively short line, stored almost one year's supply of wood in sheds along the line; this amounted to 12,900 cords valued at more than $28,000 in 1853. This wood, stacked in a row one cord high (4 feet), would have stretched 19½ miles.[8] Until 1872, most Erie locomotives wore balloon or diamond smoke stacks to catch live sparks before they set fire to bridges, tracks or the surrounding countryside. Experiments with coal in 1858 eventually led to a total conversion to bituminous when that coal became a major item hauled by the railroad.

In 1851 the New York & Erie Railroad had been completed, cutting across the Genesee Country on its diagonal route to Lake Erie (see figure 12-8). The March 18, 1852 *Livingston Republican* reported the effect of the railroad both on markets and on the price of farmland:

> Penn Yan, instead of being dull in winter, is unusually lively, and our own village is evidently the gainer. Markets and business are kept more uniform, and the facilities are such that the farmers of our county find as good a market at Penn Yan and Dundee in winter as they can at Rochester, Geneva, or Jefferson (Watkins Glen). The influence of the railroad has caused an increased de-

FIGURE 12-4. Stereo view of the Erie Railroad at Avon, New York, in the early 1870s when some engines still burned wood and the rails were broad-gauged. The south-bound, diamond-stacked locomotive in the foreground is a coal burner, while in the distance the balloon-stacked locomotive in front of the freight house is a wood-burner. View by T. D. Tooker.

FIGURE 12-5. The Erie Railroad trestle at Portage. At the time this was the largest wooden trestle in the United States. Circa 1865. Walker stereograph.

> mand for farming lands, which are every where held at prices ranging from $2 to $10 an acre higher than they were a year ago.

Of course with the sale of new farms came the need for new barns. Of these new designs the first was the gambrel roofed barn (Chapter XIV) followed by the truss-framed barn (Chapter XX).

Curiously, the Erie Railroad's state charter stipulated that it must never connect with a railroad leading to another state, necessitating something other than standard spacing of the rails. The Erie's executives decided on a six-foot gauge, making it the widest track in the United States. This lack of a standard rail gauge was a peculiar feature of the early short railroads. But if they were to be a real boon to shippers there ultimately had to be agreement on one gauge so that through passage could be accomplished; otherwise the costs of transferring goods from the cars on one line to those of another would become prohibitive.

Finally, 4 feet 8½ inches became the standard. This was the gauge once used by Roman chariots on roads grooved for their passage, and the gauge most used in England. President Lincoln had prompted conversion to 4 feet 8½ inches when signing the 1862 bill authorizing the construction of the Pacific Railroad, which was to use this standard gauge. Figure 12-4, a stereo view made by T. D. Tooker in the early 1870s, shows a broad-gauged mixed train at the station in Avon. This widest of United States tracks was in use until 1880, when conversion to standard gauge was completed.

This conversion is itself an interesting story: It was begun in 1876 by the laying of a third rail, the new inside rail being laid to the now standard gauge. This three-rail arrangement permitted operation of both wide-gauge and standard-gauge equipment over the line. Standardizing required $25 million, many times the cost of the original construction. Most of the broad-gauge rail was removed in 1880 when as a publicity stunt gandy dancers (rail gangs) removed the extra rail over the entire Buffalo Division in one work day. Stories are still told of that "conversion day."

Among the significant engineering feats pc formed in building the Erie Railroad were the valley-spanning masonry Starrucca Viaduct, at Lanesboro Pennsylvania, and the world's then-largest wooden railroad bridge, an 800-foot long, 234-foot high bridge carrying the railroad across the Genesee River at Portage (see figure 12-5). Two years of work and some 1,600,000 linear feet of pine timber were used in the construction of the Portage Bridge. Its construction required some 106,820 pounds of iron. All the materials were brought to the location by canal or by road because the railroad itself had not reached this spot when the bridge was begun in 1850. This photograph, published by L. E. Walker of Warsaw as a stereo view, shows the completed bridge with a four-car train spotted on the bridge for the photographer. To provide an additional feeling of scale, several people were carefully placed in the right side of the view.

In 1878 the Genesee Valley Canal was finally abandoned. The wooden bridge itself was destroyed by fire in 1875 and replaced by a wrought iron bridge in the same year (see figure 12-6). Today one can see the third bridge at this location, constructed of steel in 1903. Eventually a railroad was built on the tow path and near Portage the railroad was even built on the canal bed itself, adding insult to injury. Figure 12-7 shows a northbound freight working its way along the high banks of the Genesee River gorge on track laid over the Genesee Valley Canal bed at the very site shown earlier (Chapter IV) in the stereo view by Charles Bierstadt.

FIGURE 12-6. Testing the new iron bridge at Portage, New York. In the traditional method of bridge-testing, the bridge was "loaded" using locomotives as the weights. The theory was that if the bridge could support the concentrated weight of these locomotives it would certainly support the weight of one or two, along with the lighter cars of the train. 1875 Stereo view by Walker.

By mid-century the Genesee Country was crossed by many shortline railroads. An 1855 map from the 1857 *Rail-Road Commission Report* shows both canals and railroads in New York State (see figure 12-8).

The Rochester & Genesee Valley Canal Railroad Company, built on the former canal, was incorporated in 1879 (see figure 12-9). In 1882 the Lackawanna Railroad, hauler of Pennsylvania coal, was building in western New York and eventually became known as the Delaware, Lack-

FIGURE 12-7. The Rochester & Genesee Valley Canal Railroad Company was later taken over by the Pennsylvania Railroad. Undated, anonymous photograph from the Clark Rice collection.

FIGURE 12-8. Map of canals and railroads in the Genesee Country, 1855. The railroads at this time combined nicely with the Genesee Valley Canal to provide access for farm products to various markets. 1857 Rail-Road Commission Report.

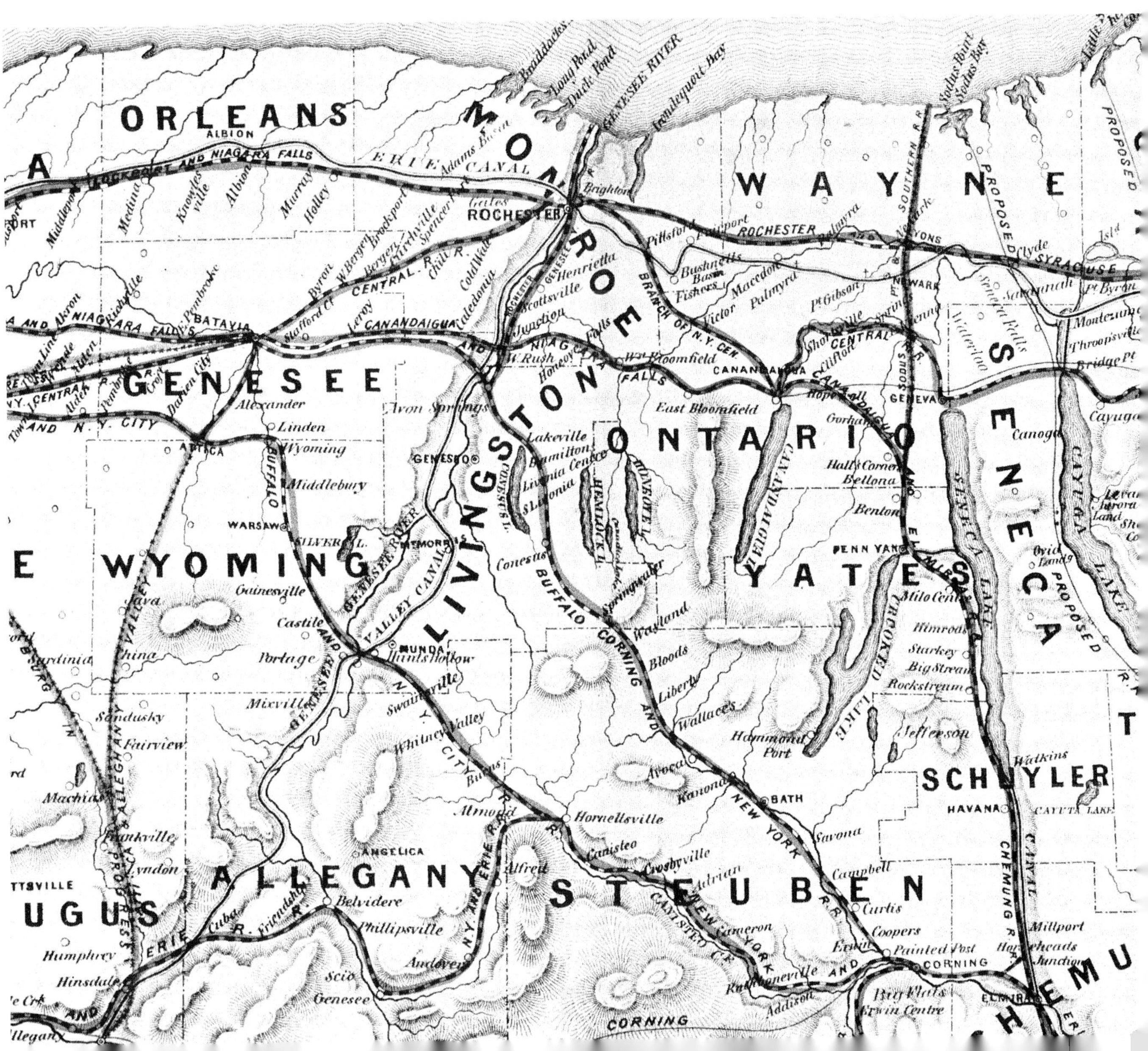

awanna & Western Railroad. The Buffalo, Rochester & Pittsburgh followed in 1887; the same year the Lehigh Valley Railroad, connecting Buffalo and New York City via Newark and Jersey City, was approved for construction through western New York.

The major positive influences of the railroads were the more reliable and speedy access to markets and the increased availability of goods from other parts of the country (see figure 12-10). Livestock and bulk items such as lumber, plaster (gypsum), coal, wheat and salt were all carried on the network of railroads (see figure 12-11). In 1857 cattle could move by train to New York City, including the watering and feed stop near Albany, in about 20 hours. Not only were railroads faster than canals or roads, but in some instances the cost was also significantly lower. Taking advantage of this speed, dairy farmers began moving milk about 1843.[9]

One of the oddities of early railroading in the region was a state law forbidding railroads that paralleled the Erie Canal to haul freight unless the canal tolls were first paid. The law did not apply when the canals were inoperable due to winter ice (see figure 12-12). By 1851 the last of these freight-limiting laws was eliminated.

Figure 12-13, a highly detailed lithograph,

FIGURE 12-9. Genesee Valley Canal Railroad circa 1895 in the Letchworth Park/Genesee River gorge area. Mt. Morris Historical Society.

ERIE **RAILROAD**

TICKETS To All Points EAST or WEST.

Train leaving AVON daily at 5:38 P.M the best and most convenient train to New York, arriving at 6:57 A.M. No extra charge for quick time and superior service.

Geneseo to Buffalo, Rochester, Corning.

7:18 A.M except Sunday, No. 109. For Buffalo and Rochester connecting at Avon for way stations to Corning.

4:40 P.M except Sunday, No. 117. For Buffalo and Rochester, connecting at Avon for Corning, New York and Philadelphia.

Geneseo to Mt. Morris.

10:32 A.M except Sunday, No. 118. Stopping at all stations to Mt. Morris.

5:42 P.M except Sunday, No. 130. Stopping at all stations to Mt. Morris.

Further information from Erie Agents or

H. T. JAEGER, Gen. Agt. Pass. Dpt., Buffalo N. Y. D. I. ROBERTS, Gen. Pass. Agt. New York

FIGURE 12-10. Erie Schedule. 1886.

The last 600 tons

OF

Pittston AND Wilksbarre

COAL!

that I have received were shipped in

COVERED CARS!

It is dry and can be well screaned. No snow, ice or frozen dust. A full supply on hand in anticipation of a

SNOW BLOCKADE

I attribute my increasing sales to my efforts to secure to my customers the best COAL in the best possible condition, and I thank them for their patronage.

JOSIAH CURTIS.

Geneseo, Dec. 6, 1883.

FIGURE 12-11. Notice of coal availability, 1883.

FIGURE 12-12. Southbound Erie passenger special, Avon, New York. 1890s. Anonymous. Negative in Clark Rice collection.

shows the juxtaposition of the warehouse-lined Erie Canal and railroad facilities. The roundhouses indicate a large and thriving rail industry. Improved transportation was vital to successful farming and the growth of cities such as Rochester. This lithograph was one stone from a series of six completed by William Henry Robinson. The drawings alone required two years of work, 1867 and 1868. A fire wiped out both drawings and stones before the projected run of 1,000 could be made. This illustration was made from a rare surviving proof print in the University of Rochester Library collection.

An example of how grains from the Genesee Valley reached distant markets by rail via New York City is presented in figure 12-14, a December 22, 1877 illustration from *Harper's Weekly.* Freight trains carried the grain to

FIGURE 12-13. Bird's-eye view of railroad terminal and canals, Rochester, New York. 1868. A formal park immediately borders the completely domed locomotive roundhouses on the left. University of Rochester, Rush Rhees Rare Book Collection, Rochester, New York.

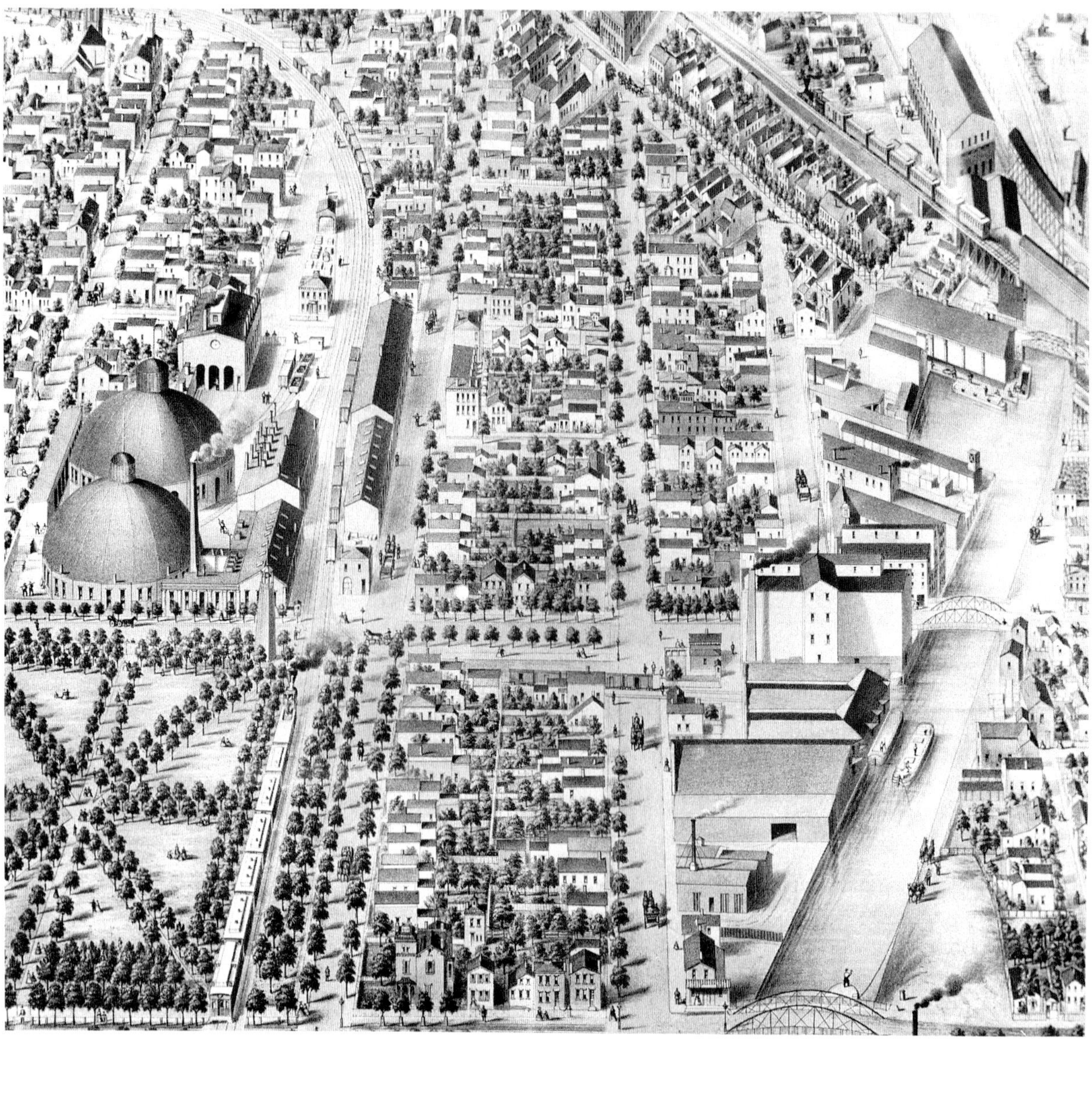

Rochester and thence to New York City by the New York Central and Hudson River Railroad to the elevator at 60th Street on the Hudson River. This 150-foot-tall elevator was powered by steam equipment, a very efficient operation. The box cars were unloaded into the bins (the total capacity of the building was over one million bushels) and the bins were emptied into waiting ships and barges for shipment to both domestic and international markets.

After the Civil War, the Genesee Country economy was adversely affected by freight rates favoring out-of-state shippers. In the mid-1870s, for example, the New York Central charged more to haul agricultural products from Rochester to New York City than it did to ship the same items from Chicago to New York City.[10] This injustice upset area farmers, and the New York Grange organized to fight the freight charges, which created hard times in the Genesee Country by giving the economic advantage to the prairie farmer and to Canadian imports. Fortunately, the formation of the National Anti-Monopoly League in 1881 ultimately led to the 1887 Interstate Commerce Act, which helped relieve the situation.

As these economic factors in rail transportation were resolved and an increasing railroad network provided access to and movement throughout the Genesee Country, the area was not left behind but continued to participate in the growth of the nation. Technological changes in every phase of life, stimulated by the Civil War, changed farming in the direction of greater efficiency.

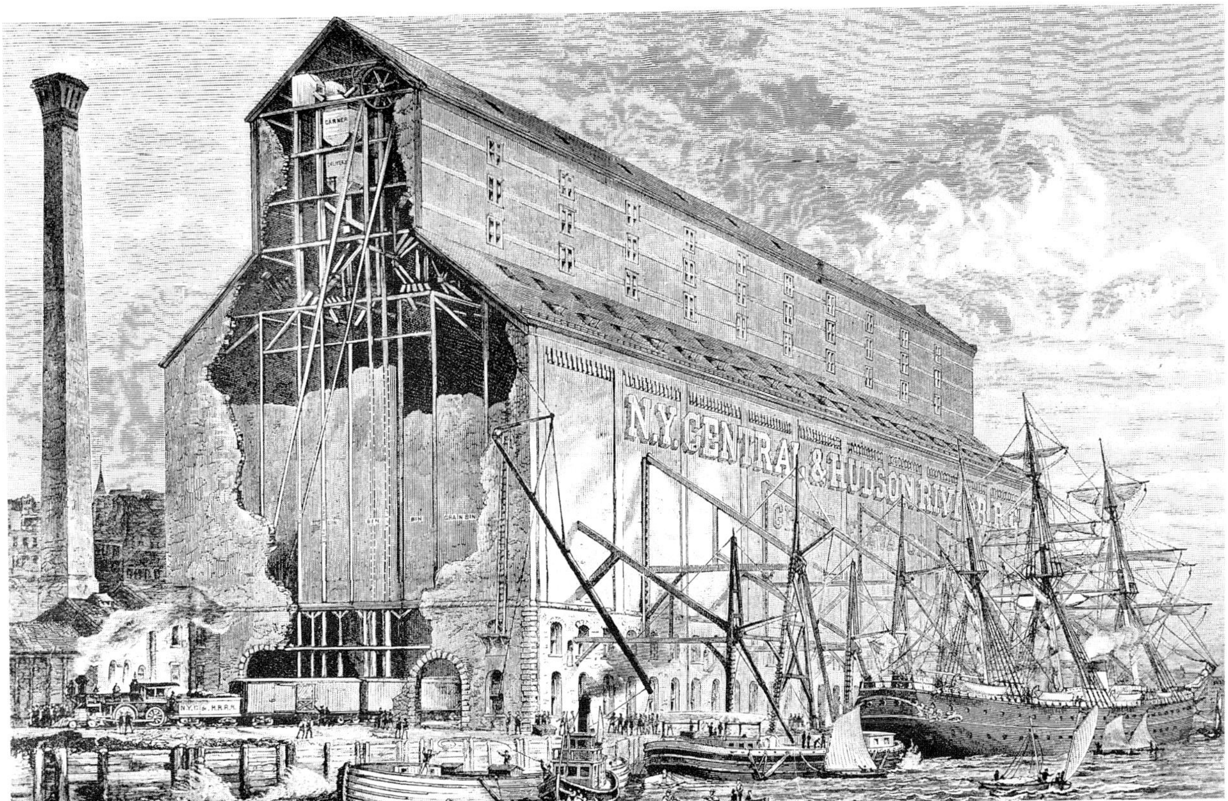

FIGURE 12-14. New York Central Hudson River Railroad grain elevator in New York City. With the elevator standing over 15 stories high, steam-powered equipment was used to move the grain into storage. The cutaway portion exposes the belting and the line shafts from the machinery. The power house stack stands to the left. Box cars are being unloaded in the elevator and the elevator itself is being unloaded into ships. Print from John Grafton's New York in the Nineteenth Century, Dover, 1980.

CHAPTER XIII

Nineteenth Century Agricultural Implement Manufacturing in the Genesee Country

A SONG TO THE GOOD OLD PLOW
By James Orr
Grange Melodies, 1910

A song to the plow, the brave old plow,
that has ruled the wide world over,
For life and good fare on his strong steel share,
shall depend for evermore;
There is strength in his beam, as the toiling team
turns the furrows so long and deep.[1]

FOR A FARM to be commercially successful it must keep up with the changes of the time. Farmers in the Genesee Country went from subsistence farming, between 1790 and 1810, to local self sufficiency after 1810, to commercial farming beginning in the 1820s. As has been shown, machinery changed farming practices while improved transportation opened new markets and speeded transport to old ones; the effects of the Industrial Revolution were being felt throughout agriculture. Within the Genesee Country farming implements were not only adopted, they were also invented, improved and manufactured. In response to increasing mechanization and production, new barns were generally built larger and framed differently. Before examining the new barn designs, a brief look at implement manufacturing in the Genesee Country is in order. So many different operations existed, ranging from grain cradle manufacture to mechanized binder production, that only a few samples can be given.

Plows have always been essential to efficient farming: the start of each growing season begins with breaking and furrowing the earth. The more efficient the plow, the greater the land

Opposite: Mowers and self raking reapers dominate this display of agricultural implements in a nineteenth century engraving. An agricultural steam engine is shown in the background.

area that one person can farm. There was a tremendous evolution in plow design during the nineteenth century and plows were manufactured at many different locations throughout the Genesee Country. The many manufactories of agricultural implements ranged from the single blacksmith making a product that he believed superior or that suited a local need, to the factory employing many hundreds of workers. Some of the descendants of these early manufactories, such as International Harvester and Massey-Harris/Massey-Ferguson, have been prominent suppliers to agriculture right into the 1980s.

To the casual observer, walking plows did not change much during the nineteenth century (see figures 13-1 and 13-2). Teams of oxen and horses still pulled most plows at the end of the century just as they had at its beginning, and the plow itself still usually had a single bottom. But the capabilities and durability of the plows were markedly improved by the century's end.

In 1800 most plows were made entirely of wood, even though Charles Newbold had patented a cast iron plow in 1797. Jethro Wood of White Creek, north of Hoosick Falls, obtained a patent in 1819 for a plow with a cast-iron moldboard (the curved plow board that turns over the earth) but as with Newbold's design it was slow to gain acceptance (see figure 13-3). To prove that his plow was superior to the traditional all-wood plow, Wood resorted to giving his plows away. After the plows gained acceptance, however, he encountered patent violators. Some plows were simply direct copies, while others had signficant modifications, such as a wrought iron share. Because this share was less brittle and more durable than cast-iron, these "patent-violators" believed that the substitution entitled them to manufacture a similar plow.

Thomas Wiard, a blacksmith by trade, moved in 1803 to Geneseo, where he made wrought iron plow points for local farmers. He was one of the blacksmiths who helped improve the all-wood plow by covering the face of the moldboard with strips of iron. In 1815, Wiard and Jethro Wood started making cast-iron plows in Wood's foundry, and in 1817 Wiard built his own foundry in East Avon, having moved there in 1804. After Wood obtained his patent in 1819 he fully expected Wiard to pay royalties but he didn't; however, in Wiard's case Wood relented. The *Livingston Republican,* of February 11, 1874 described the Wiard Plow Works as the "oldest plow works in the world"; there surely were objections to this statement, but they were not printed (see figure 13-4). The newspaper claimed that more than 100 establishments were making the Wiard plow, "not all of them legally," and remarked on the plow's durability, noting that a single Wiard plow had been used for more than 20 years on 100 acres. The Wiard plow used Salisbury iron, which could take a high polish, permitting heavy draft with less clogging, although it cost twice as much as ordinary iron. The castings were chilled to harden the plow for better durability. Seasoned white oak was used

PLOWS.

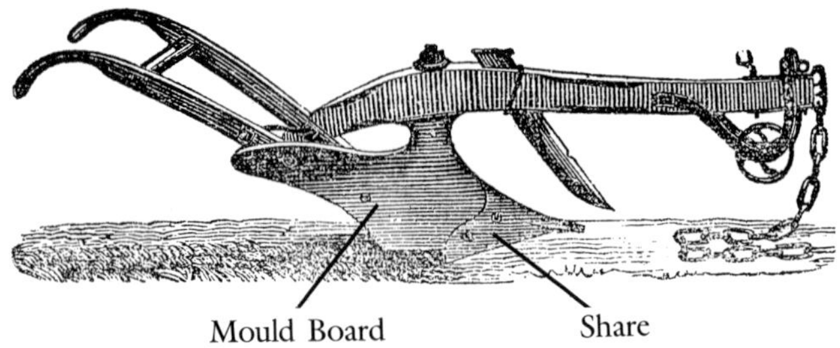

FIGURE 13-1. *Early nineteenth century walking plow made almost entirely of wood.*

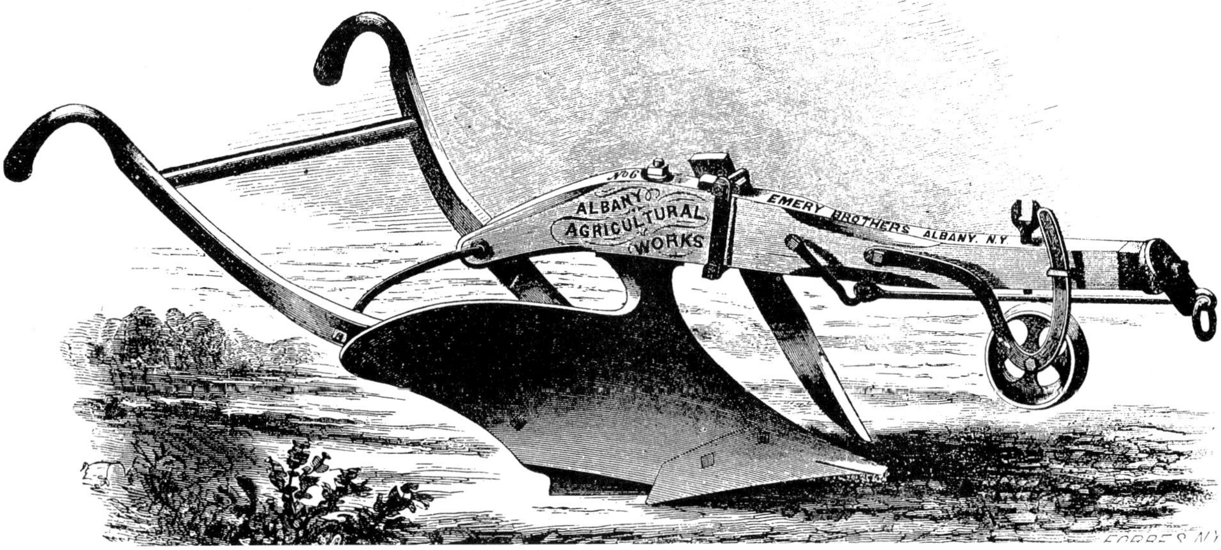

FIGURE 13-2. *1876 single bottom plow, with many cast iron and steel parts, as displayed at the 1876 Philadelphia exhibition. Emery Brothers, Albany.*

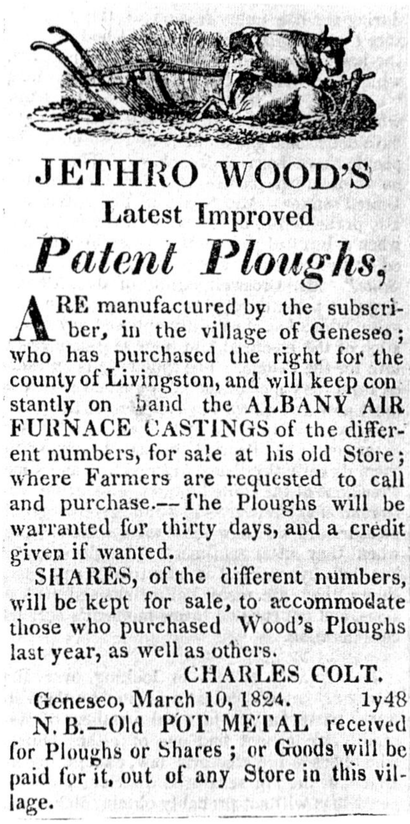

JETHRO WOOD'S
Latest Improved
Patent Ploughs,

ARE manufactured by the subscriber, in the village of Geneseo; who has purchased the right for the county of Livingston, and will keep constantly on hand the ALBANY AIR FURNACE CASTINGS of the different numbers, for sale at his old Store; where Farmers are requested to call and purchase.--The Ploughs will be warranted for thirty days, and a credit given if wanted.

SHARES, of the different numbers, will be kept for sale, to accommodate those who purchased Wood's Ploughs last year, as well as others.

CHARLES COLT.

Geneseo, March 10, 1824. 1y48

N. B.—Old POT METAL received for Ploughs or Shares; or Goods will be paid for it, out of any Store in this village.

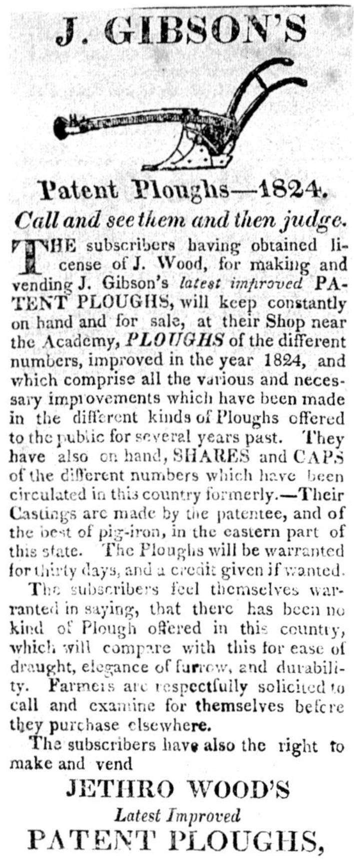

J. GIBSON'S

Patent Ploughs—1824.

Call and see them and then judge.

THE subscribers having obtained license of J. Wood, for making and vending J. Gibson's *latest improved* PATENT PLOUGHS, will keep constantly on hand and for sale, at their Shop near the Academy, *PLOUGHS* of the different numbers, improved in the year 1824, and which comprise all the various and necessary improvements which have been made in the different kinds of Ploughs offered to the public for several years past. They have also on hand, SHARES and CAPS of the different numbers which have been circulated in this country formerly.—Their Castings are made by the patentee, and of the best of pig-iron, in the eastern part of this state. The Ploughs will be warranted for thirty days, and a credit given if wanted.

The subscribers feel themselves warranted in saying, that there has been no kind of Plough offered in this country, which will compare with this for ease of draught, elegance of furrow, and durability. Farmers are respectfully solicited to call and examine for themselves before they purchase elsewhere.

The subscribers have also the right to make and vend

JETHRO WOOD'S
Latest Improved
PATENT PLOUGHS,

FIGURE 13-3. Jethro Wood Patent Ploughs in 1824 newspaper advertisements. The Livingston Journal.

FIGURE 13-4. Wiard Plow Works circa 1885, East Avon, New York. Collection of the Livingston County Historian, Geneseo, New York.

for the frame, and buyers of other plows were warned to watch for manufacturers who painted rather than varnished their wooden parts to hide beech or maple, both inferior woods for a plow (see figure 13-5). In 1921 Massey-Harris was still selling nationally "The Famous 'Wiard' Implements," including several Wiard walking plows.

Elsewhere, John Deere patented a steel moldboard plow in 1837 and James Oliver patented a plow in 1873 with interchangeable parts, permitting easy replacement of worn or broken parts. Plows were constructed almost entirely of iron and steel by the end of the century, except for some with wooden handles (see figure 13-6).

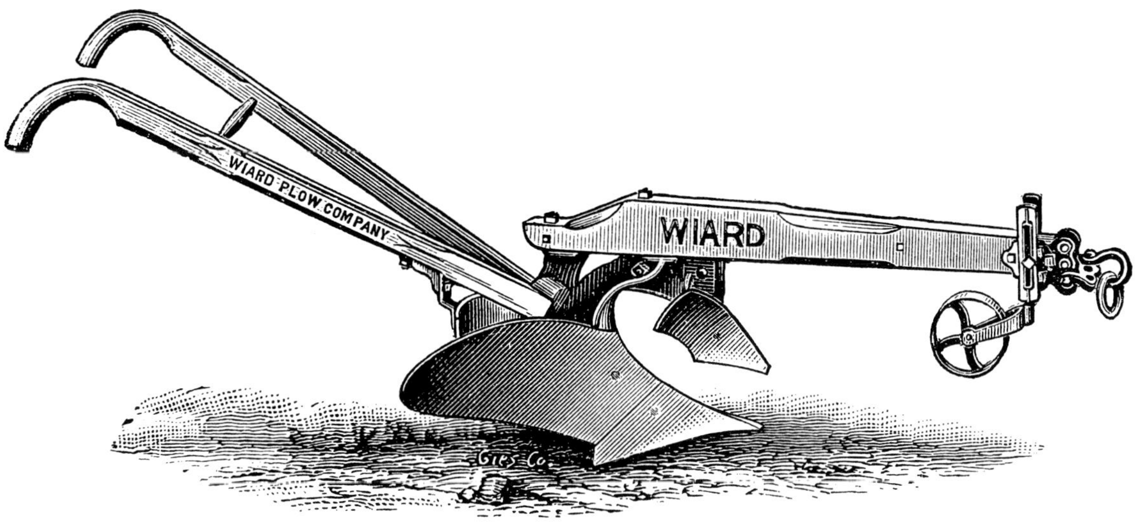

FIGURE 13-5. 1921 advertisement for the Wiard Plow (Massey-Harris).

FIGURE 13-6. Single bottom walking plow demonstration, 1984. Estes Brothers Farm, Caledonia, New York.

FIGURE 13-7. Demonstration of plowing with a sulky plow, 1984. Estes Brothers Farm, Caledonia, New York.

With a two-horse team, a single-bottom plow and proper rest for the horses, a farmer could plow a single acre in about five hours in 1915.

Experimenting with plows led to the sulky plow in the 1860s, similar in appearance to the large-wheeled sulkies used in harness racing. These wheeled plows were much faster, cutting plowing time in half while requiring less effort by the plowman. Most who could afford them made the transition to the sulky plow by the end of the century (see figure 13-7).

The Casaday Sulky Plow, manufactured by the Oliver Chilled Plow Works in South Bend, Indiana and sold locally, had one wheel angled—in fact, it looked broken—so that it rode in the bottom of the newly made furrow, guiding the plow (see figure 13-8).

A poem written in April, 1883 by one C.A.J. of Lima sings the praises of this plow:

What I wrote to you just now,
Is all about the sulky plow;
I have no time to tell a long story
But think this plow should have the glory.

Of all the plows I ever had
The Casaday Sulky beats them bad.
I plow with a pair of horses gray,
A little over two acres a day.

Some stop their teams and come to me,
And say you've broken your axletree,
I tell them no, they bung their eyes;
In that bent wheel the secret lies.

I tell them why it's made that way,
And to each other they will say:
"Well now ain't that a mighty wonder,
I do declare it does beat thunder!"

Put on manure six inches thick
I'll plow it under clean and slick
If thistles or weeds are high as a kite
I'll plow them under out of sight.

It is the plow, I tell you what,
There is no better to be got;
I sit and sing with my arms folded
My furrows look as if they're molded.

But I must write no more just now,
About the Casaday Sulky Plow,
I'll tell the truth anyhow,
It draws easier than the handle plow.

FIGURE 13-8. Casaday Sulky Plow from an 1884 newspaper advertisement by Kneeland & Son, Geneseo, New York.

A Sampler of Reaper Manufacture in the Genesee Valley

The significant increase in the speed of harvesting offered by reapers caused many companies, both large and small, to undertake their manufacture in the Genesee Country. The McCormick reaper was manufactured under patent license (for a patent fee of $20.00 per machine) in many locations throughout the United States; locally, it was made by Seymour & Morgan and the Johnston Harvester Company, both located in Brockport.

The Bacchus & Fitch Company of Brockport also manufactured the McCormick reaper. After several name changes the company became the Johnston Harvester Company in 1870. As with Seymour & Morgan, the Johnston Harvester Company abandoned the McCormick design in favor of one of its own. Its work force ultimately exceeded 500 and annual production rose to more than 5,000 machines per year at the end of the 1870s (see figure 13-9).[2] Figure 13-10 shows the Brockport plant in 1877. Following a fire in 1882 the company moved to Batavia for better

THE JOHNSTON HARVESTER CO.,
Manufacturers of
Self Raking Reapers, Combined Machines, and Mowers,
Send for Catalogue. **Brockport, N. Y.**

FIGURE 13-9. 1877 advertisement from the American Agriculturist. The Johnston Harvester Co.

access to rail transportation, which was considered essential to national sales. There more than 600 people were employed in manufacturing mowers, binders, reapers, rakes, harrows and other farm equipment. This company even had an office in Paris, France. Eventually, the firm was taken over by Massey-Ferguson in 1910.

Seymour & Morgan's production of the McCormick reaper has already been partially described in Chapter VI. By 1849 Seymour had invented several improvements for self-raking machines, including the "quadrant" platform; thus the firm could offer a machine that did not seem to infringe on the McCormick patent. A model of this "New York Reaper" used by company salesmen has survived and is on display in the Morgan-Manning House in Brockport (see figure 13-11). McCormick felt otherwise about just how different the New York Reaper was, and the legal battles went on for years with no definite resolution. (A young Illinois lawyer by the name of Abraham Lincoln was involved in the McCormick/Seymour & Morgan litigation.) In turn, patents awarded to Seymour required other manufacturers to pay royalties to his company when these important improvements were incorporated in their own basic designs. Seymour & Morgan introduced its most improved model, the "Triumph," in 1874, achieving an annual output of more than 3,000 machines by 1877.[3] By the end of the 1870s, 8,000 reapers per year were being produced by the various companies in Brockport alone.

Johnson & Company, located next to the James Wadsworth office on Main Street in Geneseo, began the manufacture of Brinckerhoff's Self Raking Reaper and Mower about 1861, employ-

FIGURE 13-10. Johnston Harvester Works, Brockport, New York. McIntosh, History of Monroe County, 1877.

FIGURE 13-11. This fully operational Seymour & Morgan salesman's model, complete with specially cast parts, provides a good idea of how the mechanical rake swept the quadrant clear while the reaper continued. One less person was thus needed in the harvesting process. Collection of the Western Monroe Historical Society, Morgan-Manning House, Brockport, New York.

ing about thirty people. The ability of the Brinckeroffs to cut close to the ground and their heavy, durable construction made them most desirable; in 1866 an order was placed from Buenos Aires, Argentina, for twenty-five Brinckerhoff machines.

The Agricultural-Machine Works of Dow & Fowler included eight buildings in Fowlerville, where the "Macedon Improved Thresher" and "The Yankee Mower" were produced, along with separators and farm steam engines (see figure 13-12).

Dow & Fowler succeeded the Livingston Agricultural Works, which had occupied the location from 1835 to 1865. In 1866, sixty-five Dow & Fowler employees turned out 111 threshers ($600 each) and 400 mowers ($125 each). An 1875 Dow & Fowler thresher-separator with a thirty-three-inch threshing cylinder can be found among the agricultural implements on display at the Henry Ford Museum in Dearborn, Michigan. In the off season the firm manufactured coffins and furniture.

Dow & Fowler became B. F. Dow & Co. in 1868. Following a total rebuilding after a fire in 1879, the firm did a large business in the manufacture of portable steam engines, threshers, separators and other farm machinery. In 1881 the firm was lured to Peru, Indiana through a cash endowment of $10,000.[4]

FIGURE 13-12. Livingston Works, Dow & Fowler, Fowlerville, New York, manufacturer of mowers, separators, and farm steam engines.

Among many other manufacturers of reapers located in the Genesee Country was the Wyckoff, Tuttle & Olin Company (see figure 13-13), once located on the Main Street of Perry in a cobblestone structure, now razed.

FIGURE 13-13. Wyckoff, Tuttle & Olin Co., Perry, New York. The plant operated between 1875 and 1894 when the business relocated in Jamestown. The Royce Reaper, its first machine, enjoyed an excellent reputation.

Binders

Mechanical binders or "harvesters" were the next logical addition to the reaper. A machine that dispensed with the work of the four or five workers who walked behind the reaper binding the cut grain was clearly desirable. In some early machines, workers rode the reaper tying bundles before they were raked to the ground. These machines were hard on the horses and it was difficult to maintain the proper cutting speed without burying the bundlers in cut grain.

Chief among the many problems in developing a binder was the need for a reliable knot-tying mechanism. The first successful binders used wire, but when the wire was chewed up in threshing and became part of the flour it became clear that twine or straw would be preferable. (Machines that could handle both wire and twine were eventually perfected.) John Heath of Warren, Ohio received a patent for a twine binder in 1850, but the knot had to be tied by hand. Many worked to make the process completely automatic, and by 1877 J. Appleby of Whitewater, Wisconsin developed a twine binder that he attached to a Marsh harvester. Appleby and William Deering of Chicago joined forces to produce some 3,000 twine binders in 1880. A year later the McCormick firm produced a binder (harvester) employing an Appleby twine binder mechanism (see figure 13-14), and in three years sold 15,000 of them. The next major step would be the combine.[5]

Almost every reaper manufacturer in the Genesee Country produced twine binders; their usefulness was apparent to all. Now all the grain could be harvested at just the right time, neither too green nor too ripe. The machine-tied bundles could be shocked in a timely manner with eight or ten bundles to the shock, heads up, set to withstand the wind, finished with a cap bundle to shed rain. One major later development was a device that kept bundles on the binder until enough were produced to make a shock, thus

FIGURE 13-14. A folding McCormick twine binder traveling north on the Avon-Geneseo road at Ashantee near the five-arch Erie Railroad bridge. In a folding binder the reel folded to fit narrow doorways or gates. Anonymous. Collection, Livingston County Historian's Office, Geneseo, New York.

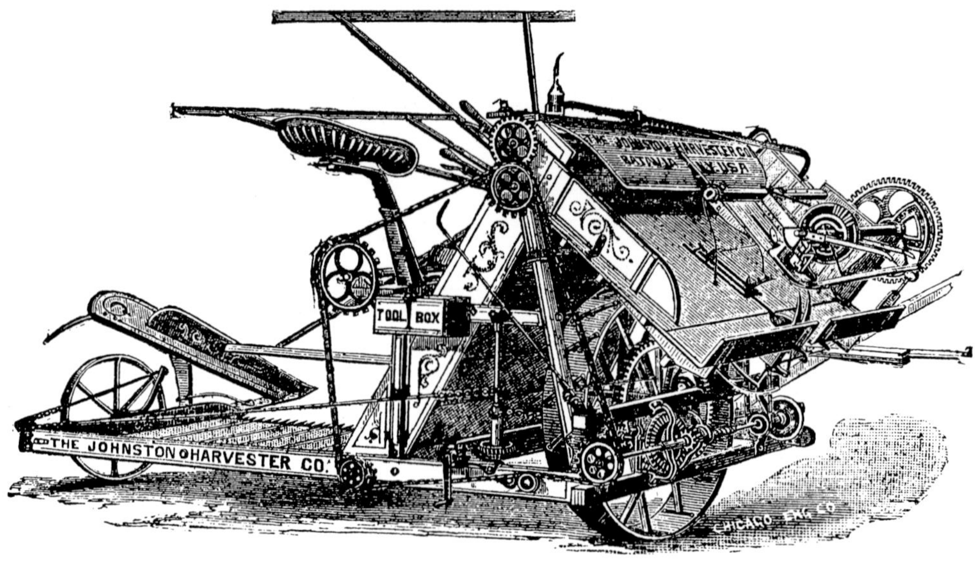

FIGURE 13-15. Johnston self-tying binder, Brockport, New York. The design, once perfected, was to remain fairly constant for some forty years.

saving much walking and stooping for the shocker. Figure 13-15 shows a Johnston twine binder. A comparison of the Johnston machine with the Aultman Buckeye Twine Binder, made in Canton, Ohio, shows many similarities (see figure 13-16). In 1888 the Seymour & Morgan Company offered its "Triumph Number 8," a steel frame folding binder that Morgan claimed was the only practical folding binder.

FIGURE 13-16. 1882 Buckeye Twine Binder built by Aultman & Co., Canton, Ohio. Scientific American, June 10, 1882.

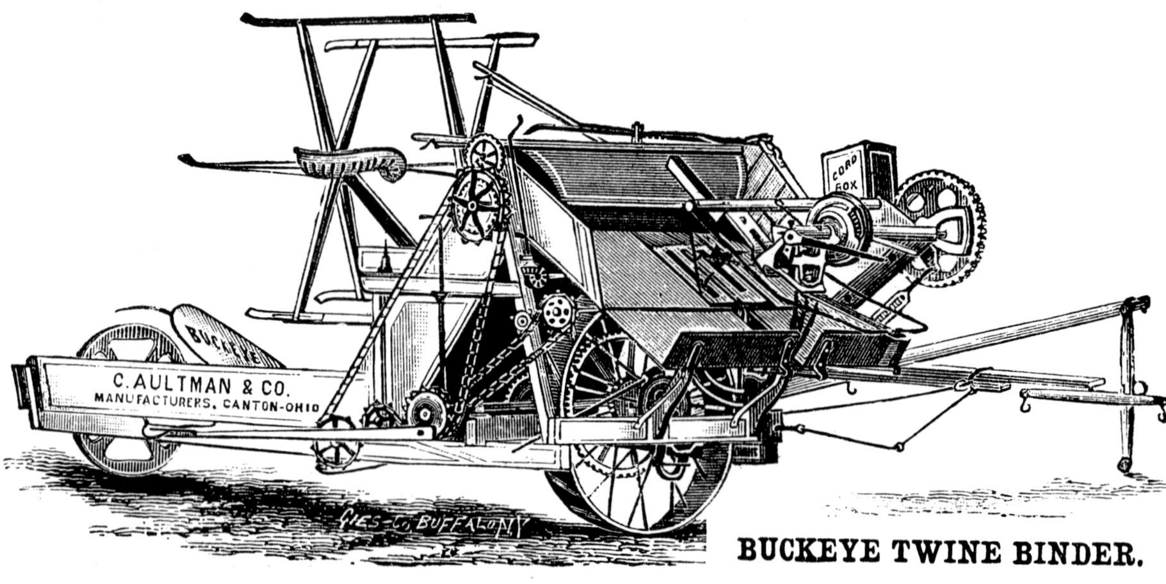

Grain Drills

Grain drills for seeding were first introduced about 1840 but not patented until 1851, when Foster, Jessup & Brown, manufacturers of grain drills in Palmyra, obtained a patent for a force-feed mechanism.[6] In 1854 the Brown portion of the company relocated in Shortsville, where the Empire Grain Drill Company was incorporated (see figure 13-17). Seed drills provided effective, orderly, mechanical seeding, which became more and more necessary as various kinds of mechanical harvesters were developed. Careful planting in even rows became especially important for large stands of corn in view of the need for mechanical cultivation between the widely-spaced rows and the mechanical harvesting required for ensilage (see Chapter X). At the end of the century The Genesee Valley Manufacturing Company of Mount Morris was selling its "Missouri" Grain and Fertilizer Drill throughout the world. The company's catalog was printed in both English and Spanish.

The invention, design improvements and extensive manufacture of farm implements in the Genesee Country had a national impact on farming. Today, traces of the region's once-extensive farm implement production are somewhat limited, but that does not diminish its importance, which reflects the vital importance of agriculture to the area. Consolidation of farm implement manufacturing into fewer and fewer companies, a trend begun in the nineteenth century, continues today and has tended to obscure the contributions of the original local manufacturers. Additional manufacturers are discussed elsewhere throughout the book.

Improved farm tools and machines helped expand crop production, thus increasing the need for larger barns to store the larger yields. In the wake of the Civil War the gambrel-roofed barn, the most common barn type in the Genesee Country, helped fill the need for increased storage capacity.

FIGURE 13-17. The Empire Grain Drill Works, Shortsville, Ontario County, New York. Production commenced here in 1854.

CHAPTER XIV
Gambrel-Roofed Barns

LOG OR TIMBER-FRAMED, most early barns were gable-ended. However, the gambrel roof, with its shallow-pitched upper slope, provided significantly more storage space for sheaves of grain or loose hay (see figure 14-1). The gambrel-roofed barn began to appear in the region soon after the mid-nineteenth century, and today it is the most common barn profile in the Genesee Country. Another important factor in the widespread acceptance of the gambrel roof was the increased availability of power-sawn lumber, which was used in its construction.

That the balloon-framed gambrel roof design was considered superior can be seen where a gable-roofed barn has been converted to one with a gambrel roof. Figure 14-2 shows a barn on Rte. 15 between Wayland and Springwater in which the original gable roof has been modified to a gambrel roof. This barn is unusual in that the conversion can be seen from the outside; in most such cases the exterior no longer reveals the change. The increase in the storage capacity of this barn has been calculated to be on the order of fifty percent, and was accomplished without changing the foundation size. Although many other examples of externally evident gable-to-gambrel conversion exist in the Genesee Country, the number presented here must be limited. Figure 14-3 shows another such barn; here the major external key to the alteration is the diamond window. Originally placed in the gable, with the new gambrel roof the window now appears at an unusual position. Of course, the boarding of the gambrel section also hints at the conversion, but one must be aware of situations which seem to reveal both roof shapes but which instead are the result of removing a gable barn connected to a gambrel barn. The remaining markings at first might seem similar.

A truly extraordinary set of barn moving photographs made by Herman Wright in 1916 illustrates how an older structure could be made to serve modern needs, especially if modified with a gambrel roof. Herman Wright was not only a farmer but an avid amateur photographer who photographed the daily events as well as extraordinary events of his life. This documentation through a series of photographs was an innovative approach for even advanced amateur photographers of the time.[1]

The seventy-foot basement barn on the Herman J. Wright farm in Lyons burned during the last week of October, 1916. Figure 14-4 shows the Wright farm before the fire with the original barn on the lower left.

Wright decided to move an old English barn, even though it was smaller and shorter, from its hilltop location above the farmstead (see figure 14-4) to the foundation of the burned barn. Will Rush, experienced in such work, was hired to move the barn. For power Rush relied on a winch

Opposite: After raising the Shearing/Krenzer barn frame the builders pause for their photograph. Rush, New York, circa 1907. Collection of the Town of Rush Historian.

FIGURE 14-1. *Gambrel-roofed barns (also called curb-roofed) abound in the Genesee Country. In this view the barns are hard by a meander in the Genesee River, on the flats between Mount Morris and Geneseo. Behind the barns the valley is seen rising gently to the west.*

FIGURE 14-2. *In this gable-roofed barn, converted to one with a gambrel roof, the storage space has increased by 50 per cent. West side of Route 15 south of Springwater, New York.*

FIGURE 14-3. The diamond window in the gable of the original structure now seems unusually placed in this gambrel-roofed converted barn. The barn is located on Route 36 two miles north of Arkport, Steuben County, New York.

and a single horse, the same horse that drew his buggy to the site (see figure 14-5).

The English barn was jacked up and beams were placed under the sills. Eight wooden wheels and stout wooden rollers were used to move the barn down the hill, across the road on rails supported by cribbing, and onto the stone foundation where it was turned into its final position (see figures 14-6, 14-7 and 14-8).

The entire moving operation, including preparation of barn, road and foundation, required nearly six weeks. Stout ropes were used to draw the barn and others to prevent its running away. The lines were fastened to trees along the hillside route and were shifted as the moving progressed (see figure 14-7). After arriving at the site the barn was moved onto the old foundation and turned in the course of one day. The turning had to be done very carefully so the barn did not tip over or knock out one of the foundation walls (see figure 14-9). Once in place, it was decided to remove the gable roof and replace it with a more capacious gambrel roof, but this work was put off until the following spring. With the increased mow capacity provided by the new roof the originally smaller barn came close to providing storage capacity similar to that of the original longer barn (see figure 14-10).

FIGURE 14-4. The barn complex at the Herman Wright farm on Middle Sodus Road north of Lyons, New York sometime before October, 1916. The long, gable-roofed basement barn was replaced by the barn just barely visible on the hilltop above the complex. Collection of Mrs. Herman Facer.

FIGURE 14-5. William Rush's horse and buggy stand in front of the English barn, which has already been jacked up and fitted with moving beams and wheels. This horse also provided the power for the moving. Collection of Mrs. Herman Facer.

FIGURE 14-6. A moment is stolen from the downhill moving by Herman Wright's camera. The taut ropes restraining the barn can be clearly seen, as can the wheel assembly under the moving beam. William Rush, in the white shirt, is standing to the left in the photograph. It seems apparent that boards were removed to permit communication from one side of the barn to the other during moving. Collection of Mrs. Herman Facer.

FIGURE 14-7. The winch used in the moving process is in the right foreground and all the downhill cables are slack as things are arranged for the next stage of the barn moving. Planks were laid down under the wheels to even out the ground and to prevent the wheels from sinking into the soft earth. Collection of Mrs. Herman Facer.

FIGURE 14-8. A dramatic moment is reached when the barn begins to traverse the planks laid on the cribbing across the road. On the downhill side the wheels remain in use while on the upside rollers are being used (three can be seen in this photograph). Collection of Mrs. Herman Facer.

FIGURE 14-9. In less than one working day the barn was moved from its position adjacent to the foundation up onto the foundation and then turned into its final position. Cribbing was used for intermediate support and the wheels were entirely dispensed with in favor of the rollers. The season is late, as is evidenced by the dusting of snow on the ground. The moved barn is obviously too short for the old foundation, and after a new end wall was constructed the unused original part was allowed to slowly crumble away. Collection of Mrs. Herman Facer.

FIGURE 14-10. The new gambrel framing is being installed. Collection of Mrs. Herman Facer.

The January, 1884 *American Agriculturist* discusses the origin of the term "gambrel," saying that it refers to the stick put between the gam (or gamb, or gambrel) tendon and the bone of the hind legs of a hog or steer to suspend it for dressing (see figure 14-11). Later, the editor quotes the dictionary definition of "gambrel" as "the hind leg of a horse," noting the obvious similarity in shape between the hock and the barn roof (see figure 14-12).

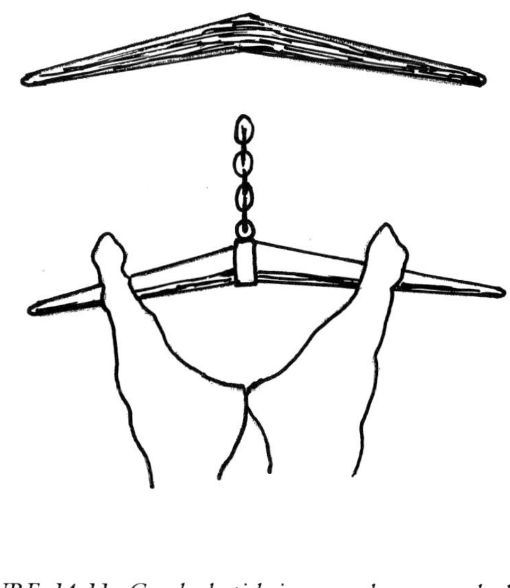

FIGURE 14-11. Gambrel stick in use; also, a gambrel stick shown by itself revealing its similarity to a gambrel roof design.

In England the gambrel roof form can be traced at least to the early sixteenth century, and in American domestic architecture the gambrel roof appears at least as early as the late 1670s in the Peter Tufts house in Medford, Massachusetts.[2] The roof beam on the Tufts house is shaped like a gambrel stick providing support for the two upper surfaces of the roof. The re-

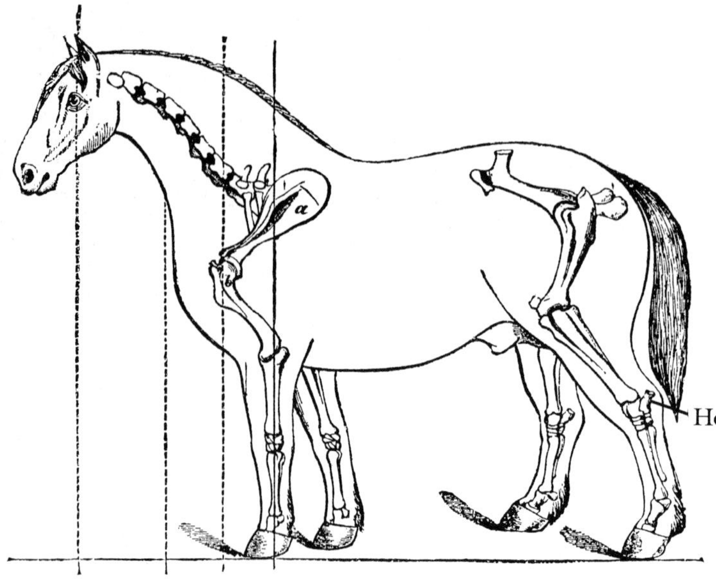

FIGURE 14-12. The hind leg of a horse.

semblance is remarkable; even the grain of the wood follows the "curve" of the Tufts' upper roof beams as one finds the grain following the "curve" in a gambrel stick. The earliest gambrel-roofed barns had relatively narrow upper roof surfaces, strongly reminiscent of the shape and angle of a gambrel stick. The gambrel roof seems to have been used by Dutch and English settlers on houses, mills and other storage buildings in the seventeenth and eighteenth centuries, but it was less often used on barns. The widespread appearance of the gambrel roof on the American farming scene seems to be associated with the increased availability of power sawn lumber and the need that was generated by increased crop production made possible with improved agricultural machinery. Sketches made in the late '30s and early '40s for Barber and Howe's historical books on New York and New Jersey show only two gambrel barns in some 304 illustrations; the rest of the barns are gable-roofed.

References to designs for gambrel-roofed barns began to appear in agricultural periodicals of the 1860s.[3] The appearance of gambrel roofs on Genesee Country barns can be roughly dated by a check of the various county histories. For example, out of 172 barns in McIntosh's 1877 *History of Monroe County,* only eleven were gambrel-roofed. W. W. Clayton's 1879 *History of Steuben County* similarly shows that only ten out of the sixty-five barns included were of gambrel design. The 1879 Beer's *History of Allegany County* shows a higher ratio, fifty-eight gable to fifteen gambrel. These surveys give a general rather than an exact ratio because of sampling bias; nevertheless the results are significant and useful.

Balloon framing of the gambrel roof was common and allowed a large unrestricted open space which can be readily appreciated in an interior photograph. Traditional post and beam framing would have required cross beams, which even in the gambrel design would have interfered with the efficient loading of the mows with a hay track. The gambrel-roofed barn on the Estes Brothers Honey Locust Farm near Caledonia is a good example of a bank barn in which the original hewn frame was later modified with sawn timbers to produce the gambrel roof. In this instance the collar beams between the purlin posts and plates

FIGURE 14-13. Interior of the gambrel barn on the Maxwell/Estes Brothers Honey Locust Farm, Maxwell Station Rd., near Caledonia, New York. The newer lumber used in the gambrel roof contrasts with the original timber frame. Collar beams connecting the purlin beams are only seen in barns without a hay track.

hark back to earlier European gambrel designs. Collar beams could be used because no hay track was installed (see figure 14-13); it probably was considered unnecessary, since the mows on both sides of the driveway are open all the way to ground level. The mows were easy to fill by pitching down but difficult to empty when it came to threshing (the thresher itself was located on the drive-through).

On the Little Road in the Town of Avon stands the barn complex of the Wade Farm, its first barns built by Herbert Wadsworth in 1888. The newer gambrel-roofed barn and the older gable-roofed barn form an "L" (see figure 14-14).

FIGURE 14-14. Gambrel barn in the Wade complex Little Road, Town of Avon, New York.

A side-by-side comparison of the two Wade barn interiors demonstrates some of the advantages of the gambrel-roofed barn. It should be noted that both structures were built to accommodate a mechanical hay track. In the gable-ended barn a center aisle is formed by purlin posts and lateral beams (see figure 14-15). In the newer, gambrel-roofed barn there is clearly much more space for hay storage, unimpeded by purlin posts (see figure 14-16). The only restrictions inside are those of the diagonal braces which stabilize the structure against the outward thrust of loose hay. The barn gains its strength from its balloon framing. closely-spaced roof rafters and roof boards pinned together with plenty of nails and spikes.

At the outset of the use of balloon construction, many were skeptical of its ultimate strength, so in many cases heavier timber framing, reminiscent of earlier European gambrel-roofed structures, was combined with the lighter-weight materials of the newer technique. When these two techniques are combined, one is reminded of the man who has worn suspenders all his life and is confronted with the gift of a belt; he solves the problem of whether to trust the belt by wearing both belt and suspenders at the same time. In gambrel barn construction timbers continued to be employed in the main frame while balloon framing was used in the critical roof rafter construction.

On the Robert McClellan Farm in Geneseo, the gambrel-roofed barn in the center of the U arrangement originally had a gable roof (see figure 14-18). The interior view reveals an interesting combination of hewn posts with sawn rafters and gussets. The conversion to a gambrel roof and removal of the original crossbeams resulted in a structural problem, indicated by the presence of replacement beams and tie rods in addition to the diagonal bracing. While the increase in space offered by the new roof must have made the risk seem worthwhile, problems of this kind were not unusual in converted barns (see figure 14-19).

Two glass plate images recently discovered in Livingston County show two stages of construction of a gambrel-roofed barn next to a gable-roofed barn. Figure 14-20 shows the sawn timber frame with its heavy, two-story diagonal braces set in place on the foundation. In figure 14-21, six men continue construction, installing the rafters of the balloon-framed gambrel roof and vertical side sheathing boards prior to shingling. Gussets provide stability at the break in the slope of the roof and also along the ridge.

Figures 14-20 and 14-21 may be compared with figure 14-22; the Shearing/Krenzer barn in the Town of Rush has a much heavier timber frame, which provides a central access different from the lighter barn. In addition, the interior purlin posts in the Shearing/Krenzer barn are from an earlier design. The unidentified, undated barn (figures 14-20 and 14-21) is clearly a newer design, but this does not necessarily mean more recent construction than the Shearing/Krenzer barn. Such is the manner in which barns were built: Some builders simply retained older designs, even though advanced designs were available (see figure 14-23).

A group of photographic postcards reveals several scenes of gambrel-roofed barn construction. Figure 14-24 shows the raising of the last bent; because of the stability provided by securing the previous bents, this was probably the easiest bent to raise. The characteristic roof form of the gambrel design has taken shape in figure 14-25.

The cooperation of barn raising and the associated preparation of food for the raisers were important social occasions. Food preparation required a great deal of work, as shown in figure 14-26. In spite of the dangerous and difficult work of the raising, the memories of those lunches and dinners are precious to those who attended. The raising and the meal were traditionally capped by cigars for the raisers after the day's work was completed. If the barn wasn't filled immediately a barn dance might be held as a way of celebrating the success of the raising. The

FIGURE 14-15. Wade gable barn interior. This essentially timber-framed structure was designed with large cross section purlin posts and well-braced cross beams, which permitted an open aisle for use of the hay track.

FIGURE 14-16. Wade gambrel barn interior. The balloon framed rafters impart strength without the need for the purlin posts in 14-15, making the mowing away of loose hay much easier.

FIGURE 14-17. Gambrel-roofed barns eventually became dominant in the Genesee Country. Mendon Center Road, Pittsford, Monroe County, New York. The Thornell/Bruner Barn on the right is also seen in Figures 21-10 and 21-11.

FIGURE 14-18. McClellan Farm barn complex, Reservoir Road, Geneseo, New York. The barn converted from gable to gambrel is in the center of the photograph.

FIGURE 14-19. This interior view shows both the new lumber used for bracing and the tie rods in the modified barn. McClellan Farm, Reservoir Road, Geneseo, Livingston County, New York.

FIGURE 14-20, Above Left. Construction of the side posts and braces on a gambrel-roofed barn. Anonymous.

FIGURE 14-21, Below Left. With the side boarding completed, work proceeds on the roof boards and purlin posts at the ends of the barn shown in figure 14-20. Anonymous.

FIGURE 14-22, Above. The hard and dangerous work of the barn raising is over. Shearing/Krenzer barn, corner of Routes 15 and 251, Town of Rush, Monroe County, New York, circa 1907. Collection of the Town of Rush Historian.

FIGURE 14-23, Below. The finished gambrel-roofed Shearing/Krenzer barn as it appeared seventy-seven years later, in 1984.

FIGURE 14-24. Anonymous photographic postcard view of the raising of the end bent in a gambrel-roof barn. The lines from the rigging used to raise this bent are revealed. Also, the difference in configuration of the bents can be seen in this view. Collection of Hans Tanner.

FIGURE 14-25. A second anonymous photographic postcard. Details suggest this barn was constructed by the same builder as that of the barn in figure 14-24; it exhibits the characteristic roofline of the gambrel barn after the installation of the purlins and the roof rafters. Indeed an occasion for a photograph. Collection of Hans Tanner.

FIGURE 14-26. Shearing/Krenzer raising day dinner, circa 1907. Town of Rush Historian.

dances were usually held when the barn was nearly completed.

Barn raising photographs are scarce and most were made after the frame was already raised, but each existing photograph adds its own dimension to the story of barn building. About 1913 Dan Sy raised a gambrel-roofed barn on the Fred Pafk farm in the town of Alabama.[4] In figure 14-27, the seated man in the center of the front row is holding a boring machine, which in some cases replaced the hand-guided auger for starting mortises and boring holes for treenails. Less skill was required to maintain proper angles with the boring machine, hence its widespread adoption. Unfortunately, the Pafk barn burned in 1930.

Many variations exist on the basic gambrel roof design. Figure 14-28, from an 1876 lithograph showing the Taylor Farm of Elba, reveals variations of both proportion and pitch in the gambrel barn roofs. In other instances in the Genesee Country the lower roof surfaces are pitched at 60 to 70 degrees. Figure 14-29 shows a barn on the Caledonia-York road with an unusually high-pitched gambrel roof.

Another interesting variation is the Dutch gambrel, with its slight flare in the roof near the eaves, reminding one of the starched caps that women once wore in the Netherlands (see figure 4-30). This flare is visible in the barns shown earlier in this chapter and will be quite evident in Chapter XXI, which deals with the Wells barn. The flare keeps rain water away from the siding and foundation and also prevents avalanches of any heavy snow that might accumulate on the barn roof. In days when the barn doors were hinged to swing outward, snow piled against

FIGURE 14-27. Portrait of the raisers of the Pafk barn, Feeder Rd., Alabama, Genesee County, New York, circa 1913. Thirty-five of those who gathered together to raise the Pafk barn pose for a group portrait after the sawn timber frame is safely up and in place. Much work remains, including rafter-raising, boarding and shingling. To animate the photograph the raisers all heft a tool of some kind such as a square, saw, hammer or adze. Collection of the Alabama Town Historian, Alabama, Genesee County, New York.

FIGURE 14-28. Complex of gambrel-roofed barns exhibiting varying roofs, Taylor Farm, Elba, New York. Genesee County Atlas, 1876.

them might have pinned them shut until it could be shoveled away. The hanging of barn doors on rollers seems to date back to the early days of railroad car construction in the 1830s, when box car doors were hung with patented rolling hardware. The tone of the 1866 *Cultivator and Country Gentleman* suggests an impatience with the fact that more farmers have not adopted rolling barn doors by that time.

Masonry Dutch Colonial buildings of the seventeenth century also had a flaring, or "overshoot," designed to protect the relatively weak mortar of the period from rain water. A direct connection between these seventeenth-century masonry structures and nineteenth-century gambrel-roofed barns has not been shown, but builders with a practical bent probably had no difficulty adopting this obviously functional form.

Some area barns built in the 1880s possess almost horizontal flaring, which permitted the construction of box gutters piped to cisterns (see figure 21-10 for a later box gutter installation). In still later construction, eaves troughs were preferred to the box gutter because the suspended eaves trough poses less threat to the roof during freezing and thawing cycles.

FIGURE 14-29. An example of a gambrel-roofed barn in which the pitch of the two lower roof surfaces is exceptionally steep. Caledonia-York Road (Route 36), Livingston County, New York.

FIGURE 14-30. Dutch Gambrel roof—icicles on the left help demonstrate the effect of the offset eaves.

The illustrations from McIntosh's 1876 *History of Ontario County, New York* reveal interesting aspects of nineteenth century barns and farming. The large gambrel barn of H. Padelford in Canandaigua is depicted receiving a load of loose hay (see figure 14-31). The Padelford barn has a full masonry basement but is not built into the side of a hill; therefore, a ramp is needed for the wagon. The ramp wharf or ramp abutment, as often is the case, does not bear against the foundation lest it be destabilized during periods of freezing and thawing or movement; a short bridge connects the ramp to the barn. There are two main barn doors and the ends and cupola are louvered for ventilation. The entire structure is vertical-boarded, typical of the region. Padelford clearly had easy access to needed transportation.

The gambrel-roofed hay barn on the Fall Brook Farm just south of Geneseo is unique because of its open sides (see figure 14-32). This particular barn is used to store feed for wintering

FIGURE 14-31. H. Padelford Farm, Canandaigua, Ontario County, New York. McIntosh, History of Ontario County, New York, 1876.

livestock kept in the barnyard. The open sides provide the ventilation needed to prevent the fires experienced with some other barn designs when hay is improperly cured. An additional but not unique feature of this barn is the hood over the extension of the hay track. In most barns the hood exists to protect a hoisting beam to which the track is fastened. Where these hoods appear, the barn usually has an enormous door allowing entry of the loose hay. The doors are variously hinged or tracked to permit opening and closing.

Many years have passed since the operation of the Underground Railroad helped slaves escape to freedom in the northern states and Canada. Several such routes to Canada converged in the Genesee Country, one passing through Rochester and another, more southerly, leading to Lake Erie, Buffalo and St. Catharines. Time has changed or eliminated the "depots," the inconspicuous hiding places along the trackless route of the Underground. The need for secrecy made identification of specific hiding places difficult even during the main operating years of the Underground Railroad; occasionally, however, through the secret mists of the past emerge the remnants of an Underground Railroad station, a hiding place revealed only in the timbers and foundations of a house or barn.

A period map of Henrietta shows a structure on or near the present gambrel-roofed barn lo-

FIGURE 14-32. Fall Brook barn. Geneseo-Mt. Morris Road, Livingston County, New York.

cated north of Calkins Road on East Henrietta Road (see figure 14-33). If existing maps are correct, this is the location of a barn used by David Richardson, a man of enormous size who assisted blacks seeking freedom by hiding them in the original structure. An examination of the framing of the present barn indicates that this side-hill barn was enlarged and modified sometime in the past. Although the roof is relatively new, portions of the frame and foundation are clearly old

FIGURE 14-33. Richardson/Michel gambrel-roofed barn probably used for a station on the Underground Railroad, East Henrietta Road, Rochester, Monroe County, New York. The crescent marks on the barn were caused by a tree, since cut down, whose wind-blown branches rubbed back and forth on the boards.

enough to have been part of a barn that served the Underground Railroad. Today, there is no secret space, equipment or other remaining scrap of physical evidence indicating that this barn was itself used as a station, but a historical marker between the barn and the highway identifies it as a "depot" location.

Richardson was not nearly as well known as

Frederick Douglass, who used every possible means to move slaves on to free territory. Known internationally for his efforts, he made Rochester his home between 1847 and 1872; here he published the influential abolitionist newspaper *The North Star* (later called the *Frederick Douglass Paper*).[5] It required many people to create a system of hiding places along the various routes to freedom, and great courage from them all because of the 1850 Fugitive Recovery Bill. This was serious business: A fine of $1,000 or six months imprisonment was the penalty for assisting a runaway slave. Little wonder, then, that many of the rural "stations" were not known at the time or easily identified today. There certainly are many other barns and farm outbuildings which were used, and in all probability the subject is broad enough to support an entire study.

Decorated Gambrel Barns

The main horse barns of the Wadsworth Homestead Farm in Geneseo included two large, horizontal-sided gambrel-roof barns with interesting surface decoration, although all the barns had fancy decorations in a Renaissance mode. The smallish classical structure in the middle of the barn complex was the stableboy's house; this structure included a classical entablature complete with pediment supported by rusticated pilasters (see figure 14-34). The barn on the left had an interesting and rather large rectangular cupola, gracefully decorated by louvered arcading. The cupola not only provided the required ventilation but also was appropriate for a period in which towers and cupolas, not all of them functional, were a general part of accepted popular architectual design. The use of classical detailing is further apparent in the corner quoins. These barns had clapboard siding similar to that of the cow barn (already examined in Chapter IX). Another similarity was the use of circular windows in the gable end, double in the horse barns and single in the cow barn. The gable-ended barn on the right and the gambrel-roofed barn on the left had stalls on both sides of the lower level with mows above. Fire claimed the entire complex in 1972; the new barn has a similar feeling but with quite different detailing.

An additional gambrel-roofed barn of distinguished history, located in Caledonia, should not be overlooked even though its purpose is quite different from that of barns already examined. This structure was built as a fish hatchery barn. Fish cultivation as a rural industry was the lead article in the July, 1853 edition of *The Genesee Farmer.* 31 years later the 1884 *American Agriculturist* described the development of the New York Fishery Commission, established in 1869, for the express purpose of the artificial culture of fish. The first superintendent of fisheries, Seth Green, in association with his brother Monroe, developed many new techniques of fish handling and hatching.[6]

Seth Green was born in Rochester in 1817 and as a youth studied the fish of the Genesee River. By the age of twenty he had evolved a method of artificial propagation of fish. A photograph, now lost, showed Seth Green "stripping" or removing the eggs from a fish into a pan.[7]

By 1884 the original hatchery buildings at Caledonia had been replaced by a gambrel-roofed fish-hatching barn complete with dormers and cupola. This new barn contained hatching tanks and laboratory equipment. Interior lighting requirements mandated the numerous windows,

FIGURE 14-34. Wadsworth Homestead barns, Geneseo, Livingston County, New York. 1967 photograph by Elizabeth Williams.

including the windowed dormers. A strong overlay of Gothic is found in the steep pitch of the dormers, the hoods over some of the gable windows, and the scroll work of the verge boards (see figure 14-35).

Robert Roosevelt, writing in the 1884 *American Agriculturist* says of the hatchery,

> . . . now that the ancient time-eaten, bug-bored and weather stained shanty has been replaced by a reputable building, with some pretentions to architectural attractions, and the ponds have been rebuilt with walls which will hold water and trout securely, and in spite of the fact that the State owns no individual right of entrance, and not half enough ground for ponds, there is no establishment in the world where as good work is done as at the hatchery at Caledonia. . . . Farmers should bear in mind that water-culture can be made a part of agriculture . . . These [fish] will increase rapidly, and take care of their own young, [and] . . . will supply food for the table. . . .[8]

The gambrel-roofed barn became dominant in the latter quarter of the nineteenth century in the Genesee Country. The major differences among gambrel-roofed barns are found in their various framing designs; some of these will be examined in later chapters.

FIGURE 14-35. Fish Hatchery barns, 1884. Caledonia, Livingston County, New York.

FLORAL HALL
AGRICULTURAL HALL
SMITHS DRIVING PARK
& VALLEY POINT FAIR GROUND

CHAPTER XV

Agricultural Information in Almanacs, Periodicals, Manuals, at Colleges and Fairs

IN THE EARLY NINETEENTH CENTURY, many farmers relied heavily on almanacs as important sources of information and new ideas about farming. These almanacs provided recommendations as to when to plant and harvest and they also included predictions of weather conditions throughout the year. Almanacs are still published, but although they still contain weather predictions and planting recommendations, the farmer has generally shifted his attention to agricultural periodicals.

The need for agricultural information required more than once-a-year almanacs. One of the earliest publications in New York State was *The Plough Boy*, first published in Albany in 1819. Another early publication was *The New York Farmer And Horticultural Repository*, which began publication in New York City in 1828. According to noted agricultural historian Ulysses P. Hedrick, the first really useful journal was *The Genesee Farmer (And Gardener's Journal)*, edited and published by Luther Tucker in Rochester beginning in 1831 (see figure 15-1). While the title might suggest otherwise, *The Genesee Farmer* became a national publication. In fact, in his 1835 book recording his trip to America, the Scotch farmer Patrick Shirreff notes that he received copies of *The Genesee Farmer* at his home in Scotland.[1] Tucker changed the name in 1840 when he moved to Albany to begin publication of *The Cultivator: A Consolidation of Buel's Cultivator and The Genesee Farmer*. In 1853 the consolidated publication became a weekly and was renamed *The Country Gentleman*. A "second series" *Genesee Farmer* appeared in 1840 with a succession of publishers and editors. In 1856 Joseph Harris, founder of Harris Seed Company, undertook its publication and continued as editor and publisher until he sold the publication to the *American Agriculturist* in 1865.

The Genesee Farmer featured practical information, interchange of experience through letters, and the introduction of new ideas, including barn plans. Many excerpts from it and other nineteenth century periodicals have been used throughout this work. Articles were often quickly followed by letters to the editor or by other articles in subsequent issues on the same subject,

Opposite: Period lithograph of various events at the 1868 Allegany County Fair.

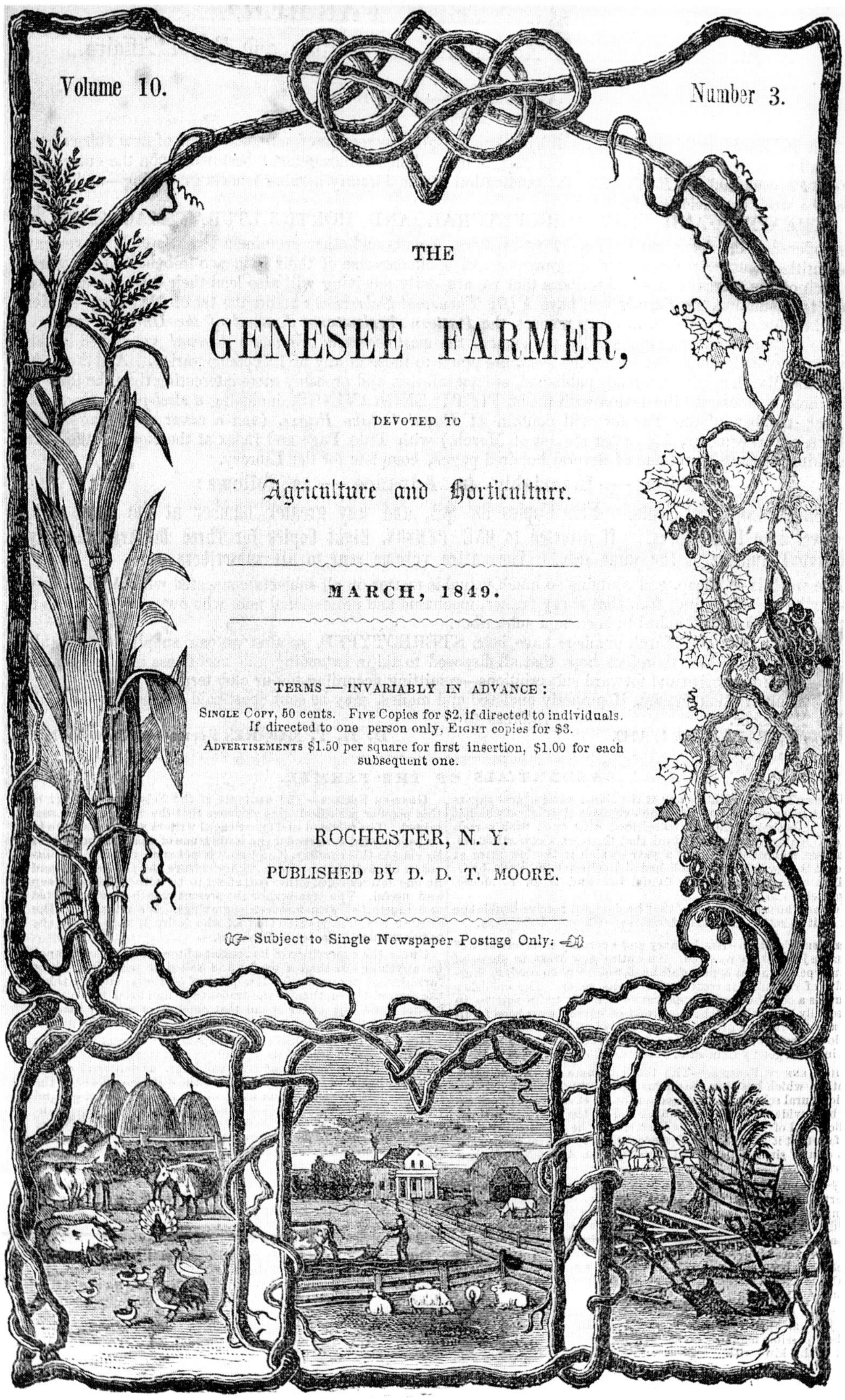

Volume 10. Number 3.

THE

GENESEE FARMER,

DEVOTED TO

Agriculture and Horticulture.

MARCH, 1849.

TERMS — INVARIABLY IN ADVANCE:
SINGLE COPY, 50 cents. FIVE Copies for $2, if directed to individuals.
If directed to one person only, EIGHT copies for $3.
ADVERTISEMENTS $1.50 per square for first insertion, $1.00 for each subsequent one.

ROCHESTER, N. Y.

PUBLISHED BY D. D. T. MOORE.

Subject to Single Newspaper Postage Only.

FIGURE 15-1. The Genesee Farmer title page, March, 1849 (decorative title page from second series).

presenting another method or point of view—all signs of a vigorous publication. For example, during 1851 *The Genesee Farmer* published this exchange concerning the relative merits of the flail and the threshing machine:

I wish to say one word to the farmers of this

region, and to all whom it may concern, about the threshing of oats with a machine. If the farmer has cattle, it is much the best way to thresh oats with a flail, and the cheapest. I will give the cost, in this town, of threshing 600 bushels, calling it a day's work for a threshing machine:

COST OF THRESHING WITH A MACHINE

Cost of Threshing, $2.00 per hundred	$12.00
Four horses,	2.00
Four hands,	3.50
Board of seven hands	2.00
Board of eight horses	2.00
Total	$21.50

COST OF THRESHING WITH A FLAIL

Hand, at $13.00 per month, 30 days,	$15.00
Board four weeks,	8.00
Total	$23.00

Any farmer can tell whether he had rather have oats threshed with a flail, or machine, by the trial; and the improved condition of the cattle will testify in favor of the flail; and unless the thresher be uncommonly wasteful, the farmer will have more bushels of oats, and his straw will be worth double.—Think of it, brother farmers, figure it up, and see how it stands. Artel.—Hornby, N.Y., Dec., 1850.[2]

These figures and figuring were too much for H.L.E. of Albany, whose letter to the editor was published in the very next issue:

Having seen in your February number a communication signed "ARTEL," of Hornby, N.Y., containing an estimate of cost attending the threshing of oats with the ordinary large "Itinerant Threshing Machines," making the cost, including board of men and team, upwards of three and a half cents per bushel, (a pretty heavy per cent. upon their value,) and comparing the same with the cost and advantages of threshing with the flail, I beg the privilege of going a little further, first acknowledging the latter mode the better of the two described by him. The farmer's grain not always being in a condition to allow long delay, and the quantity too large to admit of the slow process, as well as the fluctuation of the market at different seasons, together with the exposure and waste from various causes, he is in most cases compelled to submit to this extravagant tax, levied on him by the "Goths and Vandals" of the surrounding country, in the form of the large *eight* and *ten* horse powers.

Having for many years devoted the greater portion of my time and attention to the wants of the agricultural public in the way of labor-saving implements and machinery, and been engaged in the manufacture and introduction of the same, I may perhaps be allowed to go a step further than "ARTEL" and see if a better mode can not be adopted, by which the farmer may be placed in a more independent position in all respects, save much time and expense, and better suit his own convenience at the same time. As the Railroad Horse-Powers, together with the small spiked cylinders now used for the threshing machines, have been so simplified as to do away the prejudice heretofore so justly felt against all endless-chain horse-powers, (on account of injuring horses,) I will compare the cost, &c., of threshing an equal amount with the R.R. machines, (viz., 600 bushels,) rating the capacity of the machines at the average amount they will do, (and which the warranty of responsible manufacturers makes a condition of sale, the purchaser being his own arbiter, after three months use and trial.)

With a two-horse power and the other parts to match; with two horses which shall together weigh not less than two thousand pounds, on an elevation of 1½ inches to the foot, and a travel of 2¼ miles per hour, either with or without harness; and with four men, four hundred bushels of oats can be threshed per day for months together, without change or injury to horses. Taking his rate of charges, we arrive at the following result, assuming the farmer to own his own machine:

Four men 1½ days each,	$5.25
Two horses 1½ days each	1.50
Board of four men 1½ days	1.14
Board of two horses 1½	.75
Cleaning up for market with fan-mill	1.75
	$10.39

making a fraction over one cent and seven mills per bushel, and having at the same time the advantage of doing his own work with his own farm help and team, and at such times as the market, the weather, or his own convenience is best accommodated. So much for the two horse machines in 1½ days.

As a majority of farmers pursue a mixed farming, and have a variety of grains and grasses, with a power and a machine of the improved kinds now in use, he can do all his threshing and secure his own grass and clover seed, if not a surplus for market. The power will be found of sufficient use, in sawing wood and for various mechanical purposes, to more than pay the interest on the investment and ordinary repairs from wear, &c. Should the farmer prefer to thresh his oats day by, at the rate of twenty or thirty bushels, with his monthly laborer he can do so with the assistance of a boy in from one to one and a half hours, with but one horse, leaving the balance of the day to be devoted to making and repairing implements and attending to the thousand and one wants always about the farm, buildings, &c., &c. When the farm is small, and the quantity of grain small, a saving of some $25 to $30 is made by the purchaser in buying a one instead of a two-horse power, with which about one-half the amount can be done as with the double power when each are worked up to their full capabilities. H.L.E.—Albany, N.Y., 1851.[3]

For the most part the exchanges were civil and designed to point out superior methods of doing things. These periodicals were popular among non-farmers as well and did much to promote the intense interest in horticulture which developed at mid-century.

Additional periodicals include the highly regarded *American Agriculturist,* still in print and now headquartered in Ithaca, New York, which began publication in New York City in 1842. The *American Agriculturist* provided important primary material for this study. Other significant periodicals include: *American Farmer and Farm News,* established 1881; *The Country Gentleman,* established 1853; *The Practical Farmer,* established 1855; and the *Rural New Yorker,* established in Rochester, 1850. In 1853 *The Genesee Farmer's* editor noted that he received some sixty "agricultural papers" printed in the United States.

Agriculture Education

Probably the most common way of learning to farm was from one's family through practical experience. In the United States in the mid-nineteenth century, however, agriculture was becoming more scientific and complex, and the belief grew that a person should have an opportunity to study agriculture at a college. After a considerable period of argument in Congress, Land Grant Colleges were established under the provisions of the Morrill Act of 1862. Agriculture, engineering and home economics were the prime offerings of these new colleges. Financing for the new college in New York State came from the sale of 999,000 acres of federal lands. The money from the sale was first intended to support the "People's College" at Havana (Montour Falls, New York), but eventually the Havana project was abandoned in favor of Cornell University in Ithaca in 1865.

Partially endowed by Ezra Cornell, the school was also supported by both federal and state governments. In 1904 the Cornell College of Agriculture became a state institution.

After the Civil War many farm manuals and books were published which were designed to make the reader aware of what was required to run a successful farming operation. Some of these manuals are still used as references by families whose grandparents and great-grandparents purchased the books when they were first published. By the early 1880s, the New York State Agricultural Station at Geneva had been established, soon to be a part of Cornell University. The Station provided information acquired through scientific experimentation and disseminated through the new manuals.

Agricultural fairs sponsored by agricultural societies have also been a means of disseminating ideas on farming, livestock, economics and so on. Fairs have a European heritage old enough to be lost in the mists of time. In the Genesee Country, as a part of his land promotion scheme, Charles Williamson conducted his first fair near Williamsburgh, in 1794 (see Chapter III). County Agricultural Fairs, similar to those which still exist today, were formed on a model evolved by Elkanah Watson in Pittsfield, Massachusetts around 1810. Watson attempted to form an agricultural society which would conduct annual fairs and offer premiums for worthy displays of animals, produce and agricultural techniques. Watson hoped that awarding prizes for excellence would encourage farmers to their best efforts. Subsequently ideas could be exchanged as a result of these competitions. The 1811 Berkshire, Massachusetts County Fair was successful to the extent that additional fairs were held and localities elsewhere became interested in holding similar fairs.

Watson moved to Albany in 1816, where he was approached by the people of Cooperstown, Otsego County, to assist them in conducting a county fair that same year. Watson then went on to help other eastern counties to establish fairs.[4] In 1819 a New York State Board of Agriculture was established which made provisions for agricultural societies and fairs.

A few samples of agricultural fairs in the Genesee Country are worth mentioning: The Monroe Agricultural Society held its first meeting for a "cattle-show and fair" in 1823 with rules established for the conduct of the fair. McIntosh records at length the winners of premiums at this first fair in the county. Best horse was presented by Elias Proser of Clarkson, best sheep by William Garbutt, Ezekiel Mose, Whiting May and Joseph Colt. Best hogs were presented by two Chili men, Henry Widner and Thomas Sheets. Horace Bush won the prize for the best acre of wheat with 59½ bushels per acre, and John Fargo's prize winning corn yield was 144 bushels per acre. Mrs. Mary Smart of Mendon won a prize for her lace, "one of the best specimens ever produced in our county." Among the implements exhibited was a fanning mill shown by Colonel Abner Hubbard of Rochester.[5]

The Wyoming County Agricultural Society, formed in 1843, held its fairs in Warsaw, using tents (a common practice) until permanent buildings were constructed in 1874. Almost all fairs had driving parks or driving tracks. How else could the fastest horse be determined with certainty? The original one-third mile driving track at Warsaw was expanded to a half mile in 1866.[6]

Figure 15-2 depicts the Allegany County Fair at Cuba in 1868. The peaceful Genesee Valley Canal contrasts with the throngs of fairgoers come to see the new agriculture implements and ride the four-basket ferris wheel.

Figure 15-3 shows an exhibition of farm equipment including mowers and disk harrow. The signs from other exhibitors fill the background. In addition to providing space for displays of machinery, the fairs sponsored competitions of all sorts, with plowing contests extremely popular. Sometimes manufacturers

FIGURE 15-2. Engraving of the Allegany County Fair at Cuba, New York, 1868. Mowers, plows and separators can be seen to the right and behind the exhibition buildings. Included among the buildings are the Floral Hall, the Agricultural Hall, the Mechanical Hall and the Poultry Hall. Beers, History of Allegany County, N.Y., 1879.

FIGURE 15-3. Anonymous photograph from a stereo view showing new equipment at an agricultural fair; probably 1870s. Printed advertisements were useful, but seeing the equipment and talking with others who used it were important to farmers in deciding on what to buy.

held trials such as the Hussey/McCormick reaper competition.

The Livingston County Agricultural Society, organized in 1841, held a fair in Geneseo in October of that same year. The premiums offered stimulated interest in improving stock and agricultural products. Premiums were presented for the best yearling colt, best bull two years and older, best cow, best stallion, best pair of working oxen, best pair of fat oxen and best pair of matched horses. The highest premium, $10.00, was paid for the best cultivated farm of fifty acres or more. The level of interest and pride taken in fairs are evident in an 1851 portrait of a prize cow and calf, plus ribbons from years of exhibiting at state and local fairs (see figure 15-4).

FIGURE 15-4. Fair ribbons and an 1851 painting of a prize cow and calf by Henri Delattre. Wadsworth Homestead Office, Geneseo, New York.

The first state fair was held for two days in Syracuse in 1841 with more than 10,000 attending from all over the state. New farming implements, produce and stock were on exhibit. Various competitions were held, the plowing contest proving to be the most popular. The state fair moved each year from city to city until 1890 when it was permanently located at Syracuse. The fair was held in Rochester in 1843 and 1851, with the latter setting a record attendance of 100,000 people. Thayer Gauss mentions the 1843 State Fair in his diary. Figure 15-5, a woodcut from the October, 1851 *Genesee Farmer,* provides an idea of the appearance of the 1851 State Fair at Rochester. E. P. Prentice of Rochester took first prize at that fair with "Dundee II," his one-year-old Ayrshire bull. Mr. Ayrault's Hereford bull won another first premium and Mr. Wadsworth won the first premium on working oxen; both gentlemen were from Geneseo. Devon stock on display numbered some 350 head while neat cattle were set at 700 head. Oh yes, Mr. Alex Rumsey of Ogden took the plowing match. Ellwanger and Barry, the nationally-known Rochester nurserymen, exhibited more than 100 varieties of pears in addition to their apples. Apparently many sheep were exhibited at the 1851 fair with Mr. D. Hillman of Avon winning the first premium for Merino bucks and Mr. Church of Vernon taking the first premium for Saxon bucks.

The Genesee Farmer observes:

> In no other department was improvement so conspicuous as in that of Agricultural Implements. It would fill a volume to describe those which were really meritorious. Mr. McCormick was present with his world-renowned Reaper; Mr. Ketchum, of Buffalo, with his admirable Grass-cutter. Steam engines were travelling about on wheels, over rough ground, as steadily as ox-carts; and wheeled cultivators, gang plows, seed drills, clod crushers, grain threshers and separators, straw cutters, draining tile and pipe machines, harrows, plows, shovels, axes, hoes, ox-yokes, and farming tools of every description, were shown in almost endless variety.[7]

FIGURE 15-5. 1851 State Fair held at Rochester. Woodcut from The Genesee Farmer, 1851.

The Grange

Formed in 1867 by Oliver Kelley of Washington, D.C., the Grange was at once an educational, fraternal and political force for improving the lot of farmers. The Grange, whose members are called the Patrons of Husbandry, was the outgrowth of the economic problems following the close of the Civil War. Throughout the years since its founding the organization has been a powerful influence on both the state and national level; it is well to remember that at the time of its founding almost half of all Americans were directly involved in farming. Organized against abusive railroad tariffs and corporate monopolies, the Grange brought seven cases to the Supreme Court in 1876. The two major acts to emerge from these cases are the 1876 Interstate Commerce Act and the 1890 Sherman Antitrust Act.[8] The Grange remains active in many areas today.

Agricultural periodicals and fairs of the nineteenth century dealt with all manner of things of interest and use to the farmer. Changing crop patterns have already been noted; the next chapter will contain a brief look at livestock in the Genesee Country during the nineteenth century.

PEASE. DEL ET SC ALBANY.

CHAPTER XVI
Livestock

THE KEEPING OF LIVESTOCK changed radically from the early days of settlement in the Genesee Country to the period of World War I. Early accounts indicate that pioneer settlers brought little livestock with them. The Genesee Country is noted today for its horses, but until mechanized farming equipment required them, beginning with the reaper, horses were kept in smaller numbers than oxen. McNall gives the ratio of one horse to four oxen for the period prior to the Erie Canal. Except where speed was required the ox was superior for many farming purposes, since it could be turned out to browse in the woods and turned into beef when grown old. Initially, milk cows were not common except for the immediate needs of a family. Cattle kept by those who had pasturage were chiefly for beef.

Early visitors to the Genesee Country were unimpressed by the quality of beef cattle, a situation made worse by marketing practices. For most farmers, getting cash for their cattle meant selling to the drovers who bought at each farm, picking the best of each herd until sufficient cattle had been gathered for a drive. Selling the best and breeding the rest unfortunately does not improve the breed. The condition of cattle in winter was also a problem during pioneer days: Many accounts describe only a simple shed or no shelter at all, and the cattle were often left to browse by themselves, winter and summer. It should be noted that in the 1830s, while wheat was still the major agricultural product, attempts were made to improve the cattle in the region. The Wadsworth brothers, for example, arranged to bring improved breeds from the Boston area, but in the 1840s critical observers were still not impressed, even complaining that no distinct breeds could be identified.

By 1848 there had been a definite improvement in the area's beef cattle: *The Genesee Farmer* described the twin steers raised by Judge Allen Ayrault of Geneseo, as "Certified by over 100 Butchers to be the best pair of Fat Cattle ever exhibited in New York City" (see figure 16-1).[1] These seven-year-old Durham steers had a combined live weight of 5,522 pounds and a dressed weight of 4,376 pounds. The Durham, a predecessor of the Shorthorn, was raised because of its superior size. Until the mid-nineteenth century, breeding ever-larger steers was something of a fad, the largest on record being some 4,365 pounds (see figure 16-2)[2].

Transportation costs were a continous problem in the marketing of stock. Canal boat operators did not like to take livestock (they preferred their beef in a barrel) and railroads were fast but high-priced. There was an ebb and flow in cattle raising in the Genesee Country in the second half of the nineteenth century brought about by changing access to markets. In the same period many Genesee Country farmers were purchasing young beef cattle in Buffalo, which were shipped by rail to the nearest siding and driven to the farms to be

Opposite: One of Mr. Ayrault's record breaking five-year-old twin Shorthorn steers. Geneseo, Livingston County, New York, 1847.

FIGURE 16-1. Genesee Valley twin steers owned by Allen Ayrault, Geneseo, New York, 1848. The Genesee Farmer, 1848.

fattened for market. When the timing was right the fattened cattle were sent to Rochester or New York City by rail. Transportation to market continued to be a great concern because even though stock trains could reach New York City in twenty hours the rail costs could make the operation financially marginal. The problem was so great before the Civil War that some farmers considered resuming the wasteful "driving" of their cattle to

FIGURE 16-2. 3,086 pound ox "Sovereign," an example of the fad for fat cattle in the first half of the nineteenth century. The Plough Boy, 1819.

THE SOVEREIGN.
LIVE WEIGHT, 3086 *POUNDS*.
Fatted by P. Fink, of Goshen, Orange county. Slaughtered by T. Gibbons, of New-York.

Comparison of several extraordinary oxen.

SOVEREIGN.	
Live weight,	3086 lbs.
Height,	5 ft. 9 in.
Length,	9 2
Girt,	10 3
Rough tallow	295 lbs.
SOVEREIGN'S MATE.	
Live weight,	2800 lbs.
Height,	5 ft. 10 in.
Length,	9
Girt,	9 6
Rough tallow,	307 lbs.
COLUMBUS.	
Live weight,	2962 lbs.
Height,	5 ft. 9 in.
Length,	9 1
Rough tallow	213 lbs.

SOVEREIGN and his MATE, were raised in this state; fed by Mr. Fink, of Orange county, where they obtained a premium from the Agricultural Society. Sold to Thomas Gibbons, butcher of the city of New-York. The mate not weighed, but supposed to overrun the estimate.

COLUMBUS and DELAWARE Ox, both slaughtered in Baltimore; an account of them has been published in the *Plough Boy*.

BRIGHTON Ox, exhibited in 1819, obtained the premium from the Massachusetts Agricultural Society, being a few pounds heavier than any before exhibited at Brighton.

STRAYED,

FROM the subscriber, living on the west bank of Conesus Lake, in the town of Geneseo, about the 20th ult. a small red OX, 8 years of age, with a white face, short tail and high horns—some white hairs on the top of his neck, occasioned by a yoke.—Whoever will return said Ox to the subscriber, or secure him and give information where he may be found, shall be generously rewarded.

BOAZ BATEMAN.

Geneseo, Aug. 5, 1826. 3w18

STRAYED

FROM the subscriber, living five miles north of Hosmer's Hotel, in the town of Rush, Monroe co. on the 21st July last, a bay HORSE, of middling size, shod all round, and having a small white spot on his face.—*Also*—A likely 2 years old COLT, both having switch tails.

FIGURE 16-3. *Portrait of Wadsworth steer. Wadsworth Homestead Office, Geneseo, New York. Circa 1850.*

market. The passage of the 1876 Interstate Commerce Commission Act did much to level rail transportation costs.

By 1881 Livingston County historian James Smith could claim that everyone possessed some choice stock, in contrast to the period before the Civil War when few but the common sort were to be found.[3] Raising beef in the Genesee Country became an important business. Perhaps the increasing importance of cattle breeding can be seen by the portraits of prize animals, which were painted just as one might commission a portrait of a family member. Several portraits of bulls and steers hang in the Wadsworth Homestead office in Geneseo (see figure 16-3 and figure 15-4).

Because the Wadsworth family seems to have been the target of ridicule in the writings of some who observed their stock in the 1820s to 1840s, it is important to note their progress in cattle breeding. The 1871 *American Rural Home* characterized the stock of Craig, Charles and James W. Wadsworth as "thorough-bred," adding that many of their 6,000 acres were given over to "stock purposes."

In some areas of the Genesee Country during pioneer days livestock were pastured on a "common green," according to the European model, or were left to forage on their own. Later, fenced pastures were preferred to these practices, even though the commonly used rail fences were expensive to make and maintain. Stump fences were less expensive where the materials for their "construction" could still be obtained. Articles on new and improved varieties of fencing became common in agricultural periodicals; the invention of barbed wire in 1884 caused a major change in fencing practices even though many were against the new wire.

One interesting outgrowth of successful livestock keeping was experimentation with the by-products of slaughter. A use was sought for everything except the "moo," to paraphrase a famous saying. Bones, horn and hooves could be boiled down to make gelatin but the result was tasteless, odorless, transparent and of no real use at first except for glue and sizing. Gelatin ultimately was used in large quantities in the production of the emulsions for photographic film, plates and paper. Gelatin emulsions of the sort that first appeared commercially in the late 1870s are still used today.

Although unflavored gelatin was (and is) used in cooking, experiments to make a truly good-tasting gelatin food succeeded in 1897, when Pearl Wait, a LeRoy resident, produced a palatable, fruit-flavored, sweetened gelatin. Wait's wife called the new product "Jell-O." Orator T. Woodward, owner of the Genesee Pure Food Company of LeRoy, New York bought the recipe in 1899 from the Waits for $450; by 1906 the sales of Jell-O had reached a million dollars.

Hogs

Hogs were important almost from the outset of settlement (see figure 16-4). At first the number of hogs kept was generally limited to those used by each farm family, but eventually, in some Genesee Country areas, salt pork and sausages were produced in sufficient quantities to sell or trade. As with cattle, the number of hogs increased as transportation systems improved. In the early days of settlement hogs ran loose in the woods, foraging for themselves; that is, wherever the number of predators was low enough to permit their survival. Later on in the nineteenth century pure-bred hogs were raised; figure 16-5 shows an advertisement from the June, 1877 issue of the *American Agriculturist* featuring the Magie or Poland China breed. Figure 16-6 shows hogs being driven toward the barnyard on the Daniel Brown farm, in a scene taken from *The Combination Atlas Map of Genesee County, N.Y.*, 1876.

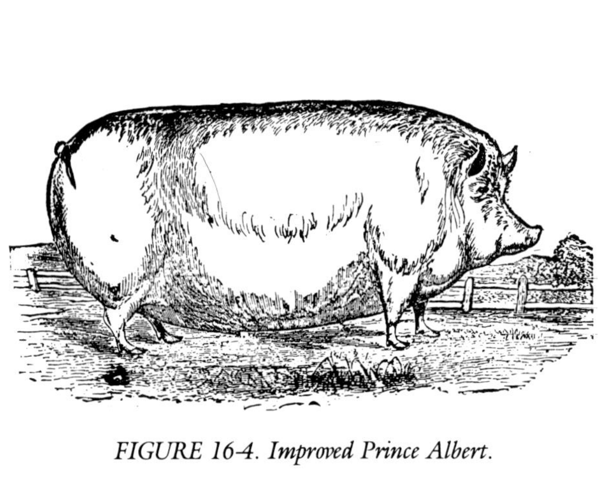

FIGURE 16-4. *Improved Prince Albert.*

FIGURE 16-5. *Magie or Poland China Hogs. June, 1877 American Agriculturist.*

FIGURE 16-6. A hog drive on the Brown Farm, West Bethany, Genesee County. Combination Atlas Map of Genesee County, 1876.

Sheep

In the late eighteenth century conditions in the Genesee Country were generally unfavorable for raising sheep, but small flocks were kept nevertheless for both wool and meat. Difficulties included wolves, poor feed and the heavily wooded conditions in areas where the topography was good for sheep but not for general farming. It was not uncommon for the first business conducted by new local governments to involve the setting of cash bounties on wolves and the establishment of laws governing the building of proper sheep pens. McIntosh mentions that sheep pens "sixteen rails" high were called for in Ontario County.[4] Bounties paid on wolves varied considerably, but $10 to $15 seems to be the general range throughout the region. As this represented quite a bit of cash, in some areas the bounty was only to be paid to residents. Sheep from various farms were allowed to run loose and ear marks were used for identification, recorded locally so sheep could be sorted out according to owner when wintering or shearing time arrived. Figure 16-7 shows the first page of ear marks from the 1792 Painted Post Town Meeting ledger.[5]

As an example of how ear marks were used, George Williamson recorded in the July 19, 1794 Painted Post Town Meeting ledger:

> I hereby certify that there came in September 1793 two stray sheep to my Plantation: one of which was a Ram the other a Ewe: The Ram mark'd a Halfpenny the under side the left Ear. The Ewe marked a half crop off the right ear.—The Ram goared by an Ox February 20, 1794. The Ewe was killed by the wolves May 21st a lamb of the Ewes still remains—[6]

Of the many breeds of sheep, the hardy Me-

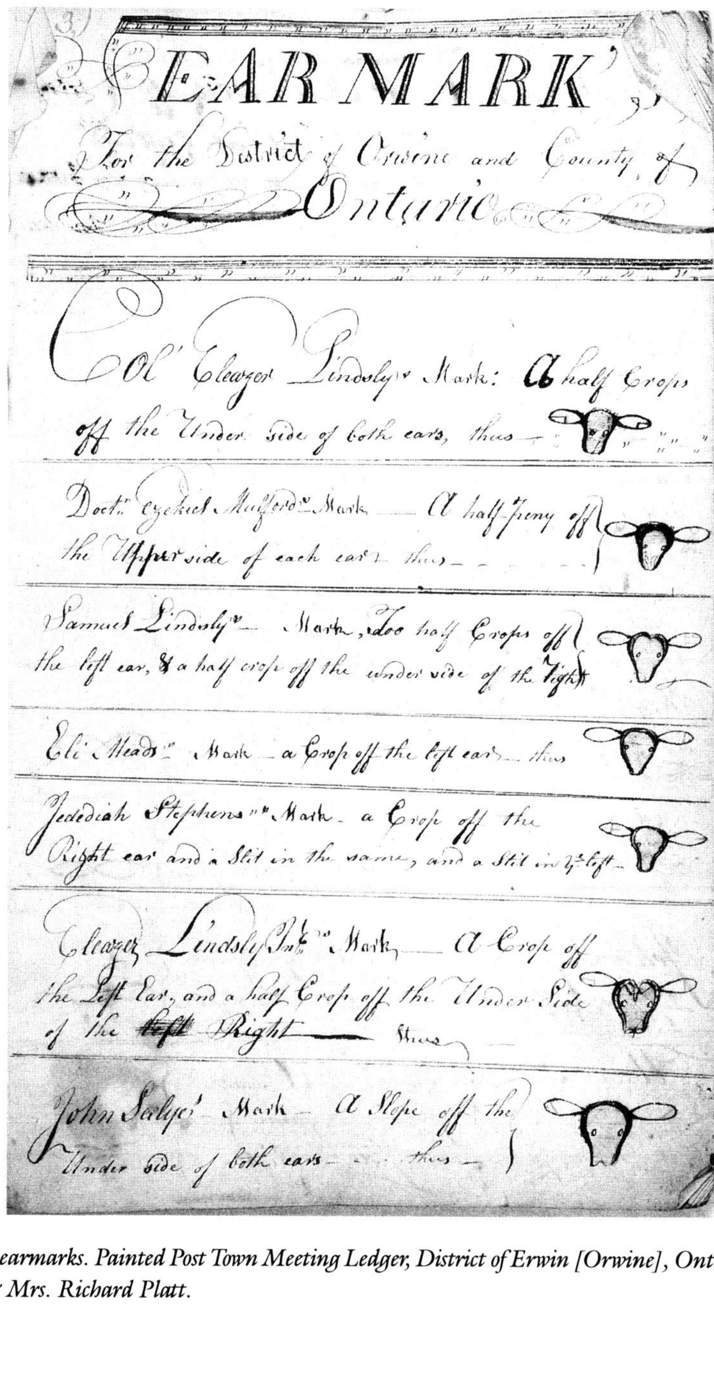

EAR MARK

For the District of Orwine and County of Ontario

Col. Eleazer Lindsly's Mark: A half Crop off the Under Side of both ears, thus

Doct.r Ezekiel Mulford's Mark — A half peny off the Upper side of each ear thus

Samuel Lindsly's Mark, Two half Crops off the left ear, & a half crop off the under side of the Right

Eli Mead's Mark a Crop off the left ear thus

Jedediah Stephens's Mark a Crop off the Right ear and a Slit in the same, and a Slit in the left

Eleazer Lindsly Jun.r's Mark — A Crop off the Left Ear, and a half Crop off the Under Side of the ~~left~~ Right thus

John Seelye's Mark A Slope off the Under side of both ears thus

FIGURE 16-7. 1792 sheep earmarks. Painted Post Town Meeting Ledger, District of Erwin [Orwine], Ontario County (now Steuben), 1792. Collection of Mrs. & Mrs. Richard Platt.

FIGURE 16-8. A prize Merino ram.

rino was preferred because of the high quality and yield of its wool. In 1809 William Wadsworth contacted Robert Livingston, the first importer of Spanish Merinos, to ask about starting a Merino flock. In 1811 DeWitt Clinton described in his private journal an advertisement he saw from Geneseo concerning William Wadsworth's Merino sheep:

> In the tavern there was an advertisement of William Wadsworth, dated Geneseo. He proposes to let out half-blooded Merino rams, to be delivered on the 1st of September, each ram to be put to fifty ewes, and no more, before the first of October, and to be returned on 1st of June, unsheared. All the ram progeny to be returned, and he is to have all the ewe lambs except two (from each ram), for each of which he is to pay eight shillings cash, on the 1st September, 1811. He charges nothing for the use of the rams.[7]

According to the 1877 *American Agriculturist,*

> The Merino sheep produce the heaviest fleece, some of them yielding 18 to 30 pounds of wool. The Cotswold yield the next heaviest fleeces, these being from 11 to 18 pounds. The best market lambs are a cross of Southdown with Merino, and the heaviest lambs a cross of Cotswold with Merino.[8]

At first Merinos were very difficult to import from Spain; and smuggling seems almost to have been a necessity. After the problems of Spanish export restrictions were solved Merinos became almost a fad until the wool market softened in the late 1840s (see figure 16-8). At first, most flocks had just twenty to fifty sheep, but when the market was strong flock sizes increased dramatically. A Wadsworth account book lists almost 2600 sheep sheared in 1839. Over a million pounds of wool were shipped from Rochester by canal in

1845, probably the peak of interest in Merinos by the wool industry (see figure 16-9). The fact that Merinos could be grazed in such numbers is an indication of how the farming landscape in the Genesee Country had changed in the early nineteenth century.

MERINO SHEEP.

The subscriber offers for sale **600** MERINO EWES AND **200** MERINO LAMBS, on a liberal credit.—They will only be sold in large lots.

WM. WADSWORTH.

Geneseo, 29th Sept. 1829.

MERINO BUCKS.

Saxony Buck Lambs.

EIGHTY BUCKS, FROM FULL TO HALF BLOOD MERINOS.

Also—Some half blooded Saxony Buck Lambs, will be offered for sale on Monday the 28th inst. so many of the Bucks as may not be disposed of on that day will be sent out of the County. Any persons wishing Bucks and are not able to attend on the 28th to select them will please send a description of such Bucks as they may wish, and if the subscriber has them, they will be selected and delivered on Saturday the first day of November, the risk of the animal to be the purchasers, from the time of selecting him. To save trouble, the price will be from three to seven dollars

Wm. H. SPENCER.

Sept. 15th, 18[illegible]

FIGURE 16-9. Late 1820s newspaper advertisements for Merino sheep.

FIGURE 16-10. Sheep on the Clark Farm, Covington, Wyoming County, New York. Beers, History of Wyoming County, 1880.

The Genesee Farmer described the statewide decline in the sheep industry in its April, 1853 issue:

> In 1845, New York contained 6,443,865 sheep; in 1850 the number was reduced to 3,453,241; decrease in five years 2,990,624. The number of sheep slaughtered in the five years named, was much larger than the difference between the larger and smaller sums, otherwise the natural increase would have prevented so great a reduction. It was necessity that cut their throats. To replace three millions of sheep by dairy stock, at least four hundred thousand cows would be required. We shall take another occasion to demonstrate beyond all dispute, that not less than eight and a half of the twelve and a half million acres of land under improvement in New York is suffering deterioration at the rate of three dollars a year by parting with its elements of fertility.[9]

Sheep continue to be raised today throughout the Genesee Country, although the emphasis is on breeds that produce superior meat rather than wool. While special sheep folds are sometimes encountered, most barns can easily be adapted to sheep raising even if not originally constructed with them in mind (see figure 16-10).

FIGURE 16-11. A copper weathervane, symbol of the dairy.

Dairy Farming

Specialized buildings, improved barns and better methods of product preservation and transportation were needed for an increased dairy industry. Many are surprised to learn that a milk cow was rare on the average farm at the beginning of settlement in the Genesee Country. Henry Conklin remembers that buying a milk cow for his mother and large family of brothers and sisters was a very special event in 1839.[10] Early attempts at dairying meant keeping just a few cows because of the difficulty of processing, storing, and transporting milk products. Where wheat was impossible to grow, especially in the southern portion of the Genesee Country, cheese making took an early hold. While dairying was slow to develop as an industry, successive improvements in transportation, from the canals to the faster railroads, opened new markets for butter and cheese. Dairies at mid-nineteenth century averaged some ten to fifty cows. Over the years greater and greater attention was paid to the dairy industry on an international level, with the result that herds grew in size and the cows themselves were treated much better than they had been. Special attention was given to the design of larger, efficient dairy barns: again and again agricultural periodicals showed dairy barn designs offering improved ventilation, feed storage and handling, and so on.

The April, 1853 *Genesee Farmer* describes the growth of the dairy industry in New York State:

> It had a slow growth for twenty years, during which several millions of acres of native forests were cleared and brought under the plow, and paid a larger immediate profit in grain culture, than could be realized from milk, butter, and cheese. As the virgin soil parted with its elements of crops by tillage, and failed to produce remunerating harvests, farmers extended their pastures and meadows, and diminished their grain fields, until New York has become the largest stock growing and dairy State in the Union.[11]

According to the 1850 United States Census, New York led both Ohio and Pennsylvania in the value of livestock kept. The same *Genesee Farmer* went on to advise:

> The dairyman should keep no poor cows, and as few good ones as will consume the food to be converted into milk. Provide, by the aid of corn grown for forage, by manuring pastures and meadows, and by root culture, a plenty of food for whatever cows are kept for dairy purposes. Have yards and stables so arranged and littered with straw, muck, forest leaves, or some other absorbent, as to preserve from waste or loss all the droppings of domestic animals.[12]

FIGURE 16-12. A nineteenth-century farm cartoon from Periam: "Shall I move the barn or the manure pile?" The problem was not a small one. Periam, Home & Farm Manual, 1884.

Jonathan Periam's *Home & Farm Manual* of 1881 includes a cartoon addressed to the question of what to do with accumulated manure (see figure 16-12). The solution may seem obvious to us today, but apparently many did not see the advantage of animal manure as a fertilizer in the 1880s.

New dairy barns were designed to provide proper shelter and ventilation; the addition of the louvered cupola was perhaps the most conspicuous evidence of the concern for ventilation essential to the well-being of the cows. Gradually even these dairy barns were redesigned to provide for the easier handling of bulk items such as milk, feed and manure. Side-hill and basement barns were constructed with basement space for cows in mind. Pitching feed down from a mow that had been filled from a wagon or hay fork on its own level was a clear labor-saving improvement, permitting larger herds to be handled (see Chapter XVII). From mid-century on, agricultural periodicals were filled with articles and letters suggesting ways of upgrading manure handling—a centuries-old problem that required new attention with the growth of dairying, where the animals are kept inside in some seasons for much of the day. Some suggested an additional level in the barn, beneath the cow floor, for the collection and efficient removal of manure. The idea was not new but the interest in practical solutions was spawned by the growth of the dairy industry.

Some of the improvements and changes in the barns themselves will be discussed further in subsequent chapters but a few of the essential requirements of milk storage will be considered here. Before the advent of the milkhouse, with its ice or artificial refrigeration, milk was stored in a springhouse or cooling cellar. The June section of Robert Thomas's 1826 *Farmer's Almanack* describes a dairy:

> One day last June I visited the dairy room of my cousin . . . It was a part of the cellar of the dwelling house, and a neater place mortal man never set eyes on. It was well ceiled with plaster to prevent the dirt's descending. The top and sides were white-washed, to increase the light, and not a hole or chink could be seen for insects to harbor in. The floor was of stones, and she told me that she often washed it in summer with the coldest water, to keep the air cool and sweet as possible. On the northern side were two windows, and on the western, one, which could be opened and shut at pleasure. In a corner was a little stove, where she sometimes in winter burned a few coals to keep the room of an equal coolness. So her milk never grows sour in hot weather, and never freezes in cold; of course there is no obstruction to the rising of all the cream. She has a little thermometer in the room, and endeavours to keep the air at about 50 or 55 degrees. The moment a spider spins his web there, it is death.

Well or spring water was used to bring the temperature of the milk from 101 degrees at milking to the 45 or 50 degrees required for relatively long-term storage. During the summer months, when the water was not cold enough and the milk had to be used within a short time, cheese and butter were often made. In a springhouse the cans were placed directly in the running water, in a specially formed trough. Where springs were not available, well water was used to cool the cans, the cooling trough often placed outside close to the pump or well. In later years the ice house became a necessity, but in months when ice was not available the milk was flowed over pipes through which cooling water ran, speeding up the cooling process.

Before the centennial year of 1876 the average daily production of a cow on pasture was about four quarts, and it was not uncommon for cows to be milked for only six months of the year. Today the average cow produces about twenty quarts per day. After the Civil War, with increased milk production on dairy farms both cheese and butter became much more than home products (see figures 16-13 and 16-14). The skimmed milk remaining from cheese and butter production was fed to pigs and calves. Butter, well washed and stored in brine, could last an entire year and cheese could last at least that long.

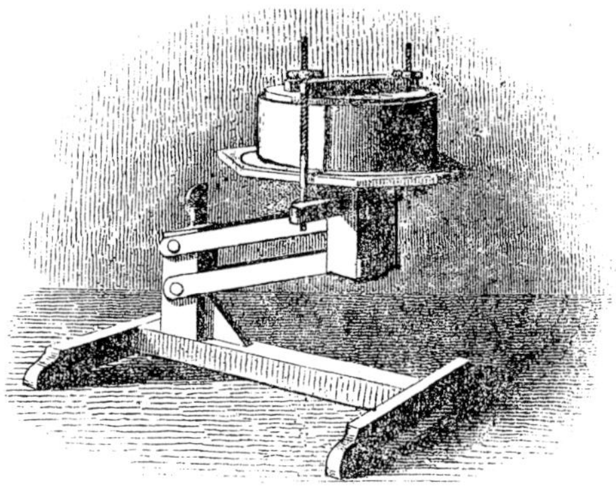

FIGURE 16-13. Collin's & Stone's Patent Cheese Press. 1849.

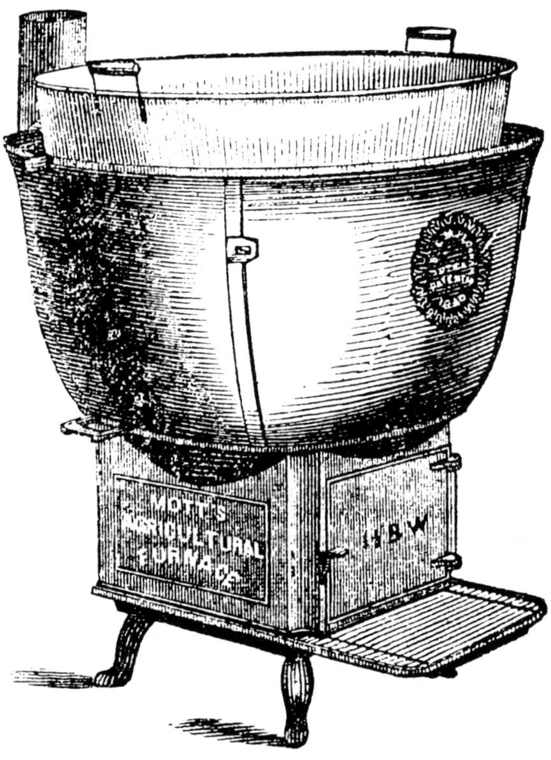

FIGURE 16-14. Mott's Patent Dairy Stove. 1849.

Factory-made cheese and butter began to replace those made on the farm about 1875, roughly the time of the first mechanical refrigeration. The Genesee Country is still known for its outstanding cheddar cheese.

Milk trains or regularly scheduled passenger trains hauled milk to processing centers if local dairies were not available or patronized. These early milk trains are not to be confused with the regularly scheduled, fast-running milk trains with their special cars operating to New York City in later years. In most instances the milk cans were hauled to the railroad in wagons or, in winter, sleighs.

Milk from the country was superior to the milk from city-kept cows, fed on distillery and brewery refuse grain and producing "swill milk." However, many apparently could not conceive of carrying milk in cans on a jolting trip over the railroad to a city processor, until Thomas Selleck succeeded in convincing a few Orange County farmers to ship milk experimentally to New York City on the Erie Railroad in the spring of 1842.[13] There were some disastrous shipments at first, but shipping milk by train eventually proved successful. Cans of all sizes were soon in use (forty-quart cans became standard) because all the milk was sent, none held for another day. Special cars were built and ice was used both at the farm and on the cars to prevent spoilage. The Erie Railroad reported in 1845 that milk transport accounted for two-fifths of its entire freight receipts. By 1899 the Erie ran a daily milk train to New York City from Hornellsville, a trip of 350 miles, with stops at nearly every station in Steuben, Chemung, Tioga, Broome, Susquehanna, Delaware and Pike Counties.[14]

The following example may serve to illustrate the process of shipping raw milk within the Genesee Country in the first decade of the twentieth century: Even though the Doty Farm was next to the Maple Tree Dairy in the Town of Geneseo, the milk from the Doty Farm was shipped by rail from Geneseo to Rochester for processing. The milk cans were taken from the milk house to the depot in Geneseo, from which the 9:05 a.m. Erie milk train took the cans to Rochester, arriving at 10:00 a.m. If ice was unavailable for cooling the milk, the whole shipment was jeopardized. John Black, son of the Maple Tree Dairy owner, remembers taking milk cans to Rochester in the early twentieth century, in a truck with solid tires. The sixty-mile round trip took up most of the daylight hours: top speed was twelve miles per hour and then only if the roads were in good shape. In contrast, the scheduled one-way trip on the Erie Railroad was less than one hour. It doesn't take much speculation to understand why the railroads were used to haul milk.

The first bottled milk appeared about 1878 and pasteurization was first practiced commercially in the mid-1890s. With these two advances the demand for fluid milk increased and the old practice of summer dairying gave way to year-round dairying as it is practiced today. The four most popular breeds in the 1890s were Holsteins, Jerseys, Guernseys and Ayrshires. In 1915 processed milk sold for six cents a quart.

Ice for cooling the milk was harvested on nearby lakes and brought by sled to the farms. Figure 16-15 shows a typical scene of hauling lake ice. Most of the smaller lakes in the Finger Lake region were better suited to ice harvesting than Lake Ontario. For example, both Conesus and Silver Lakes froze earlier than Lake Ontario and were less hazardous to the workers. Irondequoit Bay, northeast of Rochester, was also a significant source of ice because of its sheltered nature. On a smaller scale almost any pond could have been used to supply ice.

After an experimental railroad ice car proved successful in 1895, whole trainloads of Conesus

FIGURE 16-15. Hauling lake ice, 1893. Anonymous.

Lake ice were shipped from Lakeville to Rochester.[15] In addition to hauling these ice trains, railroads were heavy users of ice themselves, requiring large quantities for refrigerated cars. Harvesting began as soon as the ice was thick enough (see figure 16-16). Sawed with coarse-toothed icesaws, man or horse powered, the cut blocks were floated, through a channel left open for that purpose, to the loading conveyor at the icehouse (see figures 16-17 and 16-18). Commercial ice houses were often quite large: For example, R. W. Sanborn's round Lakeville icehouse on Conesus Lake was 100 feet in diameter and the icehouse at Silver Lake, 350 feet long, held 60,000 tons. The ice was either packed away in insulating sawdust or sold for use in private icehouses.

A typical design for an above-ground farm icehouse included wooden walls and door eight to twelve inches thick, framed and sheathed to provide "dead air" space insulating the interior from elevated exterior temperatures. The ice was further insulated by a foot or more of sawdust on the sides and top to help retard melting (see figure 16-19). The floor had to be drained and usually some provision was made to ventilate the upper portion. The main purpose of the ice, of course, was to cool milk to prevent souring. A dairy farmer tried very hard to avoid the predicament of running out of ice, since the older methods of cooling were much less reliable and there was no electricity on most farms until 1910. Also, milk and butter cooled with ice were said to taste better.

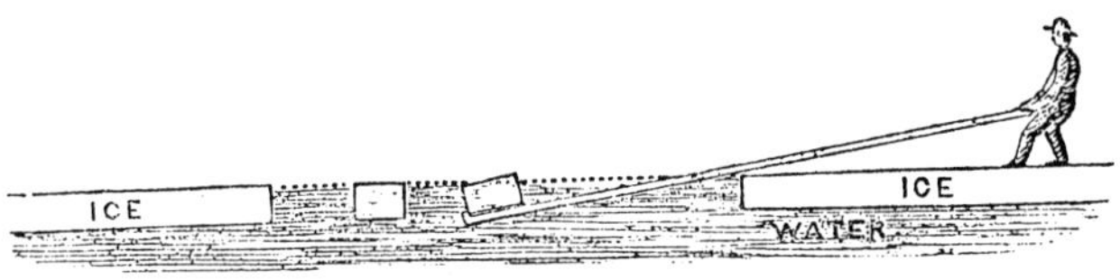

FIGURE 16-16. Drawing ice with a lever or ice grapple was a relatively safe method of securing ice from a local pond for an icehouse.

FIGURE 16-17. A lineup of teams, sleds and saws on Conesus Lake, Lakeville, New York, prior to the ice harvest. Circa 1910.

FIGURE 16-18. Ice harvesting and storage on a large scale. The ice is sawed into cakes and floated to the conveyors, where it is taken to the proper level of the ice house for storage.

FIGURE 16-19. Above ground icehouse. Country Gentleman, December 17, 1861.

The Milk House

Most farm operations with a herd of milking cows during the late nineteenth and early twentieth centuries also had a building known as the "milk house," in many cases the enlarged successor of the spring or pump house. The milk house usually stood close by the milking shed or milking place in the barn. The size of the milk house depended largely on the size of the herd served, early examples tending to be small, perhaps eight by ten feet. They supplied space sufficient for a cooling trough, milk cans, possibly a cream separator and work area; larger operations sometimes included butter and cheese making rooms. Milk houses were usually frame buildings, but the Doty Farm in the town of Geneseo had one built of textured concrete blocks to match the adjacent barn; today concrete block milk houses are quite common (see figure 22-14; the milk house is the small building on the right side of the photograph). The milk house on the McNinch Farm is a small gable-roofed building on the left of the main barn (see figure 19-27). Milk houses may not have been significant from the standpoint of architectural design, but they did signal a major change in farming techniques, techniques which were themselves destined to change the design of newly built barns.

DIARY

Jennie Mather, Livonia, Livingston County, 1867

These selections from Jennie Mather's Diary add to our understanding of changing conditions of farm life in the nineteenth century, this time from a young woman's point of view.
(from the collection of Mrs. Patricia Murphy, a direct descendant of Jennie Mather)

From 1867 Diary of Jennie Mather, Livonia, New York, age 23.

1867

Friday, February 1:	*It thawed all night and all day. Today the snow is almost gone. The wind blew our straw stacks over and it fell on our white cow and almost smothered her.*
Saturday, March 9:	*Cloudy, Thawed a little. We got home a little after 9. Ed came down and Henry drew him a load of straw. . . . We baked and I sewed a little but I felt rather sleepy and did not do much.*
Tuesday, April 30:	*Looked very much like rain all day. We baked and ironed this forenoon and have worked in the yard some. Pa drilled his barley in.*
Friday, May 17:	*Pleasant—Pa and Henry was going up in the hill to sew oats.*
Thursday, June 6:	*Very pleasant—I went to Julie's this morning to have her make Ma a bonnet. Caroline's was ready for her, it is a purple silk with white ties. Caroline and Jennie came a ways with ——— on the track. I got home before sundown. Pa is planting corn. C- is helping him. I am very tired.*

Wednesday, June 12:	*It rained some this morning and has been quite cool all day. I ironed this forenoon. Mrs. Patterson visited here this afternoon and Ma went home with her tonight. Pa saw a head of wheat today for the first harvesting will soon come.*
Monday, July 15:	*Cool and windy looks like rain tonight. We did a large washing and baked. I am quite tired tonight. Pa is howing corn and Henry is planting.*
Monday July 22:	*Looked very much like rain this morning. We washed and ironed some and baked. Pa went to the Station, he has sold his wheat . . .*
Monday, August 12:	*Pleasant not very warm, we washed Ma went up to Mrs. Densmore's she hurt her feet. Capt. Densmore reaped our barley for us.*
Wednesday, August 14:	*Not very warm. We quilted all the forenoon and this afternoon. Mrs. Nottingham and Mrs. Densmore visited us. Mr. D. helped Pa stack barley.*
Thursday, August 15:	*Hot and dusty. We quilted. Henry and Johnny went to Geneseo to a circus. Pa went to the station, they got home about dark*
Wednesday, August 21:	*Very comfortable. We had considerable to do this afternoon . . . Pa helped Mr. Richardson thrash.*
Thursday, August 22:	*Cloudy and looked very much like rain but it did not. We ironed and baked and quilted some. Pa and Henry helped D.H. Densmore thrash.*
Friday, August 23:	*Very hot and dusty we had considerable to do this morning we are fixing for thrashers . . . the thrashers have come.*
Saturday, August 24:	*Looked very much like rain we had to stay around some to wait on thrashers . . . they are finished and went home. Pa went to the station and brought me a letter from J.B.*
Monday, September 16:	*Pleasant this forenoon we washed. It rained this afternoon we made some beeswax. Pa is fixing to sew wheat.*
Thursday, September 26:	*Cloudy and quite cool. We baked and sewed. Today is the first day of the fair at Geneseo. Pa went to the Station . . .*
Tuesday, October 8:	*Pleasant, we ironed this morning. Pa and Henry is picking apples . . .*
Wednesday, October 9:	*Cloudy and it rained some this afternoon. Pa went to the station with a load of apples. Densmore is thrashing.*
Thursday, December 5:	*Pleasant the sun shines and it is still. Pa butchered, Mr. Nottingham helped him . . .*
Saturday, December 7:	*Pleasant—this forenoon but it was very stormy this afternoon we made our sausages and baked.*

Monday, December 16:	*Not very cold we washed and baked some. Pa went after a load of wood. D.H. Densmore is threshing clover seed.*
Saturday, December 21:	*It blows some. Pa and Henry drew ice from the lake for to wash with Monday . . .*

Perhaps it would be well to remember that when Jennie Mather kept her diary the Civil War had been over less than two years and it would be almost another two years before the first transcontinental railroad would be completed. Each of these events was destined to change life significantly on the farm in the Genesee Country and in America. Both events brought about an exchange of ideas and promoted increased mechanization in all phases of life.

FIGURE M-1. Helping to unload wheat bundles in anticipation of the threshing machine. Anonymous.

CHAPTER XVII

Barn Arrangements

THE RELATION of one farm building to another can be an interesting study. For instance, in New England it is not unusual to find continuous, interconnected farm buildings. The main house may be connected to the woodshed, the woodshed to the carriage house, the carriage house to the hay barn and so forth. Many believe that the purpose of such an arrangement was to avoid having to go out in deep snow. Others think that such complexes may have grown with each succeeding building built onto an earlier one so that one side of the new structure would be already in place. A third thought is that New England farm buildings may have been connected in an attempt to produce an orderly appearance. The most recent additional interpretation is that some of the connected structures might have housed the "cottage" industries needed to make marginal farming situations successful.[1] Connection of the main house to the rest of the farm buildings was rare in the Genesee Country, but barns and related buildings are often arranged in close proximity (see figure 17-1).

The main barn in the Genesee Country often stands against the strong western and northwestern winds that drift the snow in deep piles during the long winter months; the average annual snowfall is more than eighty inches. When the barn doors were opened, this same barn orientation, in barns built before fanning mills and separators became common, provided the winnowing wind needed during summer, fall and winter (see Chapter V). Other, smaller buildings often surrounded the barnyard providing essential protection for the animals. Such an arrangement is exemplified by the barn complex of the McClellan farm in Geneseo (see figure 17-2).

Unlike New England, one usually has to go outside in the Genesee Country to get from house to barn. Some localities forbid connection of the farm house to barns due to fire hazards. Figure 17-3, a lithograph from the 1880 atlas of Wyoming County, shows an arrangement similar to that of the McClellan barns.

A series of aerial views (figures 17-3 to 17-7) illustrates the organization of connected barns in the area. Perhaps the most common arrangement is a U shape, with the main barn providing a windbreak. The Stapley-Merrimac Farms barn complex in Geneseo, New York, illustrates the U form (see figure 17-4).

The Mulligan Farm north of East Avon has a complex organization of buildings of many different ages and types (see figure 17-5). The barns are situated in the typical arrangement to shield the barnyards against winter winds.

The Ashantee Farm (pronounced "a shanty"), built by Herbert Wadsworth south of Avon, shows one of the more enclosed arrangements (see figure 17-6).

The Crosett barns, figure 17-7, were designed from the beginning to form a U shape, whereas the other barn complexes were created through

Opposite: The wagon ramp in the Thomson/Smith barn rises from the road to the peak in the opposite end of the barn, Warsaw, New York.

FIGURE 17-1, Above. Barn complex on Jones Bridge and Crossett Roads, Geneseo, Livingston County, New York.

FIGURE 17-2, Below. The McClellan barn complex, looking west. Long Point Road, Town of Geneseo, Livingston County, New York.

FIGURE 17-3, Above Right. E.C. Sherman Farm, Middlebury, Wyoming County, New York. Beers, Wyoming County Atlas, 1880.

FIGURE 17-4, Below Right. The barnyard of the Stapley-Merrimac barn complex, Groveland Road, Geneseo, Livingston County, New York, opens to the east.

FIGURE 17-5. The Mulligan Farm barn complex with structures ranging from the 1850s to the present. The west is to the top right of the photograph. Barber Road, Avon, Livingston County, New York.

FIGURE 17-6, Above Right. Four buildings surround the Herbert Wadsworth Ashantee Farm barnyard to the left (south); the buildings protect a second barnyard from winter winds, permitting exercise of the animals. The west is to the top right in the photograph. Martha B. Wadsworth increased the number of stables to provide the best accommodation for horses. Ashantee Farm, Avon, Livingston County, New York.

FIGURE 17-7, Below Right. Crossett Farm barn complex. West is to the right. Crossett Road, Geneseo, Livingston County, New York.

additions to an existing barn. The Crossett complex will be discussed in some detail later.

The Wells barn on the Getzinger-Bailor Farm included a pair of sliding "doors"—actually sliding extensions of the walls—which could be closed in winter to complete the U for protection of cows in the barnyard (see figure 17-8).

FIGURE 17-8. Getzinger-Bailor arrangement. The tracks for the doors that once closed off the barnyard still remain. Reservoir Road, Town of Geneseo, Livingston County, New York.

Shaker Barns At Groveland (Sonyea)

In 1834 the Shakers formed a community in the Town of Groveland, Livingston County, New York. This new community engaged in agriculture and manufacturing on a fairly large scale. Most know the Shakers for their inventions and their furniture, which combined good lines with practical durability. This same concern for practicality was evident in the Shaker barns.

The United Society of Believers in Christ's Second Appearing had its start in England during a Quaker revival; the "vulgar" name given to the sect was the "Shaking Quakers" for the marching, singing, and dancing and other body movements that characterized phases of their religious worship. In 1774 Mother Ann Lee, leader of the Shakers, immigrated to America and two years later founded a colony in Watervliet, New York, drawing many new converts. Originally the Groveland Believers owned a tract of land located partly in Sodus and partly in Huron, purchased in 1826. With the threat of encroachment from the new canal the property was sold in 1837 and more than 1600 acres were purchased in the Town of Groveland. Curiously the Sodus Canal was never constructed. The colony at Groveland (shortened from the Town of Groveland and today known as Sonyea) began with some 145 original members; at first it grew and prospered, but by 1892 the members had become aged and membership declined to the point that the Sonyea site had to be sold. The State of New York purchased the site and buildings in 1892 to be used as an epileptic colony, and it exists today as a state hospital and medium security prison.

The Society of Christian Believers practiced communal living with community sharing of possessions and equality of the sexes. The performance of work was consecrated to the common good and the glory of God. The Shakers were pacifist, celibate and generally separated from the rest of the world in their daily activities.

Henry Conklin, while staying in Mount Morris in 1854, comments:

> I stayed over Sunday with them and we went to the Shaker Meeting. I heard a good sermon and after preaching they marched and danced about an hour, men and women all together, but Cousin Ben said the men and women never spoke to each other . . . they were a true and honest sect, nobly acting out the true Christian character. Everything about their garments was clean and neat as wax and their church and dancing floor was almost white as chalk. A sort of marching of low sweet singing accompanied their dancing and I thought that was the way I would like to serve the Lord.[2]

The Groveland Shaker Community farms were considered quite successful—producing, for example, forty bushels per acre of wheat. Almost 2,000 fruit trees helped provide an income for the community and additional cash was generated by growing, packaging and selling quality seeds, as has been discussed in Chapter XI. Brooms were one of the best-known items manufactured and sold in the Groveland community. Broom corn was extensively planted and a separate building was constructed for broom manufacture. In fact, broom corn was sold as a separate item.[3]

Among the early buildings constructed at Groveland were an office, a church, and flour and saw mills. Eventually there were two groups of buildings, known as "East House" and "West House," both with rather large barns (see figures 17-9 and 17-10). The many Shaker barns at Groveland were an integral part of community organization and consequently were built in among the rest of the structures. This integration

FIGURE 17-9. In the East House group the gambrel-roofed barn (second building from left) was a horse barn. The barn behind the boiler house, with its smoking stack, was a stock barn, and the large building on the right with a cupola was the broom shop. Lithograph from J. Smith, History of Livingston County, 1881.

of the community buildings, barns included, reflected the common ownership and use of all Shaker property. To appreciate the uniqueness of Shaker organization, try to imagine barns located on the main street of secular villages. Unfortunately, none of the barns built by the Shakers at Groveland still stands, but the available illustrations reveal the organization (see figures 17-9 & 17-10).

The East House had a gambrel-roofed barn that was used to stable horses, as is suggested by the side windows (see figure 17-9). The use of the ventilating cupola was fairly common to Shaker barns. The Groveland stock barn was the largest in the complex, built for thirty-six head of cattle. It was unique because of its cupola-capped clerestory, an interesting roof design similar to that of the large Shaker barn built at New Lebanon, New York, about 1860. In addition, the complex contained a seed barn built for the drying, storage and packaging of seeds. According to the state inventory of 1892, the fodder capacity of the East House was 150 tons in the grain barn, 50 tons in the horse barn, 100 tons in the stock barn, and 60 tons in a fourth barn. The West House cow-and-grain barn had a capacity of 75 tons, while the horse barn stored 50 tons.[4] The unique flat roofs of some of the barns and other buildings were susceptible to leakage in this climate due to the effects of ice and snow, but they demonstrated the attitude of invention and experimentation so often associated with the Shakers; this experimental roof shape was not repeated elsewhere in Genesee Country barns.

FIGURE 17-10. The Groveland Shaker West House group. The flat-roofed horse barn is on the right; to the left of it is the broom house (the light structure). To the left of the broom house is the dry house where fruit was dried for preservation (see Chapter XI). On the left of the photograph are the main house (the white structure with cupola) and the office. Anonymous, circa 1881.

Interior Arrangements: The Octagon And Other Polygonal Barns

All over the Genesee Country, barns were arranged in efficient relationships to each other and to other farm buildings (or, in the case of the Shakers, community buildings). Barn designs were constantly being modified for greater interior efficiency as well; thus, the side-hill barn was an advance over the English barn, not only because it contained more space but because this space could be more efficiently used in handling feed, produce and manure.

Another approach to more efficient use of interior space at a reasonable cost involved alternative barn shapes. In fact, round and polygonal barns were under consideration by designers since at least the early nineteenth century. For example, the farming world knew of the round, 90-foot diameter Shaker barn at Hancock, Massachusetts, built in 1826 to handle fifty-two head of cattle and described in the August, 1831 *Genesee Farmer*. The multi-level structure (like the

FIGURE 17-11. Nine-sided carriage house circa 1815, Church-Bromeley Farm, Belvidere, Allegany County, New York.

later Black & White barn in the town of Groveland) allowed wagons, eight to ten at a time, to enter on the second level with stalls located on the lower level. The barn was constructed with a minimum number of posts for relatively easy maneuvering. Stalls were located on the lower level and manure could be stored below the stalls in pits for later removal. Many circular barns were built throughout the United States, but none seem to have been built (at least none that have survived) in the Genesee Country, perhaps because of resistance to the unusual shape.

On the Church-Bromeley Belvidere Farm in Belvidere, a nine-sided brick carriage barn, about ten feet on a face, was built soon after the main house was begun in 1804 (see figure 17-11). The Church family at one point had 100,000 acres of land in Allegany County. The main house was built by pioneer Philip Church, first owner and Alexander Hamilton's nephew; its design is attributed to Benjamin Latrobe, the well known Philadelphia architect. This sophisticated complex of buildings was built in a wilderness whose wildness can be measured by the fact that of twenty-four sheep driven by Philip Church to Belvidere in 1805, nineteen were killed in the fold by wolves the first night.[5] On the other hand the bricks for the polygonal carriage house were made on location and some were made to shape for the corners, indicating an advanced level of technical capability. Common brick would have required cutting for the corners. The nine sides permitted framing of the second story floor joists without posts, which might have interfered with the storage or turning of carriages. There were no stalls in the carriage house (see figure 17-12).

Orson Fowler of Cohocton advocated structures which employed an octagon plan. In the 1852 edition of *A Home For All* Fowler described the many advantages of the octagon form. Not the least of the octagon's advantages was efficient use of materials: fewer materials were required to enclose the same space as a rectangular building. While the circular form was still more efficient in that respect, Fowler believed that the rectilinear character of framing and sheathing lumber made the octagon the desirable choice. Octagon houses as well as a smaller number of octagon farm buildings were built throughout the United States.

FIGURE 17-12. Framing of the Church/Bromeley carriage house.

A small octagon barn stands on Coe Road, northeast of Conesus, New York (see figure 17-13). The sides are 15 feet long and sheathed in board and batten. Windows are centered in the walls. It is difficult to tell if the barn once had a cupola because it has been reroofed, but a cupola is an expected feature since Fowler was a strong advocate of good ventilation. With the inherent strength of the octagonal design, interior posts are not necessary in a barn of this size. Fowler describes the advantages of having all the stalls open toward the center of the barn, thus saving steps. Also he believed there would be many readily accessible spaces provided for the storage of "cattle, horses, hay, wheat, oats, straw, stalks, etc." Fowler also was pleased with the idea that the center of an octagon barn was the proper shape for threshing with horses or with a flail.

FIGURE 17-13. Coe Octagon barn. Coe Road, South Livonia, Livingston County, New York, circa 1860s.

In his book, *A Home For All,* Fowler makes additional observations dealing with barn arrangement, materials and size:

> If a farmer can find a knoll or bank, so that he can drive in on to his main floor, several feet above the foundation—and the higher the better, for it is easier to pitch down than up—so as to have a cellar or basement story, say 8 feet high, and arrange bins under his floor for carrots, beets, potatoes, turnips, etc., so that he can drive in on to the floor and *dump* them right from the cart down a hatchway into each bin, he will save half his labor in handling them. And by arranging the floor on which the stock stand a foot or two below the barn floor, the cattle can feed off of the barn floor. This plan has several advantages over a manger, one of which is, that, in turning their heads, as in feeding, they do not drop their hay or grain under their feet, but only on to the barn floor, and within their reach. In this case, since the heads of the cattle are over the barn floor, their breath is not confined, but ascends readily, not into the mow, to vitiate the smell of their fodder, but into this center, so that they get far better air—a point as important, relatively, to beast as to man. This plan gives several feet under them to receive their manure and keep it *under cover* and *in a body,* so that it retains all its original virtues; whereas, if thrown out, and especially under the eaves, or if the water from the barn-yard *runs off,* it loses a large proportion of its fertilizing elements. Manure should always be kept under cover till drawn out, and then should be spread and plowed under the very day, and, if possible, the very hour it is drawn. The bad smell from manure is caused by the escape of its nutritious elements . . .
>
> As to the cost of a frame, as compared with a concrete barn, I can not state, but as our upper story costs only $80, or less than four cents per running foot, the walls of a barn 40 by 40, and 20 feet high, at this rate will cost $68, and could be at least built for $100, with any sort of economy and management.
>
> A large barn is far preferable to a small one. Does not every farmer lack barn-room? How many things have you got, reader, which ought to be housed, and are rapidly decaying for want of it? Would not more barn-room pay many times over the interest on its cost? Then let farmers build larger barns; especially since large barns are relatively so much cheaper than small ones. To see stacks of hay and stalks standing out of doors, and cattle "run out" winters to eat them in the lots, is poor policy; for hay wastes by summer rains, wastes when fed on the bare ground, wastes by storms during its use, while the top of the stack is off, and wastes by taking a much larger quantity to keep cattle equally well out of doors than in. Nor should corn be allowed, as now, to stand out till its stalks are nearly spoiled. ..One load cured under cover is worth four cured out. Corn-stalks soon perish if out in the weather, yet make the very best of fodder if cured under cover.
>
> To facilitate this and other like ends, and give cattle sun, it will quit cost to have a part of the roof made of glass. Its cost is not much more than shingles, and allows you to have sun in your barn with which to dry potatoes, corn, hay, etc. . . .
>
> Nor can I see why two or three stories on barns are not as advantageous as on houses. A neighbor of the author drives his grain in upon his second story, where he threshes it—the horsepower being below—which allows the straw to be tumbled *down* into the yard below, instead of having to be pitched up on to a stack, and lets the grain sift down through on to the main floor, where it is cleaned, and passes from the tail of the fanning-mill right down into the grain-bin, still below, in the basement, from which it is loaded into the wagon.[6]

In octagon barns of considerable size a wagon could be turned completely around in the center of the barn.

In Red Creek is the three-story octagon bank barn, 12¼ feet on a side, built by Burgess Jenkins in the 1880s and now owned by Norman Weeks. The combination of windows on all exposed sides with the louvered cupola provides excellent ventilation (see figure 17-14). Most of the major ag-

FIGURE 17-14. 1880s octagon barn on Jenkins Road, Red Creek, Wayne County, New York.

ricultural publications carried articles and drawings in the 1860s and 1870s describing octagon barn designs.[7] But even though a number of octagon and other polygonal barns were built in the region, the design, like that of the round barn, seems to have been too radical for most, and by the time of the nation's Centennial the urge to build polygonal structures in the region had largely passed. National agricultural periodicals and farm books continued to feature round and polygonal barns into the twentieth century. Experimentation in barn design had not diminished. There were always those who felt they could design and build a more efficient barn.

Ramp Barns

As has been noted, the accommodation of the horse fork or hay track brought many changes in barn design, but not everyone liked the device. Dealers were still complaining of some farmers' reluctance to use the horse fork as late as the first decade of the twentieth century. Near Warsaw is the gambrel-roofed, side-hill barn known locally as "The Ramp Barn" (see figure 17-15). Built by Itha Thomson in 1886, the barn contains a magnificent alternative to the horse fork in the form of a ramp that permits hay wagons to go almost to the peak.

The property was originally purchased from the Holland Land Company by J. B. Thomson in 1836. When J. B.'s son Itha built a new hay barn next to the old one, he decided to employ an interior ramp similar to one in Oneida County, which would permit teams to draw hay wagons from road level up to the 65-foot peak of the new barn and there turn around for the descent (see figures 17-16 through 17-18). The ramp is 12 feet wide and 148 feet long, not including the turnaround. The ramp boards are worn from the shoes of the horses as they drew the loads up this steep incline. After a distance of 117 feet the ramp levels out and is horizontal to the end of the barn. Loose hay could be pitched into the mows on either side of the ramp throughout the length of the large barn.

The new barn cost $5,000 and when finished measured 40 by 165 by 65 feet high. Some details of its construction are recorded: for example,

FIGURE 17-15. Thomson/Smith ramp barn, near the corner of Higgins and Gay Roads west of Warsaw, Wyoming County, New York The large main barn was built in 1886. The "entrance" barn on the left has a hewn frame, indicating an earlier construction later altered to accommodate the ramp with a balloon-framed, gambrel roof.

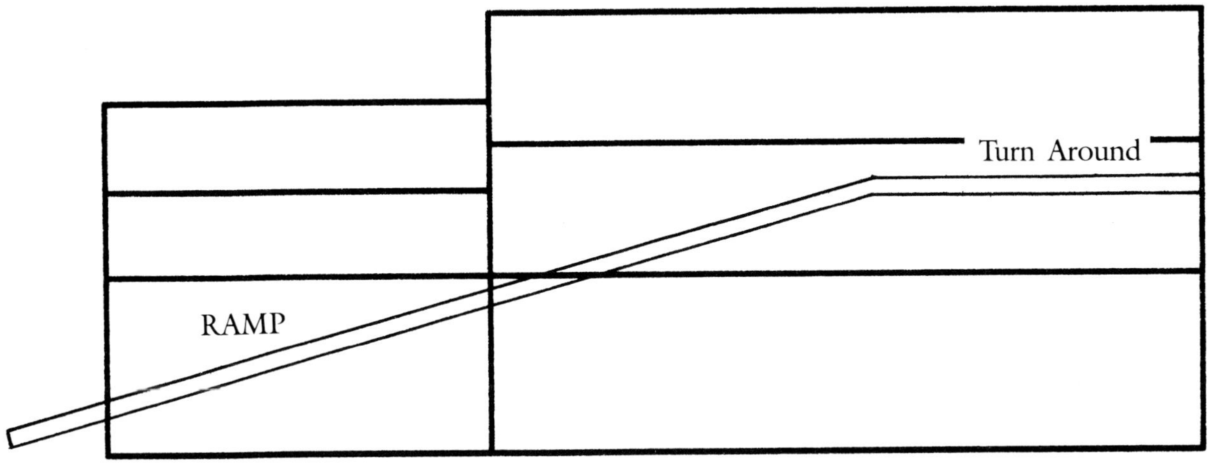

FIGURE 17-16. Elevation of the ramp. Not to scale.

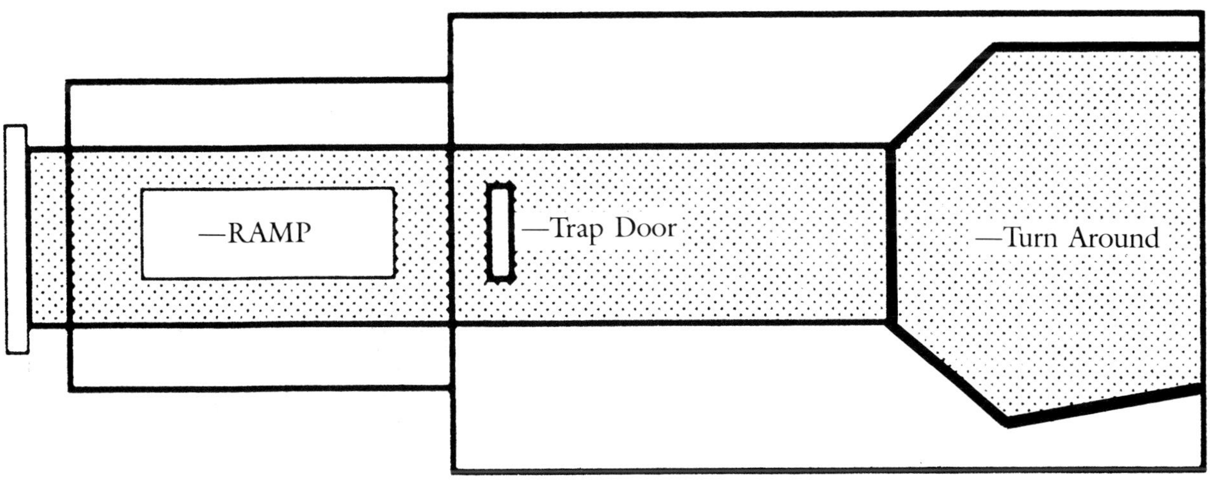

FIGURE 17-17. Plan view of the ramp. It is as tall as the peak less the height of a tall farm wagon. Original measured drawing by Eric Hepler.

FIGURE 17-18. The ramp barn from the turn around. The ramp is centered in the barn providing equal access to the mows.

Louis Higgins hauled stone all winter for the foundation and hardwood timbers for the barn were sawn on the farm. At the raising, part of the frame was pulled too far by a team in the basement, causing two men to be thrown off the bent and nearly killed. The frame cross beams and posts are 9 x 9 inches, the main cross beams 40 feet long. The basement continues to house a dairy herd and feed can be easily pitched down from the mow above. An attached side barn was once used for horses.

Partway up the ramp there is a trap door wider than a wagon box. Here, threshed grain could be unloaded into the granary below (see figure 17-19). One problem created by unloading hay from the ramp directly into the mows was the compaction caused by gravity. In addition, the weight of the loose hay in the mows compressed the hay so much that it was difficult to remove even with a hay knife. The barn is still in use although the hay is now baled.

There were other ramp barns in the area, one

FIGURE 17-19. View from the mow up toward the ramp. In the center of the photograph is the granary, which is fed from the ramp trap door.

of which survives in the Town of Junius, Seneca County. The ramp is no longer complete but its former presence is revealed on the western end of the 110 foot barn as the roofline abruptly breaks to parallel the diagonal of the ramp (see figure 17-20). The barn, easily seen from the New York Thruway, has the appearance of having a broken ridge. Charles Cosad, a New York Assemblyman, designed and built the barn in 1895. The ramp was steep enough so that horses could not draw the loads up the ramp; instead the teams were unhitched and the wagons were winched up the ramp (see figure 17-21). The remaining original portion of the barn is 90 feet long and a section of the ramp can still be seen inside the barn, now owned by Calvin Brown.

There is an outstanding example of a barn in the Genesee Country for which efficient organization played a fundamental role in the design. In the next chapter the Black and White Farm barn is examined. This barn combines the advantage of multi-level entry found in bank barns with the feed storage and handling efficiency found in round and octagon barns.

FIGURE 17-20. The Cosad/Brown ramp barn was constructed in 1895 in the Town of Junius at the corner of Route 318 and Whiskey Hill Road, Seneca County, New York. Large doors in the west gable end opened to admit the hay wagon. The walls of the barn and the shed built beneath the ramp itself were battered (inclined), producing a most unusual appearance in the region. The position of the polygonal cupola, no longer present, can be seen both in this photograph and in the interior of the barn itself. Calvin Brown Collection.

FIGURE 17-21. The doors are opened, the team unhitched and the hay-laden wagon proceeds up the ramp track for dumping into the mows. Some of the details of the original ramp framing can be seen. The original photograph was tilted, reducing the apparent angle of incline. Calvin Brown Collection.

CHAPTER XVIII

The Black and White Farm Barn

IT IS A SHAME that more nineteenth century photographs of barn building do not exist. Early photographs were mostly portraits, usually made indoors in the photographer's studio. Photography, however, matured greatly during the Civil War and in the period of western expansion that followed. The stereoscope became parlor entertainment in many homes, allowing vicarious trips to exotic places via commercially available "views." During the 1870s the cumbersome wet plate process began to give way to dry plates. George Eastman, a young bank clerk in Rochester, New York, perfected a gelatin dry plate process, partially out of exasperation with the wet plate process. Eastman first supplied these dry plates to E. and H.T. Anthony and Company, America's largest supplier of photographic materials at that time. With the introduction of the dry plate the number of photographers increased, and they began recording events and objects heretofore neglected. Instead of simple portraits of the farmer and his family, photographs of farm implements and barn construction began to appear. Unfortunately, barn interiors could not easily be photographed until the twentieth century; the process was beyond the capabilities of lenses and the sensitivity of photographic plates (see figure 18-1).

Among the photographic records of the 1880s are several albumen prints showing the construction and a finished view of the great square barn known as the Black and White Farm barn, located south of Sonyea. Joseph Cone of Geneseo, a dry goods store and land owner, had this spectacular barn built in 1884 (see figures 18-2 and 18-3). 194 by 159 feet or 706 feet around, the barn was once numbered among the largest in the state. By comparison, the large stone Shaker barn, built in 1858 at New Lebanon, New York and considered one of the largest in the state at the time, was a single rectangle 193 by 50 feet. Children who played at the Black and White barn reckoned that "less than seven times around the barn equals a mile," a calculation based on folklore that each side was 200 feet long.

The second albumen print shows the finished barn in winter (see figure 18-4). Louvered cupolas and clerestories provided ventilation for the cows and horses. The effectiveness of the ventilating system is evident to any visitor on a hot summer day. An additional interesting feature is the chimney in the southeast corner; the fire hazard involved made a barn stove a rare feature. There is no chimney today.

An aerial view of the barn puts the entire structure in perspective, showing how the buildings

Opposite: Detail of the gable and cupola on the east facade of the Black and White barn. The barn ventilation louvers clearly show in this view.

FIGURE 18-1, Above Left. Photographic documentation became possible as photography matured. Bridges, railroads, canals and barns appear more frequently in dry plate photography.

FIGURE 18-2, Below Left. Detail of an 1884 construction photograph of the Black and White barn, Dansville/Mt. Morris Road (Route 36), south of Sonyea, New York. The roofers pause for the photographer. Anonymous. Collection of Mr. & Mrs. Crawford Henderson.

FIGURE 18-3. Black and White Farm barn, 1884 construction photograph. The large centered door on the left of the photograph leads to the barnyard. Anonymous. Albumen print from the collection of Mr. and Mrs. Crawford Henderson, owners of the property from 1946-1981.

FIGURE 18-4. The finished Black and White barn, 1884. The photograph was taken looking northwest. Henderson Collection.

FIGURE 18-5. Aerial photograph of the Black and White barn. The relationship of the three new barns to the moved barns is emphasized.

surround the barnyard (see figure 18-5). What is not as apparent from the air is that this is a bank barn, allowing entry from the west on the second level for a hay wagon, which moves down the 15 foot wide drive while hay is pitched down to the cows, saving both time and effort. Holes in the floor next to the walls allowed access to the feed racks below. For many years there were forty to seventy dairy cows and 100 beef cows plus calves in the Black and White barn. This arrangement, which permitted a feed wagon to drive through 600 feet of barn without even turning around, made the feeding of a large herd much easier.

Just after the First World War the farm was sold to Craig and Mortimore Ross, who gave the name "Black and White" to the farm because of their black and white animals, including mules, cows, sheep and pigs. The barn itself was originally painted a pale yellow. The herd was mixed between beef and dairy with about 100 cows plus calves. The Rosses owned the barn until it was purchased by the Hendersons in 1946. Morris Kennedy, the owner since 1981, uses the barn in his large dairy operation.

Construction of the Black and White barn began when three hewn timber-framed barns were

FIGURE 18-6. Black and White interior, looking south, showing the hewn construction of the three barns moved to the site to form the west section. The axes of the three barns are at right angles to each other.

FIGURE 18-7. Black and White interior, sawn, 1884 construction. A portion of the floor was temporarily removed to prevent malicious trespass. Loose hay was moved to the mows with the aid of the hay track.

FIGURE 18-8. Black and White barn milking parlor.

moved to the site to form its western side. In those days moving large structures was not all that uncommon. Young Caroline Richards, who lived in Canandaigua in the late 1850s, mentions in her book, *Village Life in America,* the moving of two large buildings, one the "old court house" and the other a barn. The barn-moving in the twentieth century on the Herman Wright Farm has already been described in Chapter XIV. The axes of the moved barns are at right angles to each other, as shown in the aerial photograph. The other three barns of the Black and White barn were newly constructed on the site. The difference between the hewn timber construction (see figure 18-6) and the three new barns is startling (see figure 18-7). The regularity of the new design, with its open driveway, is perhaps most noticeable. The interior posts supporting the purlins were sawn using a circular blade, as evidenced by the saw-marks. A hay track was installed in the three new sections some time later, requiring the severing of beams in the front corners of the barn in order to make it continuous. The damage done by this alteration was repaired by Henderson, who employed both tie-rods and trusses to restore the structural integrity of the barn; strengthening was necessary because the hay continually shifted, producing too much pressure on the outside walls.

The dairy cows were milked in the forty-stanchion basement section of the barn (see figure 18-8). David Thorpe, present Livingston County Cooperative Extension agent, states that until 1966 there was no bulk milk handling; instead there was a ten can cooler holding 100 gallons.[1] In addition to the hay track on the second level there was a manure trolley on the lower level for improved handling (see figure 18-9). Although not a part of the original construction, the barn had a hexagonal silo, added sometime after 1918 and removed in 1981 (see figure 18-10).

FIGURE 18-9. Black and White barn manure trolley.

FIGURE 18-10. East side of Black and White barn showing the location of the hexagonal silo, since removed.

The silo was constructed of full 2x4s, laid flat and nailed together. Later it became necessary to add iron hoops to keep the silo from spreading (see figure 18-11). See Chapter X for a further discussion of silo construction.

The Black and White barn is a subject of much local comment and attention, primarily because of its large size. But it should not be forgotten that this barn represents an advance in bulk handling and operational efficiency, subjects which were much on the minds of farmers and barn designers in the period. Barn design simply was not a static subject.

FIGURE 18-11. Black and White barn silo detail.

FIGURE 18-12. In 1985 a 105 by 216 foot pole barn was constructed, adding to an already sizable dairy operation in which 300 cows are milked and more than 600 are kept.

CHAPTER XIX

Changing Sources of Power on the Farm

THERE WERE MULTIPLE USES for many barns, but the major traditional requirements for the grain barn included storage space for unthreshed grain and a threshing floor. In addition there probably was a granary for storing the threshed grain. Changes in methods of threshing altered the type and size of the barn required: For example, a greater threshing capacity might require a larger granary and enlarged storage space for unthreshed sheaves, while steam-powered threshers eliminated the need for a threshing floor. It therefore is important to examine some of the changes in threshing to aid in understanding changes in barn design.

Hand-cranked or horse-powered threshing machines were generally accepted in the 1820s as superior to flailing or flooring (see Chapters V and XIII); in fact, eventually some 700 different threshing machines were available to choose from. In the *Ontario Freeman* of September 2, 1823 an advertisement for the Cummings Patent Thrashing Machine claims: "Of wheat, rye or oats, one man with a boy and horse will easily thrash and clean one hundred bushels a day" (see figure 19-1).

The Case horse-powered thresher illustrated in figure 19-2 is from a J. I. Case Co. wood engraving circa 1855. This machine was said to have a capacity in excess of 200 bushels of wheat per day.

The Pitts Premium Separator (thresher), manufactured in Rochester and advertised in *The Genesee Farmer* in 1858 (see figure 19-3), had a geared mechanism that could be driven by as many as ten horses walking in a circle. Daniel Woodbury of Rochester collaborated with W. W. Dingee to patent another geared mechanism, which became the basis of most other horse powered sweeps. Figure 19-4 shows a detail of the device.

Figure 19-5 depicts twelve horses on a Case sweep power. Depending on the job and the design of a sweep, any number of horses from one to twelve might be used.

In the March, 1852 issue of the *Genesee Farmer,* Patrick Barry, the *Farmer's* correspondent at the World's Fair at the Crystal Palace, Hyde Park, London, described agricultural improvements: "Steam is *the* wonderful power of the age . . . Of such a power as this may not the agriculturist avail himself to carry on his multifarious operation? Why may he not thresh, and plow, and sow, and roll, and harrow, and mow, and stack . . . ?" Barry goes on to say that Hoard & Bradford of Watertown, New York, was advertising a three-horsepower model at $250, quickly adding, "the principal objections to the use of steam engines on the farm, are—difficulty of management, liability to explosion, and danger of fire."[1]

Opposite: For a time, everything from wash machines to water pumps to threshing machines was steam-powered. Detail of an Advance-Rumely steam traction engine used to provide power for threshing.

TO FARMERS.

CUMMINGS' PATENT

THRASHING MACHINE.

THE subscriber having obtained the right of making, using, vending, and empowering others to make, use and vend this valuable Machine, within the town of Canandaigua, for the term of fourteen years from the first of September, 1821, respectfully invites Farmers and other Gentlemen to call and examine a model of said Machine now exhibiting at E. Rowe's Inn, in the village of Canandaigua.

Some of the advantages of this machine over others are, that it is simple and cheap, and will separate perfectly from the chaff, stalk, vine or cob, every kind of grain cultivated in this part of the country. Of wheat, rye or oats, one man with a boy and horse will easily thresh and clean 100 bushels a day. JOSEPH N. BRADLEY.

April, 1823. 6

FIGURE 19-1. *1823 advertisement for the Cummings Patent Thrashing Machine. Ontario Freeman.*

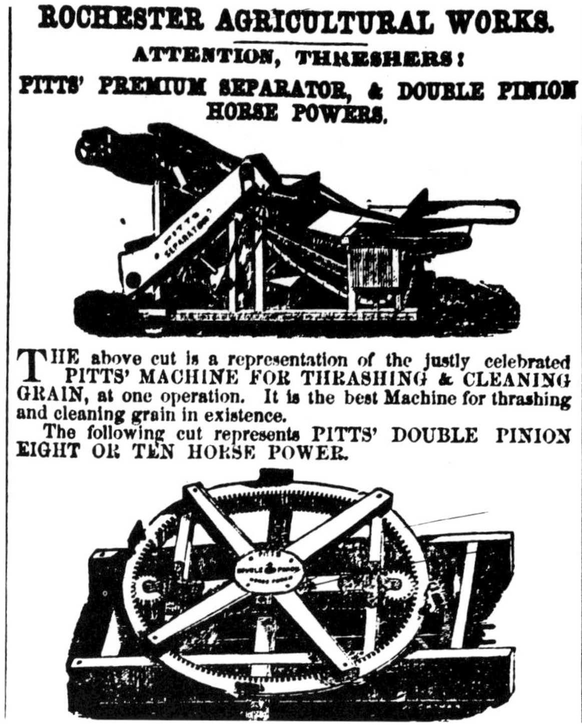

FIGURE 19-3. *Pitts Premium Separator and Double Pinion Horse Powers.*

FIGURE 19-2. *A woodcut of a Case horse-powered thresher. Circa 1855.*

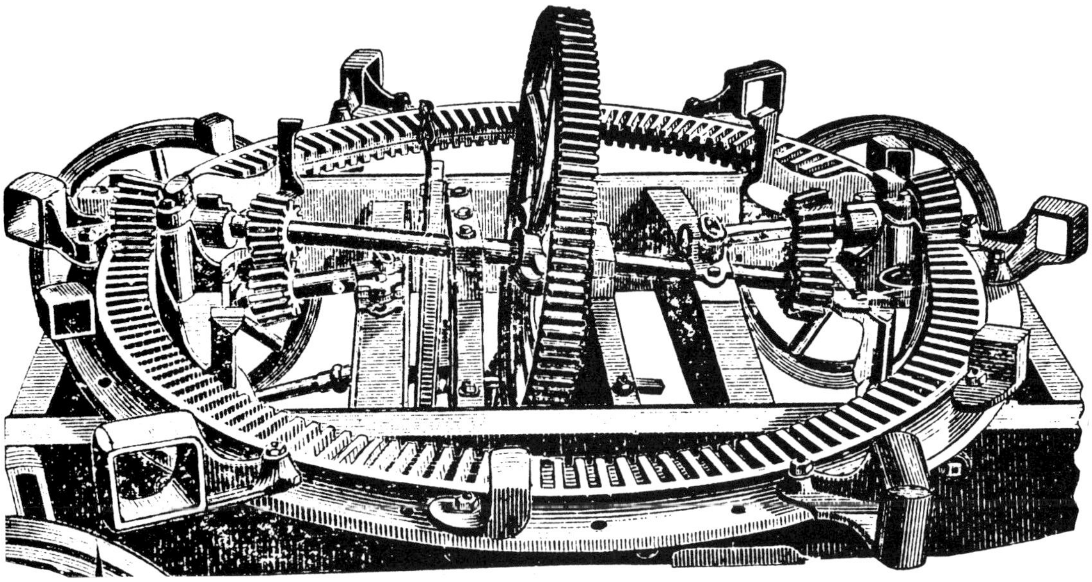

FIGURE 19-4. Closeup of the Dingee-Woodbury Sweep Power Mechanism. Woodbury's invention was introduced in the 1850s and Dingee improved it by making it stronger and less likely to break down. The shafts for the horses are omitted in this view but the shaft sockets can be clearly seen along the outer gear ring. The Dingee-Woodbury mechanism translated the round and round movement into a horizontal drive, through line shafts and universal joints, to operate the various mechanisms of a thresher.

FIGURE 19-5. Case sweep power with twelve horses. Circa 1870.

One of these "objections" is vividly described in James Disher's account of an 1873 boiler explosion in Wyandot County, Ohio:

> On the Fatal morning of July 29, an engine being used to thresh wheat blew up like the report of a cannon. Pieces of iron and steel flew in every direction. One farm worker was killed instantly, another lay bleeding and scalded and died the next day. Two more were injured. The boiler itself was blown seventy feet. More than a thousand curious folk gathered to view the awesome effect of the blast.[2]

Incidents like this help explain the natural human resistance to "new-fangled" things. However, by the 1880s most reservations were allayed and much improved and safer steam power threatened the domination of the horse and ox.

Much manual work was still required, but steam-powered one-step separating certainly was an improvement over the two-phase work of flailing and winnowing: With eight bushels considered a good day's work for one person, flailing alone could take all winter. A farmer looking at the advertisement for the Buffalo Pitts Thresher and Cleaner in 1877, which claimed some models were capable of handling from 300 to 3000 bushels of grain a day, might be inclined to think that if the machine were only half as good as the company claimed, it would still be far better than flailing and winnowing.

> Among the many distinguishing excellencies of the Buffalo Pitts Thresher and Separator, are: It threshes clean; it extracts the kernels from the head of every straw that passes through it; it does not mash or crack the berry but delivers it in a perfect state; it perfectly separates the grain from the chaff and delivers it in bags ready for market, saving entirely the labor incident to recleaning.[3]

In the spring of 1859 William Garland of Mendon purchased a Wood, Taber & Morris steam engine—according to McIntosh, the first ever introduced into Monroe County for the purpose of threshing (see figure 19-6).[4] Both the steam engine and the separator were hauled from place to place with teams; later steam threshing engines would be self propelled.

FIGURE 19-6. An 1859 Wood, Taber & Morris Traction Engine owned by William Garland, Mendon, New York. Although looking somewhat awkward at the moment, the canted stack of the engine would be righted when the rig was set up for threshing. This threshing rig was still in use in 1877. Lithograph from McIntosh, History of Monroe County, 1877.

FIGURE 19-7. Experimental plowing with steam power at Rochester, New York, 1870. Reproduced from the May-June, 1973 Iron Man Album.

An illustration from 1870 shows plowing at Rochester with a vertical boilered engine and six wooden beam walking plows (see figure 19-7). Steam plowing was tried but not practiced much in the Genesee Country.

In the June 25, 1981 *Livonia Gazette* Robert Halsted remembers details of the Halsted (originally Foote) family gambrel-roofed barn, built in 1893, which had just collapsed after years of disuse. He describes the granary and other storage spaces and also the threshing itself, powered by both steam and gas tractors.

> Half of the floor space to the left of the center aisle was taken up by the granary. This was a tightly built structure with walls of matched lumber (floors and ceilings the same) with a three and one half foot walk way through the center and the space on each side walled off into large bins. The three bins on each side of the walkway each measured perhaps six feet by six feet by seven feet high and held perhaps 150 bushels. Wheat, oats, corn, and beans were held here for use during the winter or until the crop was sold. An exterior door at the end of the walkway facing the road could be opened, and a truck or wagon drawn up in front of this door could be loaded with grain and driven away.
>
> Alongside the granary, and also on top of it (the whole granary structure was about eight feet high), was a large additional area of bulk storage space. This area was particularly necessary in the time prior to the . . . use of the grain combine—1938 or so. Up until that time, the grain (or beans) was threshed by a stationary "Threshing machine" belt—driven by a large steam engine or, later, by a gas engine tractor.
>
> Because of the large capital investment involved, there were only one or two threshing machines available in each town, the owners of which traveled from farm to farm during the later summer and fall setting up their engine and thresher at each farm and threshing out each farmer's grain as they went. This made it necessary, of course, to wait your turn until the thresher came. So, you protected your ripened crop of grain by cutting the whole thing—straw and all—with a machine that bound the cut grain into "bundles"; then, you hauled the whole crop into the barn (out of the weather) to wait for the thresher to come several weeks later.
>
> The space over and beside the granary is

where the crops were stored until threshing time. Then, the threshing machine (a big thing, six feet wide by six feet high by some twenty or so feet long), would be pulled into the center passageway of the barn (haymow to the right, grain to be threshed on the left and set up). The bundles were pitched out of storage and fed into the machine. The grain came out of the machine through a chute near the granary door and was caught in one bushel steel baskets; then, it was carried into the granary and dumped into the proper bin. The straw came out the end of the machine on a conveyor and was stacked in the middle of the barnyard to be used as animal bedding during the winter.[5]

When the timing was right, threshing could involve at least four wagons: one wagon loading in the field, one on the way to the threshing machine, one at the machine and the fourth on the way back to the field. The sheaves had been placed in round or long shocks, with one sheaf laid on top as a protective cap. If the workers were skilled and if the reaping was done at just the right time, they avoided knocking the grain from the stalk in the field during shocking.

The sheaves were pitched onto the wagon with all heads in or all heads out to make an easier job of pitching them into the separator. Either a young boy or, some say, the laziest worker was given the job of guiding the straw blower with ropes. Photographs show straw mountains many times larger than the separator itself (see figure 19-13). Halsted notes the separator was located in the barn, but in many cases the threshing was done in the barnyard or even at times in the field.

Halsted describes threshing beans:

> Beans were harvested from the field dry (but still in their pods)—vines and all—in October. These were threshed separate from the grain at a later time because a different type of threshing machine was brought in for them. The pods and vines were held inside the barn after being threshed because they were a high-quality fodder that was fed to the milking cows during the winter along with hay and some grain.[6]

A popular machine in the Genesee Country was the Farquhar Vibrating Separator which claimed performance "equally perfect in all kinds of Grain and Flax" (see figure 19-8).

The Nichols & Shepard Company advertisement in figure 19-9 illustrates the separating mechanism in some detail, including the big cylinder and grate; the "Red River Special" model is compared with other separators to show the

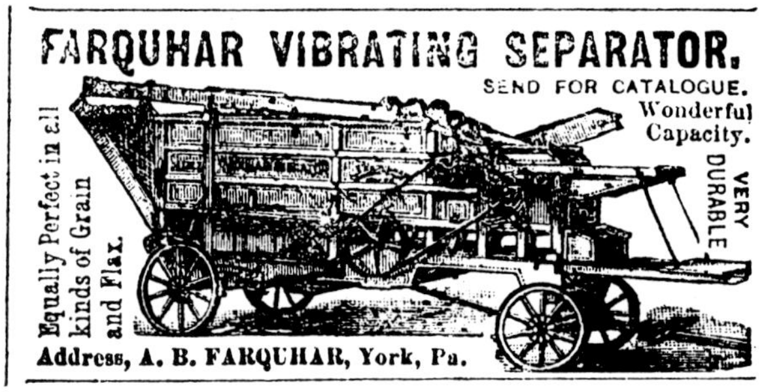

FIGURE 19-8. Farquhar Vibrating Separator advertisement. Livingston Republican, 1881.

greater efficiency of its beating action. The A. D. Baker Company compared the power of its separator to that of an enormous articulated steam locomotive (see figure 19-10). Separators manufactured by the Frick Company of Waynesboro, Pennsylvania were popular in the Genesee Country and can still be seen stored in a barn here and there (see figure 19-11).

Figure 19-12 shows the passage of the Oswaldt brothers' threshing outfit about 1912, as recorded by Herman Wright. Bill, Jake, Dan and Al Oswaldt were custom threshers from Zurich (a hamlet located between Lyons and Sodus) and were photographed along a road close to where Wright made his series of photographs documenting the moving of his barn.

Again, from an 1877 Buffalo Pitts Thresher & Cleaner advertisement:

> [If flail threshing and hand winnowing were in use in 1877] It would be simply impossible to market the present grain crops of the country, for the reason that the combined force of all the

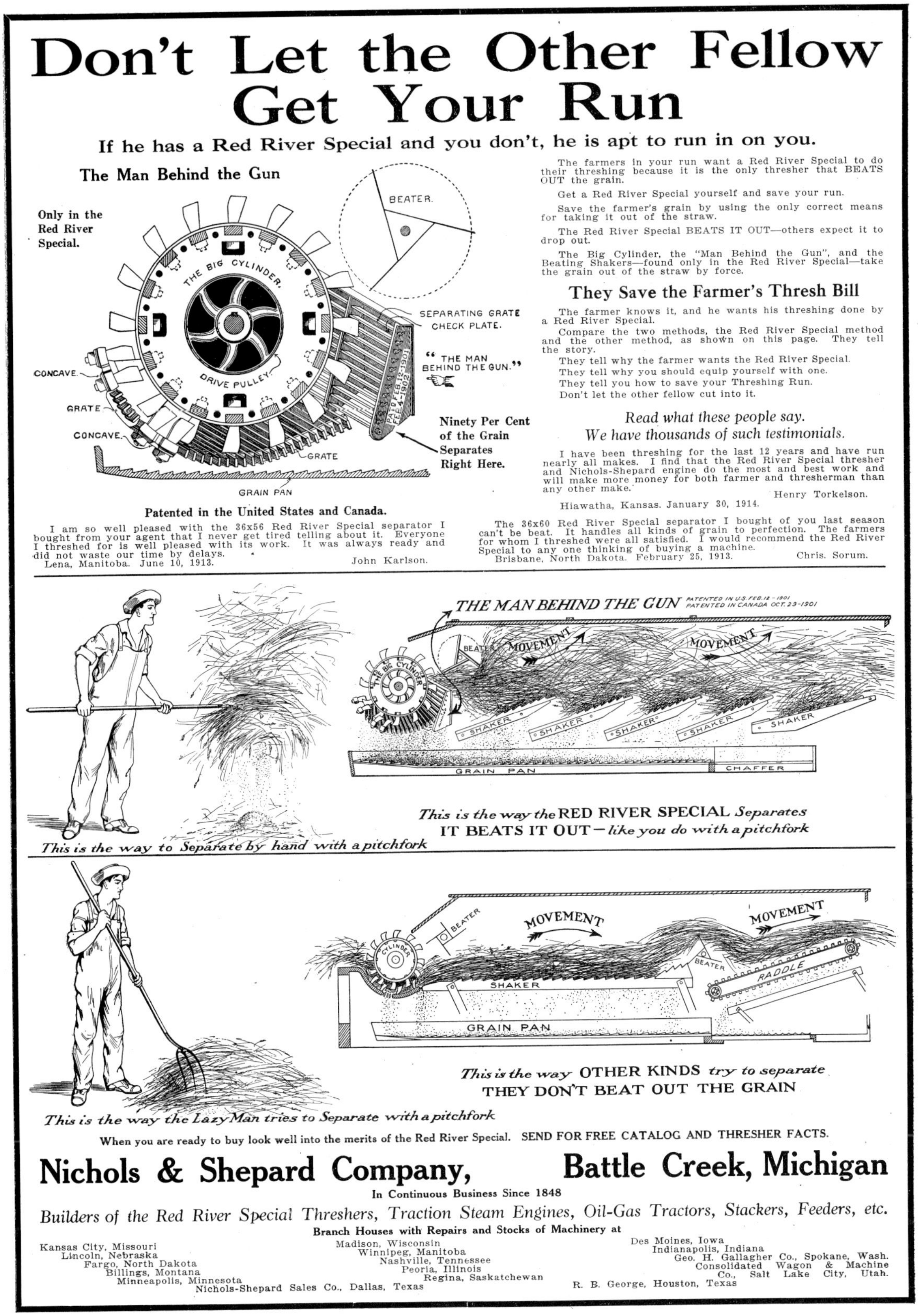

FIG. 19-9. Advertisement for Nichols & Shepard Company separator. Threshermen's Review and Power Farming, 1914.

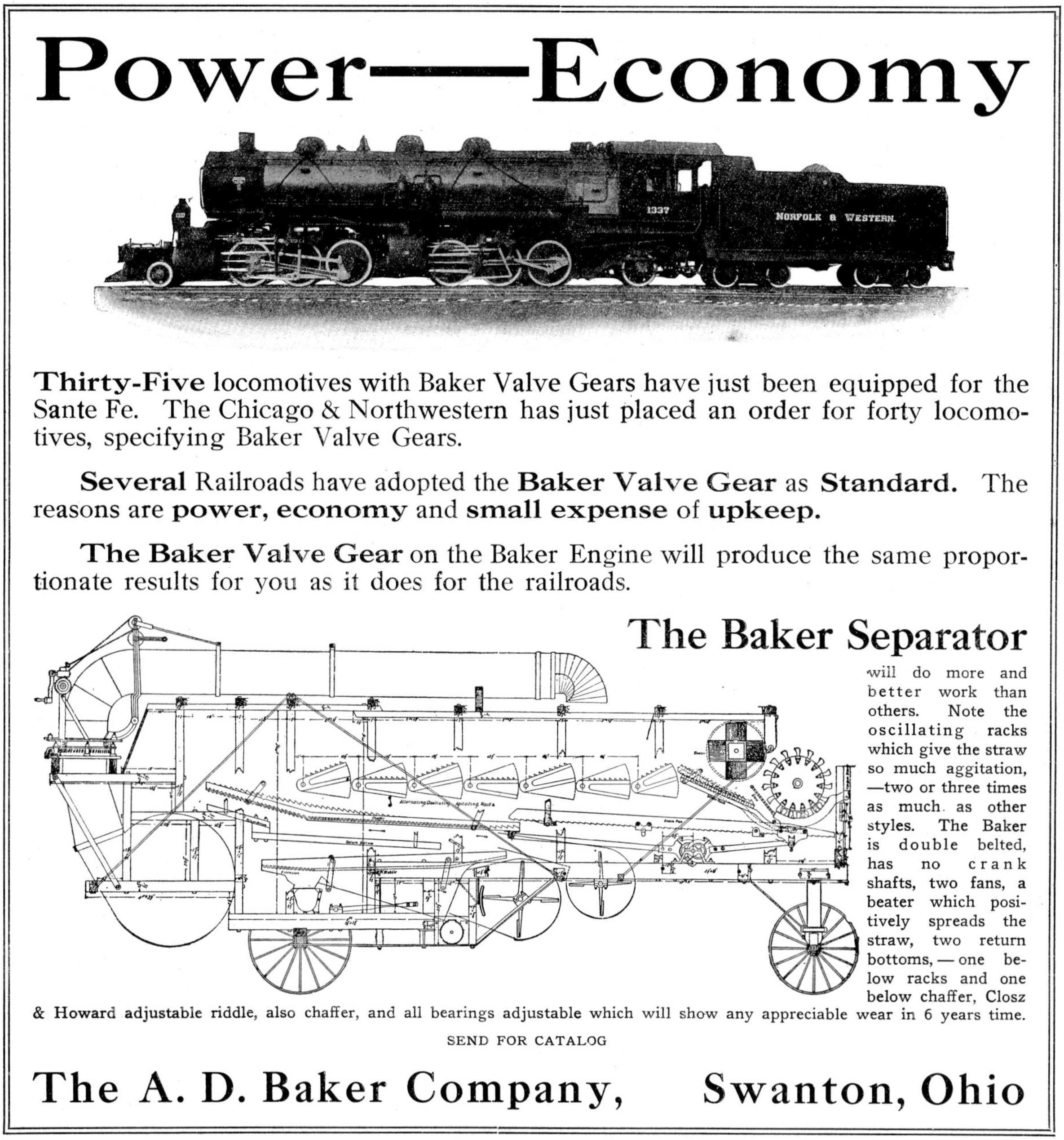

FIGURE 19-10. Advertisement for A. D. Baker Company separator. Threshermen's Review and Power Farming, 1914.

agricultural laborers in the United States, working steadily from the end of one harvest till the beginning of the next, could scarcely thresh and clean the production. In addition to the immense amount of manual labor saved by threshing machines, they effect a great saving of capital. The construction of all the barns and threshing floors on and in which they beat out the grain by hand, would involve an expenditure of many millions of dollars.

THE NEW FRICK

will eventually command your attention---why not now take notice of its merits? It will be much to your advantage to be the first to use this modern grain and time saver. Its constant satisfactory performance will reflect your good judgment---more firmly establish your prestige as a dependable operator and quickly bring you such a quantity and quality of business that you can, if you desire, spend your limited season's time serving only the more convenient and best paying.

You know what is essential to a thresher to make it better than any other. Just drop us a card today for our New Frick Thresher Circular, and you will get it by next mail. It is very complete and fully illustrated, telling in a comprehensive manner why the Closed Cylinder and Double-Faced Tooth threshes cleaner than any other and saves power and time. Why the Malleable Separating Grate and Straw Racks guarantee practically complete separation. Also shows a conclusive illustration of a separating test which establishes a new record and standing. The Gravity Cleaner and attachments for various grains is also fully illustrated and described.

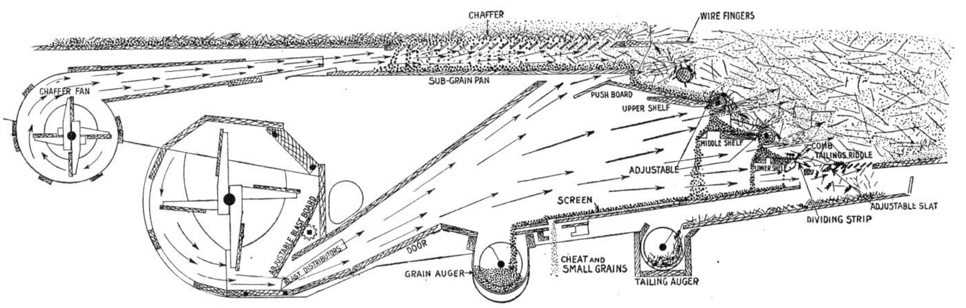

New Frick Gravity Cleaner

Importance of Ordering Now

Unless your order is filed at once there is small chance of getting one of these machines this year, but you will be posting yourself and will order for next year's delivery.

That you will want this thresher, IS CERTAIN, why not get in line by writing today?

FRICK COMPANY, Waynesboro, Pa.

Wichita, Kans. Des Moines, Iowa Bloomington, Ill. Nashville, Tenn.

BRANCHES—59 Ford St., Rochester, N. Y.; 6031-32 Jenkins Arcade Bldg., Pittsburgh, Pa.; Prior & Sallada Bldg., Williamsport, Pa.; 610 American Bldg., Baltimore, Md.; Kanawha Valley Bank Bldg., Charleston, W. Va.; Salisbury, N. C.; 2337 E. Fifth St., Knoxville, Tenn.; Box 117, Columbia, S. C.; 75 S. Tenth St., Harrisburg, Pa.

FIGURE 19-11. Advertisement for Frick Company separator. 1914. Threshermen's Review and Power Farming, 1914.

FIGURE 19-12. The Oswaldt brothers' steam threshing rig on the way to a job about 1912. The Nichols & Shepard engine trails a water wagon, followed by the separator and the team of horses used to draw the water wagon back and forth between the water supply and the engine. Photograph by Herman Wright.

FIGURE 19-13. Steam threshing on the Harry Wadsworth Orchard Farm, Groveland Road, Geneseo, New York, 1909. The photograph, taken in a Farm Census, was given to Mr. Adrian LeFeber by the State College of Agriculture at Cornell University in 1909. LeFeber is third from the right. There is a mountain of straw behind the separator. This separator has no blower but uses a straw conveyor. The engine on the belt is a Frick. Don LeFeber Collection.

Threshing

It is a distortion of fact to over-romanticize the events and labors of the past, but there was an excitement in some of the work that won't soon be forgotten. Barn raisings were special times, especially when it was a timber framed barn and workers were needed in large numbers to coax the bents into place. Threshing with a steam engine also was a special time, even at wages of fifty cents an hour for a ten-hour day that ended with a soot smear across an oft-wiped forehead and intense itching from the straw down one's shirt. All those things could not suppress the excitement of the steam tractor coming to life, with the roar of the blower encouraging the coal to almost white heat on the grates (see figure 19-13). With the pressure gauge reading 125 pounds, the engine was moved to the belt and a threshing day was begun (see figure 19-14).

Threshing involved an endless coming and going of loaded and unloaded wagons, the departure and return of the water wagon supplying the engine's thirsty boiler (see figure 19-15). Keep

FIGURE 19-14. A Case engine powers the separator. New York Steam Engine Association demonstration, Canandaigua, New York, 1980.

FIGURE 19-15. Bundled and shocked oats waiting to be drawn from the field to be threshed. Silver Springs, Wyoming County, New York, 1983.

FIGURE 19-16. Here bundles are being pitched onto the separator feed conveyor. In some locations steam engines still power separators and saw mills, but generally they can only be seen at steam shows like this one sponsored by the New York Steam Engine Association where a wide variety of steam-powered machines are operated each year. Canandaigua, Ontario County, New York. 1982.

FIGURE 19-17. Close-up of a separator made of wood and iron, complete with the many belts needed to operate the shaking grates, sieves and blowers. Western New York Gas and Steam Engine Association, Alexander, Genesee County, New York, 1977.

pitching those bundles onto the cylinder feed belt! (see figure 19-16) Enormous lunches. On some days moving the whole rig to the next farm and beginning all over. Dirty, long days, but exciting. The straw blower makes a mountain, higher and higher. The engineer's shovel clangs off the firebox door as coal fills the holes in the fire. Turn on the injector, keep the boiler water at half a glass in the gauge, don't let her pop her safety valves, wastes coal. The smell of valve oil. The music of the scene, mostly from the changing sounds of the soft exhaust, now accelerating, now slowing as the spinning ball governor controls the speed of the engine. A multitude of belts and pulleys hum and whirr as the separator rhythmically rocks in response (see figure 19-17). The working of a hot engine can only be vaguely imagined when one is looking at a cold museum specimen.

Power On The Farm

Humans, animals and water were the dominant forms of farm power until the introduction of steam in the mid-nineteenth century. The Civil War gave impetus to the use of steam power on the farm because those who fought had seen the promise of steam as it was applied to the war effort (see figure 19-18). At roughly the same time windmills, an ancient form of power, were redesigned to take advantage of newly available technology. These new windmills were especially useful for wind-powered pumps. The water could be pumped when the wind was blowing, then stored in a tank or cistern for use at any time. If the tank was located in the attic the owner could have running water with pressure at the faucet. The persistent winds of the Genesee Country made these windmills as effective in this location as they were on the prairie. Figure 19-19 shows the factory of the Tornado Windmill Company of Elba. Tornado was one of the many companies that provided farmers and village dwellers with the windmill towers and mechanisms. Farmers were the major customers because those in a rural setting rarely had access to a municipal water supply. Wind power could also be used for grinding grain: Halladay advertisements claim that each farmer could be his own miller using windmill power and a Halladay apparatus (see figure 19-20). The Robert Moody barn on Route 245 in Rushville, Ontario County, once had a windmill mounted on top of its cupola to provide power for grinding grain (see figure 19-21).

In general, farms had telephones before they had electric power. Practical electric lighting became available about 1880, but most farms used gas jets or lanterns for illumination until at least World War I. Steam power was used more extensively than electric power on the farm and many

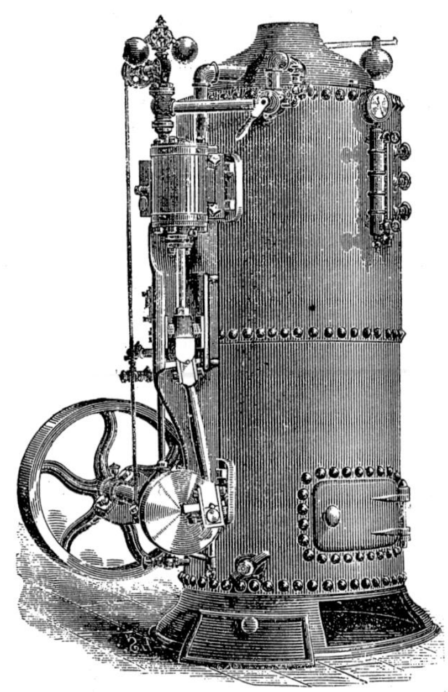

Bigelow Semi-Portable Steam Engine.

488 AMERICAN AGRICULTURIST. DECEMBER.]

TEAS REDUCED
OUR TERMS ARE THE BEST.
Send for our New Reduced Price List.
THE GREAT AMERICAN TEA COMPANY,
P. O. Box 5643. 31 and 33 Vesey St., New York City.

STEAM ENGINES

CLIPPER Vertical Engines, off and on wheels! Price on wheels, 6 Horse Power, $485. 8 Horse on wheels, Price $700. Dry Steam Horizontal Engines, 8 Horse Power Climax, on wheels, $925. Builders of Stationary Engines, 4 styles, from 6 to 200 Horse Power. Corlis's Engines, Extra Finish and cut off Engines. Crank Stationary, and Plain Finished Engines, Boilers, Circular Saw-Mills, &c. Successors to Utica Steam Engine Co., formerly Wood & Mann, Established 1857. We have the *largest* line of Engine Patterns in United States.
TAYLOR MANUFACTURING CO.,
Westminster, Maryland.
Good responsible Agents wanted.

"Eclipse"
FARM
ENGINE
[See page 240, *Agriculturist*, 1877.] Send for Centennial Judges' Report and Catalogue of Engines, Boilers, Sawmills, &c. Address
FRICK & CO.,
Waynesboro,
Franklin Co., Pa

BOOKWALTER ENGINE.
Compact, Substantial, Economical, and Easily managed. Guaranteed to work well and give full power claimed. Engine and Boiler complete, including Governor, Pump, &c., (and boxing), at the low price of

3	Horse-Power......	$242.00
4½	" "	283.50
6½	" "	343.50

☞ Put on Cars at Springfield, Ohio.
JAMES LEFFEL & CO.,
Springfield Ohio,
or 109 Liberty St., New York City.

MONEY *Made rapidly* with Stencil and Key Check Outfits. Catalogues and Samples FREE. S. M. SPENCER, 112 Washington St., Boston, Mass.

NEW ECONOMIZER BOILER AND ENGINES,
Mounted and not Mounted for Agricultural and all purposes—now taking the lead decidedly over all competitors.
The only Agricultural Engine
with RETURN FLUE BOILER in use.
Fire passes from Fire-box, which is surrounded with water—through a 12 to 25 in. flue, according to size—passing through the water in Boiler to the Chamber in rear, and then back to the front through 3 in. flues.
All sparks consumed—no danger of fires.—All wrought iron.
The Cheapest, most Economical, yet powerful Engine in the market, and is the best and safest Steaming Boiler in use. We invite competition with any first-class maker. It will make more steam from less fuel than any horizontal boiler built. Before buying, see one or send for full particulars, circulars, etc., which will be sent free by addressing
WHITMAN & BURRELL,
Little Falls, N. Y.
Manufacturers of HUBBELL & CO.'S NEW GANG CHEESE PRESS, with right to use the Bandager and all Patents Guaranteed.
Manufacturers and Dealers in New Apparatus and Furnishings for Cheese and Butter Factories.
Agents for Annattoine, Johnson's Patent Fly-proof Cheese Grease; Prof. Arnold's Liquid Extract of Rennet, etc.

SNYDER'S
LITTLE GIANT STEAM ENGINE.
FOR FARMERS, MACHINISTS, PRINTERS, AND ALL REQUIRING LIGHT POWER.
Sizes from One to Six Horse Power. Prices for Engine and Boiler Complete, from $150 to $450. We make the *Strongest Boiler* and the *Best Engine* in the country. Call at our Factory and examine, or send for free Illustrated and Descriptive Catalogue.
SNYDER BROS., 94 Fulton St., New York.

SKINNER
PORTABLE ENGINE
2½ to 10 H. P.
Made by Special tools.
All parts duplicated.
Send for circular.
SKINNER & WOOD,
Erie, Pa.

Day's $60 Automatic Incubator
Hatches and rears the young chicks! Highest possible endorsements!
Day's Fire Alarm Clocks.
Run in any position; keep good time, and give warning when there is danger of fire. Sent by mail. Price $5.
Address DAY BROS. & CO.,
92 Linden Avenue, Baltimore, Md.

FIGURE 19-18. Steam engine advertisements in the American Agriculturist, December, 1877.

different boiler and engine combinations were available. By the turn of the century the small, single-cylinder gasoline engine became important to the average farmer as an all-purpose source of power (see figure 19-22). These "hit or miss" gasoline engines were much less expensive than steam outfits and also much more portable. They pumped water, ran washing machines, sawed wood and ground feed. About this same time gasoline-powered automobiles and tractors were beginning to appear.

The problem for manufacturers of gasoline-powered equipment was that it was difficult to convert many farmers from the old ways of doing things. Public competitions, usually organized by salesmen for the new tractors, often pitted horse-drawn or steam equipment against gasoline-powered rigs (see figure 19-23 and figure 19-24). The first really successful gasoline tractor was produced by Hart-Parr in 1902; interestingly, it superficially resembled steam traction engines. The 1909 Rumely OilPull tractor, well-liked in the Genesee Country, ran on inexpensive kerosene and it too resembled steam traction engines (see figure 19-25). Gasoline tractors were eventually accepted as the standard, of course. However, as recently as the 1920s, up to forty-four horses or mules might still be used to power a single combine with lead mules selling for as much as $600. Some tractor manufacturers

FIGURE 19-19. Tornado Windmill Company, Elba, Genesee County, New York. Combination Atlas Map of Genesee County, 1876.

When not in use machines such as the above hayloader needed storage either in a shed, a new barn, or the modified interior of an old barn.

FIGURE 19-20. *The Halladay Geared Windmill. American Agriculturist, 1884.*

FIGURE 19-21. *A cupola, which was an integral part of the structure on the Robert Moody farm on Route 245 in Rushville, Ontario County, supported an Aermotor windmill whose power was used to grind grain in the barn. The windmill has been removed but the stout supporting timbers and cupola remain.*

FIGURE 19-22. International Harvester Company Gas and Gasoline Engine. Typical example of a gas engine designed for situations where electrical power was unavailable. Country Life in America, April, 1907.

catered to this reluctance to part with old ways by carrying models equipped with reins instead of steering wheels.

Manufacturers of farm machinery sometimes published booklets admonishing farmers to "meet changing conditions," meaning that they should avail themselves of the most efficient power available. Horses, after all, had to be rested during the work day and had to be cared for in the off season just as if they were working. International Harvester's 1915 booklet, called "Farm Power," argued that fully one half of a farmer's expense arose from horses and their upkeep. (In that year the Department of Agriculture listed some twenty-one million horses in use, the purchase price averaging from $150 to $250 apiece.) The arguments presented for alternate power were very persuasive: for instance, "Farm Power" claimed that a tractor rated at eight horsepower cost $750, about half the cost of eight horses. A tractor could be operated by one person, required little shed or barn space, only needed attention during periods of use, and could be shut off with the turn of a switch. It was difficult to ignore the fact that tractors could do the hard work of plowing much more quickly; some argued that tractors did it better as well, and certainly at a lower cost. International Harvester determined the 1915 cost for a horse-plowed acre at $1.25, while a tractor could do the same work for thirty cents.

As gasoline tractors became more common, some barns were altered to accommodate the new equipment operated by the tractors. In the Genesee Country one of the most common changes in barn framing was a modification of beam support that began to appear about the time of World War I. Large knee braces replaced the original beams to permit easier access to the barn floor mow for storing or retrieving hay and straw bales. Advertisements for presses used to bale hay and other materials began to appear in the 1850s and became widespread by the turn of the century. For example, the P.K. Dederick Hay Press Manufactory in Albany, established in 1854, was by the 1870s offering thirty-four different press sizes for baling hay, straw, cotton, broom corn, hemp, moss, husks, rags, hops and so forth (see figure 19-26).

FIGURE 19-23. A 1910 plowing match with a Hart-Parr 30-60 "Old Reliable" pulling at least a six bottom gang plow. Very careful attention is being paid to the quality of the work. Anonymous photograph. Numbers such as "30-60" refer to the two methods of indicating the horsepower of the engine: the first number is a measure of the wheel horsepower, measured at the drawbar; the second, the horsepower measured at the flywheel (for belt stationary work).

FIGURE 19-24. Twenty-horse combine/threshing rig on the West Coast. 1901. Anonymous stereo view.

FIGURE 19-25. A 1912 30-60 Rumely Model E OilPull tractor. To help put the size of the tractor into perspective, the driving wheel is five and one half feet high. Rumely OilPull engines burned kerosene using an ingenious patented device. Rumely restored by Jim Erdle, Canandaigua, New York, 1984.

FIGURE 19-26. An advertisement for a Hendricks Hay Press in The Rural New Yorker, July, 1905. Baled hay is packed tighter so that a given weight of hay takes up less space in the barn.

A single example will give an idea of how barn framing was modified by these developments. About 1840 a 30 by 80-foot bank barn was built for the McNinch family, early settlers near Conesus who had emigrated from Vermont. Figure 19-27 shows the southwest corner of the barn with its interesting detailed doors (the lighter toned roof section indicates the new portion of the barn). About 100 years ago a cupola was centered between the two doors but recently had to be removed. A large granary is located inside between the doors. Sheep were kept in the basement portion in the beginning as they are today. (A wool house with its smooth board interior is still a part of the farm buildings.) On the south end the mow extends all the way from basement to peak.

About 1918 Pat McNinch enlarged the barn to 30 by 115 feet. The earlier combined hewn and sawn timber framing was of conventional post and beam construction. The original north end bent was reused by moving it to the new end position (figure 19-28 shows the random width boards of the original end). Not shown but clearly visible in the barn are the places where the original lower purlin posts were located; these squares have been floored over. These original posts were replaced by large knee braces, each one 5¾ by 8½ inches. Additional "parallel" braces produced a strong truss while the opened space meant that farm wagons with their bales could now be maneuvered to almost any part of the floor. This modification was repeated many times throughout the Genesee Country. Eventually frames were constructed in new barns with the knee braces an integral part of the design.

FIGURE 19-27. McNinch barn. Springwater-Conesus Road, Conesus, New York.

FIGURE 19-28. McNinch barn interior, modified about 1918. The angle of the braces provides access for gas-powered equipment.

FIGURE 19-29, Opposite. With the replacement of horse power on the farm the harness gathers dust in the barn, symbolizing the power revolution. Farmer's Museum, Cooperstown, New York.

CHAPTER XX

Truss Framed Barns

ALTERING an existing post and beam barn frame to suit new machinery or new space requirements had been the general practice, but the demands of modern farming in the late 1860s and 1870s became such that entirely new designs for new barns made the most sense. For many builders truss frames seemed to offer the prospect of larger barns, the economical use of materials, and the use of the latest innovations in crop, feed and livestock handling.

A truss is a combination of timbers or planks (rods or bars) arranged in a series of triangles to form a rigid frame. In the Chanler sidehill barn (see Chapter VII), the sawn truss was of considerable size and weight, producing a substantial frame while at the same time keeping the center open for the use of a hay track.

As larger and larger barns were needed, economy of materials became an additional important consideration. Sawing was not free and logs big enough for an unseamed post or beam might not be available on a farmer's own woodlot, making purchase necessary. Ingenuity suggested truss framing. At first the members of the earliest trusses were of considerable size, similar to what one found in post and beam construction. Later truss frame designs replaced the heavy timbers with framing members that were either relatively short or of small cross section; strength was maintained by the rigidity of the truss design. These newer trusses were ideal for second-growth woodlots. Trusses had already proved themselves in other kinds of construction around the world, especially in bridge building and most especially in railroad bridge building with its heavy load requirements. Heavy wooden truss frames could also be seen on the enormous walking beam engines of the early steamboats.

Because trusses in a variety of designs were used in many different applications throughout the world, it is not surprising that experimentation continued in the search for the ideal barn frame truss. In 1868 the *American Agriculturist* discussed the use of truss beams in general as a way of avoiding spatial interference from posts and beams. Ten years later the same publication printed an illustration of a "Danish Barn" with a truss frame, specifically citing truss railroad bridges as a model that might be adapted to farm buildings.[1] In the Genesee Country several barn builders submitted designs to the U. S. Patent Office for approval. Award of a patent meant that a design had merit in the eyes of the office, was significantly different from other designs submitted and would be protected from unlawful copy. In 1879, David Jennings of Lyons received a patent for a remarkable truss frame, with the flexibility to be either constructed as a new barn or incorporated into a standing barn.

Jennings' design was accepted across a wide geographic area, extending from its center in Lyons at least into Ohio. The arrangement was for local builders to purchase the right to use the design. In the Lyons area the Raymer brothers built a large number of Jennings Barns, many of

Opposite: Detail of the open interior provided by a Jennings patent frame. Circa 1880. Galbraith farm, Alloway, New York.

which can be identified from the outside by the circular window over each barn door (see figure 20-1). Jennings said his frame could be used in gable, gambrel or mansard-roofed barns but most of those constructed seem to be gable-roofed. Later truss frames were commonly used in gambrel-roofed barns.

FIGURE 20-1. The Rupert Raymer Farm Jennings Barn, built by Judson and Byron Raymer, Geneva/Lyons Road (Route 14), Ontario County, New York. The characteristic round windows above each of the main barn doors reveal that a Jennings frame appears within. There are barns with round windows that do not have a Jennings frame and Jennings framed barns without round windows, but in the Lyons area the two often appear together.

The Jennings/Raymer broadside includes woodcuts that show the method of trussing plates, the elevation of the trusses and a completed frame. Equally instructive is a 30 by 40 inch model of a Jennings Barn built by John Oetzel of Ohio. The model very clearly shows how the angle of the Jennings purlin posts produces an open center area in the barn (see figure 20-3).

So sure was Jennings of his design that he offered $1,000 to anyone whose barn frame "will excell my improvement in the construction of Barns: taking three things into consideration: Cost, Strength and Convenience for which my improvement stands unequaled as hundreds can testify that have them in use."[2] The principal strength of the design derives from the canted purlin post, which was a one-piece post from purlin to sill; this large cross section purlin post is the critical member in the truss and because of its size it reduced or eliminated the need for diagonal bracing in the Jennings design. The canting of the purlin posts provided an open center to the barn, making the hay track a very useful tool (see figure 20-5). As it happens, the design was also very strong, as Jennings indicated (see Appendix II for transcript of Jennings' advertising text). In post and beam barns modified to accommodate a hay track, the beams connecting the purlin beams were eliminated, weakening the roof and often causing the sides of the barn to spread outwards. The canted purlin in the Jennings design opposed the outward movement of the purlins, providing both a substantial roof design and an open barn center.

Jennings is very specific in presenting the various mechanical details of his system and does not hesitate to list boldly the reasons for using his patent frame (see figure 20-2). He states:

> 1st, it takes less timber for the frame; 2d, it is less work to frame it; 3d, it is easier to raise and takes less time. Four reasons why these barns should be built in preference to the old plan; 1st they are stronger and more durable; 2d, they are more roomy and convenient for stowing away grain; 3d, they can be built for less money than the old plan; 4th, they are just the barn a farmer wants for a grain or hay barn.[3]

If a farmer with an old barn was convinced of the advantages of the Jennings design, he received advice for altering his standing barn (see Appendix II for complete details): "Also bear in mind that old barns may have the beams and cross-ties taken out and this improvement put in at very little expense, especially where the purlin-

ATTENTION FARMERS!

THE JENNINGS BARN is acknowledgea by every one that has seen the frame to be the cheapest, strongest and most convenient frame now in use, without any exceptions, and is endorsed by all leading carpenters throughout the United States. As there are hundreds of them now in use, it sustains the fact that this plan has only to be seen to be appreciated. Remember that there are no cross-ties on the interior, or any obstructions whatever on the floor—the building being clear from one end to the other and from the floor to the peak of the roof, so that a horse-fork can be used to very great advantage.

PATENTED JULY 29, 1879.

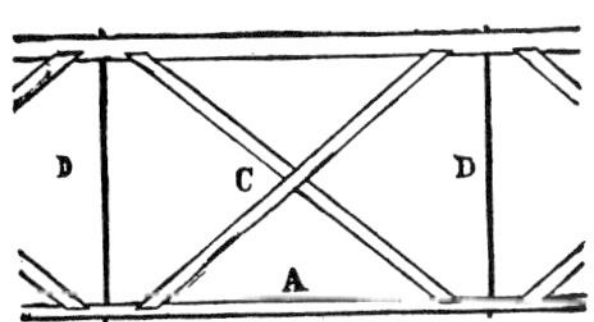

Fig. 2—Manner of Trussing Plates.'

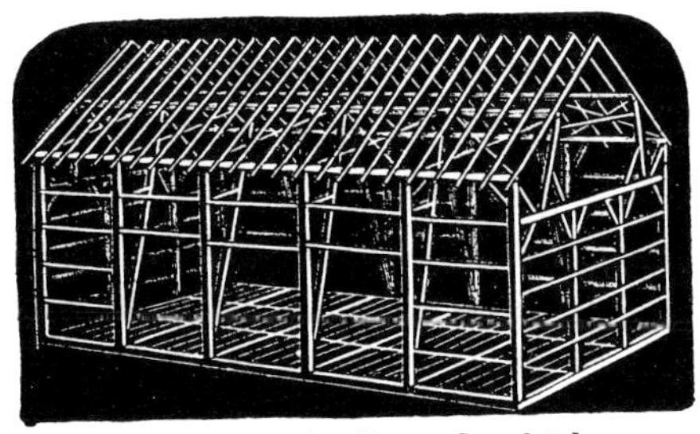

Fig. 1.—Showing Frame Completed.

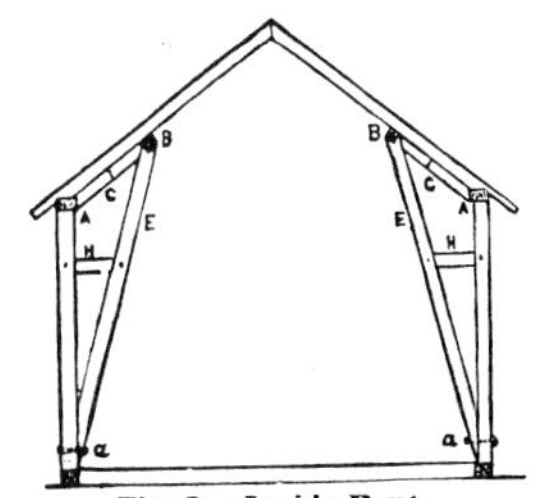

Fig. 3—Inside Bent.

In looking at the above cut, fig. 3, a person can readily see that the principle of this construction is such there is no outward pressure to the building after it is finished, but has an inward tendency, if anything; and in order for the building to spread it will not only have to break the truss-braces, purlin and main plates, but will have to lift the whole weight of the roof in moving in either direction. In raising these frames the end-bents are raised first, and staylathed, and then the main-post on the sides are raised one at a time or by pairs, putting in the girts before raising, and when raised stay-lathing them to their place; after the main posts are raised, then comes the purlin arm; bring that around in front of the post you are going to place it, put in the short girt that goes in the arm, and pin it and fasten a guy to the girt, and hand one end to a man that will climb up on the upper girt, who will hold it while the arm is drawn up by means of a pulley that is fastened to the top tenon of the main post; after the arms are raised they should be well stay-lathed; the stay-laths on main post and arms must be left on until the roof is finished, and then they can be taken away with safety, as the roof counteracts all inward pressure. The purlin-arms on end beams are set the same slant as the arms on the interior; the best way to connect the purlin-arms to purlin-plate is to notch the arm around the plate instead of a tenon; the only way they can move is sideways, and they can be spiked, which will prevent and hold them secure. Fig. 2 represents the manner of trussing or tieing main and purlin-plates together; the rafters should be well spiked to main and purlin-plates, as very great strength is derived therefrom. The X or truss-braces are gained into main and purlin-plates so that they may be dropped in after the plates are put in position; the plumb bevel being cut on foot of brace. In building a new barn I generally space off the bents equal distance apart, not exceeding twenty feet; in a twenty feet span it is not necessary to spread the X's over ten feet on plate, that is, find centre of span and measure five feet each way, that will find toe of brace. In tieing main and purlin-plates together I use a rod on every inside bent, but none on the end of building; that is, a building with six bents would require 4 rods on a side, and 4 short bolts for foot of purlin-arms.

A very quick and easy way to raise these frames is to raise the end bents first and then form a bent of one side with the main-plates on raising the whole side at one time—the end bents being leaned far enough over and stay-lathed to allow the main plate to pass when the side is raised. It is then stay-lathed and a bent then formed with the purlin-arms, with the purlin-plate placed in position. So it will be seen that these frames can be readily raised in six lifts. The ends of the main-plates are slit or mortised clear out to the end of the plate, allowing the tenon of the corner post to be the full width of the post, so that after the sides are raised the stay-laths can be taken off of the end bents—the frame being then brought together and pinned, thereby dispensing with the old way of climbing up to adjust the main and purlin-plates.

Three reasons why this barn is cheaper than the ordinary barns: 1st, it takes less timber for the frame: 2d, it is less work to frame it; 3d, it is easier to raise and takes less time. Four reasons why these barns should be built in preference to the old plan; 1st, they are stronger and more durable; 2d, they are more roomy and more convenient for stowing away grain; 3d, they can be built for less money than the old plan; 4th, they are just the barn a farmer wants for a grain or hay barn. Remember that this plan has been thoroughly tested in every particular, as regards strength, durability and economy. Also bear in mind that old barns may have the beams and cross-ties taken out and this improvement put in at very little expense, especially where the purlin-plates are already in, as is generally the case; but if not, they can be put in very easily. If the purlin-plates are already in it will only be necessary to put in the X's and the tie-rods, which can be done without disturbing the roof or siding. I generally fasten the toe or foot of these X's with spikes, as the roof being on prevents their being gained in as they would be in a new frame. It will readily be seen that by putting in the X's first and the tie-rods next, and then tightening up the rods you get the strength of the truss; then by cutting in the arm a square angle at the top, so as to fit around the corner of the purlin-plate, and working the bottom of the purlin arm, as shown in the above cut, and then by putting the arm in position as near as possible right up alongside of the beam and nailing a temporary block on the sill so as to keep the bottom of the arm even with the face of the vertical post on the inside of the barn, also to keep it from slipping while you cut out the beams and cross-ties, you can, when that is done, drive the arms in their right position by hitting them on the side with that what is generally called among carpenters a commander or maul. Thus the job is done without cutting any gains or making any mortises, except where the X's cross one another, and leaves the barn free from obstruction, and stronger than it was before.

☞ Every person should without fail see this Barn before building, and should investigate the best Barn plan in existence. They will not only save money, but will be well pleased. Anyone buying a right will be entitled to instructions that will enable any practical carpenter to build it without any trouble.

☞ FARM, TOWN AND STATE RIGHTS FOR SALE. Carpenters will do well to secure territory in time.

For further particulars apply in person or address, (enclosing stamp,)

DAVID JENNINGS, Patentee,

[Box 115.] **LYONS, WAYNE CO., N. Y.**

The undersigned have bought the right for the town of Phelps, Ontario Co., N. Y. Persons calling on them will be shown these Barns. For further particulars address

JUDSON RAYMER,
or **BYRON RAYMER,**
LYONS, WAYNE CO

FIGURE 20-2. Jennings Barn in a Judson and Byron Raymer broadside listing the qualities and particulars of this frame. Collection of Daniel La Gasse, Lyons, New York.

plates are already in, as is generally the case; but if not, they can be put in very easily."[4]

An excellent example of the Jennings Patent Barn, selected from the many that remain in the Lyons area, stands on the Richard Galbraith farm in Alloway (see figure 20-4). This 34 by 75 foot basement barn replaced a barn that burned in the late 1870s. The open interior provided by the canted purlin posts is amply demonstrated in figure 20-5. The sawn purlin posts are 6½ by 8 inches while the main frame posts are 8 by 9 inches (see figure 20-6). The trusses that connect the main plates with the purlin plates are shown in figure 20-7.

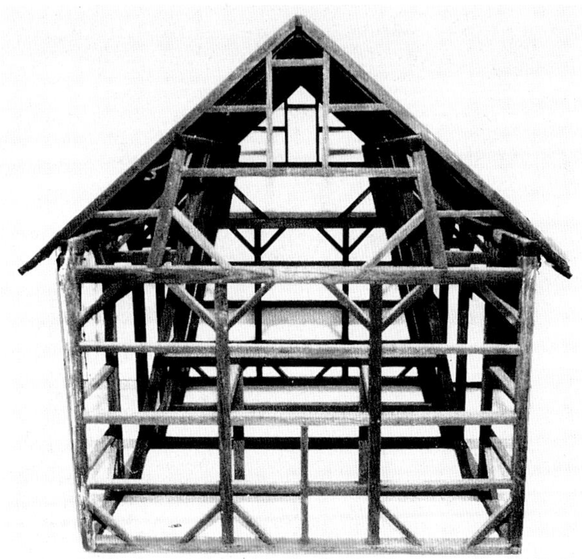

FIGURE 20-3. A 30 by 40 inch model of a Jennings Barn, constructed by John Oetzel of Canal Fulton, Ohio, now at the Milan Museum, Milan, Ohio. The white space in the center of the model indicates how open the barn center is with this patent frame. Photograph by Roy Cooley.

FIGURE 20-4. This dairy barn is typical of many Jennings Barns in the region with a gable roof, louvered rectangular ventilator in the gable end, and the circular windows centered over the barn doors. Richard Galbraith Farm Jennings Barn, circa 1880, Alloway, Wayne County, New York.

Jennings encountered patent infringement, which he confronts in one of his advertisements: "Any person willfuly infringing and concealing the same to defraud me out of my rights, will be prosecuted to the full extent of the law."[5] A person who understood building techniques probably understood how to construct a Jennings Barn after just one glimpse and would be reluctant to pay for an idea that now seemed so logical and simple. Some builders made changes, creating "new" designs free from patent obligations, and once confidence grew in truss designs based on years of experience, the concept persisted. However, it should not be assumed that post and beam construction disappeared with the advent of the truss frame, because that is clearly not the case in the period represented by this study.

Once considered the largest barn in Ontario County, the Robert Moody barn on Route 245, Rushville is a 50 by 110 foot gambrel-roofed barn with a truss frame (see figure 20-8). One of the major differences between this frame, built either in the late 1870s or the 1880s and a Jennings patent frame is the inclusion of the beams tying together the purlin posts. A truss which runs the length of the barn supports both the ridge and the hay track. Interestingly there are sawn and hewn purlin posts. The 8 by 9 inch hewn purlin posts are continuous from the sills to the purlins while the sawn posts are joined at the intermediate loft level. As with the Jennings design the center of the barn is open for easier storage of loose hay.

By the first decade of the twentieth century

FIGURE 20-5. Galbraith Farm Jennings Barn. The interior is partially filled to the purlins with baled hay for the dairy cows. The vertical rectangular structure in the center permits bales to be dropped to the level below when the barn is filled. The now unused haytrack is visible just below the peak.

FIGURE 20-6. Detail of the 6½ by 8 inch purlin "arms" or posts in the Galbraith Farm Jennings Barn. Note that the horizontal member to the right of the first post is a new addition, not required for the strength of the frame.

FIGURE 20-7. Detail of the trussing, which connects the main plate with the purlin plate on the Galbraith Farm Jennings Barn.

FIGURE 20-8. This truss barn was built for Henry M. Boardman in 1876 by Philip Wheeler, Daniel Paddock, Asahel Botsford, Frank Dicksen and Joseph Lee of Branchport. Robert E. Moody, the barn's present owner, indicates that a Van Sickle hay track manufactured in Shortsville was used in the barn. This 50 by 110 foot barn was considered the largest barn in Ontario County when built. The endmost purlin posts are continuous 8 by 9 hewn posts while the interior purlin posts are sawn and joined at the loft level. In fact the lower purlin post is set at a different angle from the upper segment. This change in angle can best be seen near the door on the lower right of the photograph. In addition the lower purlin segments are edge beveled.

major publications carried truss designs under various names. Halsted's *Barn Plans and Outbuildings* of 1881 was revised in 1903 by the *American Agriculturist's* editor, Edwin Powell, who included what he called "plank frame barns" among the new offerings.[6] K. J. T. Ekblaw's *Farm Structures* of 1917 also contained an extensive discussion, complete with formulas and diagrams, of the plank truss frame (see figure 20-9).

A comparison of these plank frames with the Jennings frame shows some changes, including the descent of the purlin post below the first floor sill and additional trussing to support the roof ridge. Powell advises farmers, because of "the scarcity of heavy timber and consequent cost," to "give study to the newer methods of framing, where no timber is thicker than two inches, and from six to eight inches wide." Powell continues:

> The plank frame which is here illustrated is the newest thing in barn framing, and at the same time is very much stronger than the old-fashioned frame made of square timber of eight to twelve inches on a side. It is about half as costly, and a first-class carpenter is not required to erect it. Where additional strength is required planks are spiked together, two or three as needed.

The tenon was the weakest part of traditional post and beam framing and therefore, Powell argues, the smaller members of the plank frame are still almost as strong as those of post and beam construction, especially in situations of tension.[7]

The bents for a plank barn are constructed on the ground with the posts located near their proper places along the sill. The last made bent, the end bent, is then raised first, set plumb, and braced. The next bent is raised, plumbed, and the side girts are spiked to the bents. The additional bents are then raised in turn. The plank design demonstrated that small members, spiked together in a truss pattern and constructed by a small crew could produce a strong, durable, economical frame. Jennings and Powell were right.

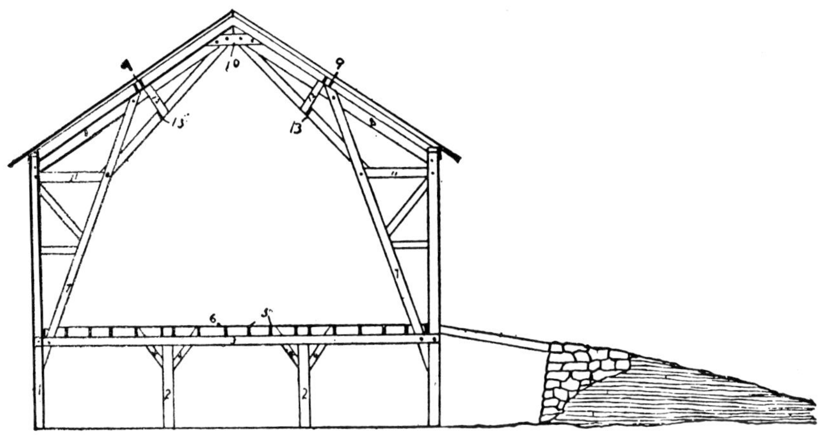

FIGURE 20-9. Interior bent of a basement barn from Halsted (Powell), Barn Plans and Outbuildings, 1903.

No plank frame exactly like the Powell model has been located, but a fairly close example exists in Livingston County on the former Merrimac farm, now owned by the International Salt Company. The truss frame of this gable-roofed barn is only slightly different from the Powell design (see figure 20-10).

Figure 20-11 reveals the diagonal bracing of the barn ends, producing almost a basket weave design as the diagonal weaves over and back around the girts. In addition to this weaving, the positioning of the end braces differs somewhat from

FIGURE 20-10. Truss-framed, Merrimac/International Salt Co., gable-roofed barn, Route 63, Geneseo-Mt. Morris road, Livingston County, New York. The truss frame is just barely visible through the open barn door.

the Powell model, yet the character of the Powell design is still maintained. The critical trussed bents are almost exact replicas of those suggested by Powell (see figure 20-12).

The roof rafters are stabilized with short collar beams, and lateral roof stability is augmented through the use of diagonals connected to the main frame plate. Stability of the structure as a whole is further increased by the use of trussed purlin plates in the two end bays. This particular barn frame design is exceptionally rigid.

Sawn timber truss-framed barns continued to be built in large numbers through the 1930s, incorporating design changes required to accommodate the increasing mechanization of the farm. Truss framing similar to the Jennings patent and the Powell plank frame was used by many who recognized the benefits of this construction. In 1912 on the Crossett Farm in Geneseo (William Crossett was one of the first settlers in the Geneseo region), a new barn was built by W. Austin Wadsworth from a wooden truss design (see figure 20-13). Architects Edward Burnett and Alfred Hopkins of New York City were engaged; remarkably, both contract and blueprints are preserved at the Wadsworth Homestead office. The barn itself is still in daily use. There is a strong similarity between this frame and the plank frame shown by Powell, but details do vary. According to his son, William, W. Austin Wadsworth examined a model of a balloon-framed barn in the Homestead office before settling on the truss-framed design.[8]

The Crossett Farm dairy complex consists of a main barn connected to two smaller ones on either side of the barnyard (see the aerial view in Chapter XVII). The main barn is a basement hay barn; the cypress shingling on this and the gables of the side cow and horse barns produces an exterior unique to the area. Interesting details include the Palladian windows in the gable ends and the cupola with its round-headed louvering. The

FIGURE 20-11. "Woven" diagonal bracing in the Merrimac/International Salt Company barn.

original cupola, shown in the illustrations, has since been destroyed by a storm. According to the contract, the trusses were to be made of Georgia pine, the roof of slate. The main barn foundations were concrete except for the masonry piers, and four-inch concrete floors were poured in the horse and cow barns. The walls of these side barns are built of "good hard burned hollow tile block 6 inches thick, ..produced by the Pennsylvania Fireproofing Co." This tile block is not only fireproof but easily cleaned, an important factor in a dairy operation.

The main barn blueprint shown in figure 20-14 (reversed to blackline for clarity) shows the trusses and figure 20-15 "translates" the blueprint into the actual barn interior. Of interest is the

FIGURE 20-12. The built-up plank frame bent in the Merrimac/International Salt Company barn is almost an exact duplicate of the Powell design.

FIGURE 20-13. Crossett hay barn. Crossett Road, Geneseo, Livingston County, New York.

method of supporting the purlins via the truss construction. The barn studs are two feet apart on center with horizontal nailing strips for the shingles.

A later example shows a typical truss construction used following World War I. This one is a gambrel-roofed barn. Violent winds in 1926 knocked down the original barn on the Wingate Farm in Livonia and in 1929 a new barn was constructed on the original foundation (see figure 20-16). The truss framing of the new barn is quite similar to that of the Crossett barn and the Powell plank frame, and rather typical of others built during the period (see figure 20-17). The Wingate Farm, long a dairy operation, had need of sizable fodder storage capacity and thus the gambrel roof was elected. The type of truss employed produced many variations in the basic proportions of gambrel roofs, variations which have combined to produce a rather visually diverse countryside.

A fuller understanding of truss barns will come about when the varieties of these designs can be more thoroughly explored. Information such as the numbers of each truss type and their geographic distribution will prove extremely useful.

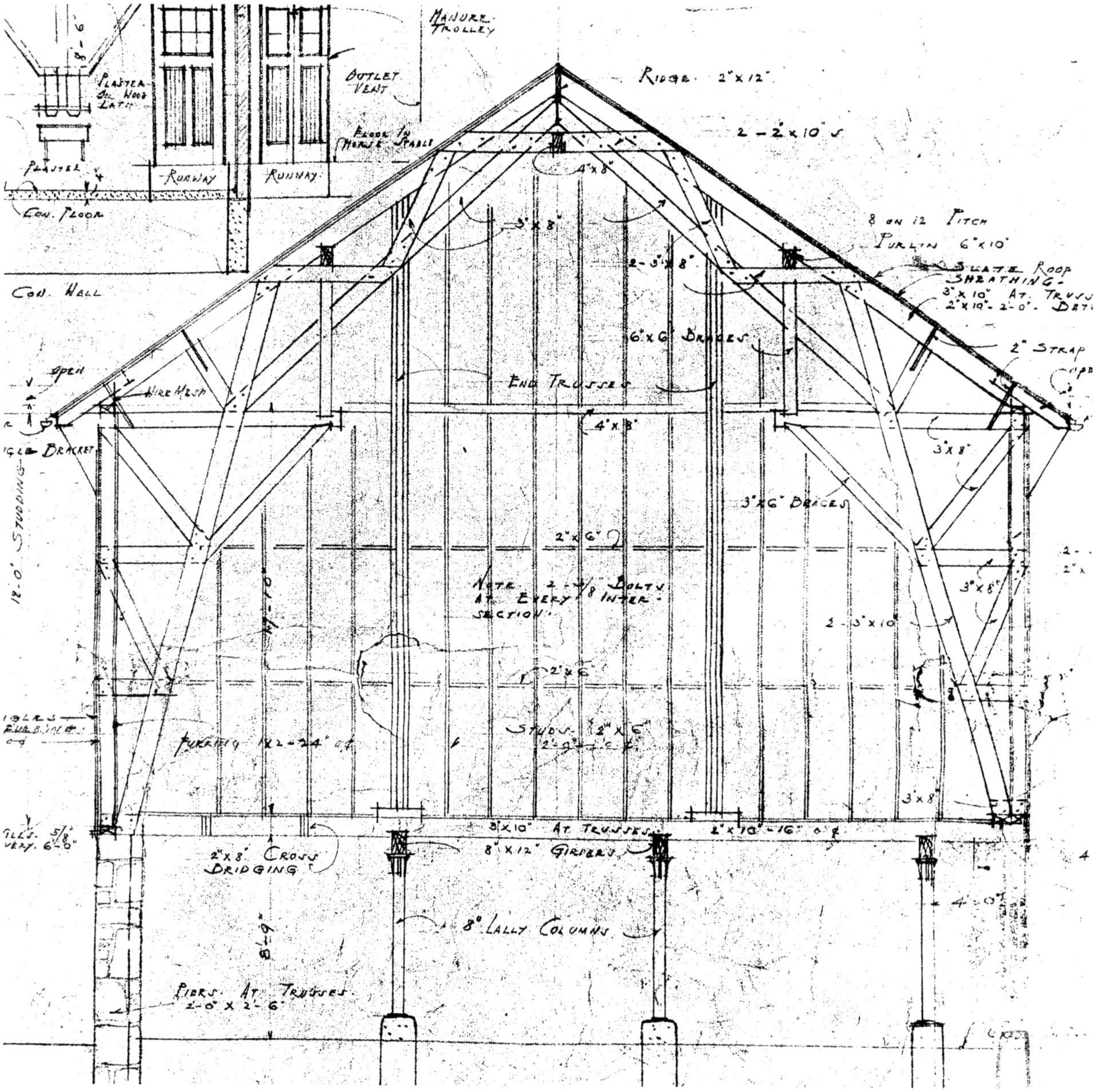

FIGURE 20-14. 1912 Crossett barn "blueprint." Wadsworth Homestead Office, Geneseo, New York.

In many instances architects begin to be associated with designing truss barns, such as Burnett and Hopkins in the case of the Crossett barn. A sampling would include the architect-designed stable with its intricate trussing system built in Lyons in 1892 for Van Rensselaer Richmond, the state engineer who had been in charge of rebuilding the Erie Canal. Wolcott Humphrey's impressively large and well preserved carriage house in Warsaw, New York was designed in 1884 by the Rochester architect A. J. Warner. The list of architect-designed barns and stables could probably fill a volume by itself.

Most examples of the Jennings frame and the other examples in this chapter were gable roofed. In the following chapter another patented truss frame, especially designed for the gambrel roof, is discussed.

FIGURE 20-15. Crossett barn interior view of trusses.

FIGURE 20-16. *George Wingate truss barn, Big Tree Road, Livonia, Livingston County, New York.*

FIGURE 20-17. *Wingate truss barn; 2 by 10 inch by thirty foot oak planks form the basis of the trusses.*

DIARY

Seneca A. Owen, Pulteney, Steuben County, N.Y., August 1908–February 1912

TURN OF THE CENTURY AGRICULTURE was becoming more and more diverse with the introduction of "new" cash crops such as grapes, beans, hops and tobacco. Grapes were the principal crop of Seneca Owen, who faithfully kept a diary from August 1908 to February 15, 1912 (there may be other undiscovered volumes). As Seneca records the seemingly endless work required to make his vineyards productive, barns of all sorts appear: Hay is constantly being moved in and out of the mows, equipment is stored in a vineyard barn, and Seneca builds a new packing house with a concrete floor.

As with the previous diaries, no attempt has been made to change the spelling or add to the punctuation. Seneca Owen wasn't big on punctuation, and he spelled phonetically; when the reader can't make sense of a word, saying it aloud will usually restore its meaning. The reward for such persistence is entry into a time when the first automobiles were still rare enought to record; when freight and people usually travelled in horse-drawn vehicles, electric trolley cars (Seneca always just refers to "the car"), or lake sidewheelers; when telephones and electric lights were largely confined to cities.

Here are some additional facts about the "Crooked Lake" (Keuka) region to help explain some of the diary entries:

About grapes—

The very first vineyards in western New York were planted in Chautauqua County in 1818. About 1830, grape slips from the Hudson Valley were brought by the Rev. Wm. Bostwick to Hammondsport, where the growing conditions were similar to those of the wine growing region along the Rhine in Germany. After Bostwick's Catawbas and Isabellas proved successful, Andrew Reisinger established the area's first commercial vineyard on two acres in Pulteney. In 1861, the first U.S. winery license was issued to the Hammondsport and Pleasant Valley Wine Company, and the Taylor license followed in 1880.

FIGURE O-1. William Carr's 1908 box-label for Lake Keuka Grapes shows a map of the lake and surrounding communities. Seneca Owen's farm is located on the far side of the lake just to the right of the "Y" in the word Pulteney. The label is from the collection of Evelyn Cole.

By the 1890s, 25,000 acres were planted to vines, and during the fall harvest the Bath and Hammondsport Railroad operated six trains a day to haul the grapes. The Pleasant Valley Wine Company shipped its first champagne in 1865 and, under the direction of the noted French champagne maker Joseph Masson, won prizes in Paris in 1867, 1889 and 1900. The company later became Great Western Producers, Inc.

About transportation—

The electric Penn Yan Keuka Park and Branchport Railway Company was organized on February 23, 1879. Business began with four open passenger trolley cars, but soon two closed cars were purchased for cold weather operation and more cars were added as business grew. The freight motor, used to haul flat and box cars, was known as the "jack car." "Even Branchport in those days furnished considerable freight business (as well as passengers), as it was not on a railroad . . . The wineries also furnished traffic, with box cars loaded with casks originating from the trolley sidings at the wineries in fall and early winter."[1] Of course a powerhouse was needed and in 1901 the line was equipped with telephones.

In 1908, Glenn Curtiss' first plane, "Red Wing," briefly flew from the ice on Keuka Lake. On May 22 of that year, Curtiss piloted a rebuilt plane from the track at the Hammondsport fairgrounds, because there was no proper airport. In November he experimented with a float plane, again due to the lack of an airport. On August 29, 1909, Curtiss won the air race at Rheims, France and in 1910 opened a flying school at Hammondsport.

Miscellaneous—

It is just about 8 miles from Pulteney, opposite Bluff Point where Keuka Lake divides, to Branchport; Hammondsport to Penn Yan is 24 miles as the crow flies; the

Lake Keuka Navigation Company was purchased by the Erie Rairoad Company in 1905.

In the vineyard, the growing vines are tied to wires strung between posts; the first tying is with willow, the second with rye straw. The vines are trimmed in winter.

Names of some of the "Crooked Lake" sidewheeled steamers operating during the time of the diary were *"Halsey," "Yates," "Mary Belle"* (which became the *"Penn Yan"* in 1905 and was considered "Queen of the Fleet"), and *"Farley Holmes."* The *"Mary Belle"* made the Hammondsport to Penn Yan trip in a scheduled one hour and fourteen minutes. She carried almost 200 electric light bulbs strung between her masts and was a marvelous sight on the lake at night.

THE MAIN PEOPLE IN THE DIARY:

SENECA OWEN
TOPSIE OWEN, his wife
HARROLD, younger son
COOK, a man for whom Seneca worked
GLENN, older son
NELLIE, Glenn's wife
Gim & Dell, hired men

FIGURE O-2. Seneca Owen's farmhouse and barn as seen from the vineyard; Keuka Lake appears in the background.

Sunday, 8/16/08:	*at home was a nice day Glenn went to willow grove This afternoon I went down to Vinyard a little while*
Saturday, 8/22/08:	*Harrold & I cut the Oats down on Ed McConnels set up part of them rain vary hard with Thunder & lightning put machine in vinyard barn & Plow in sod back of house until night Harrold Topsie & I went to town this eve,*
Monday, 8/24/08:	*I help wash in foornoon & finish shocking Oats on Eds. Glenn Plow afternoon I work at shed Glenn finish Plowing then we go to Wageners get load sand bank in bad shape cool nice day*
Friday, 8/28/08:	*Glenn & Harrold drew 10000 Handles covers & hardware for same from Italy Hill in forenoon I went to town then begin Cleaning out shop. The Boys took Crates mill home this morning also took 3 bush Timothy seed to him. in afternoon they went up to Dewit Babcocks got 5100 Baskets I worked getting shop ready for Packing sent one Bush Timothy seed up to Schyler Hibbards. by the Boys was a vary nice day cool*
Saturday 8/29/08:	*Glenn & harrold went to Italy hill & got 4900 Baskets making Totle of 10000 I churn & handle some Baskets. Afternoon all went to Ball game betwean Spaldings of Corning & the Pulteneys at Pulteney Scores 8 to 0 in favor of spaldings Glenn went to town again this eve*
Monday 8/31/08:	*was a hot day 100 degrees in afternoon I help wash in foornoon Harrold draged the lot back of house Glenn went to sullivans engage Jennie to Pack Grapes for us then He handle some Baskets. in afternoon we drew in our oats for our new ground & mow the Peas & weeds around Oats stubble*
Monday 9/07/08:	*went to shop had Tony shod Glenn helped wash. I work at shed after I get home from shop. Glenn help me in afternoon Harrold drag the Piece west of house let John Lamphire have 10 bush wheat $10.00 & 2 Bush Timothy seed $6.00 did not Pay for it was a fine day Glenn went to town this eve.*
Tuesday 9/08/08:	*Glenn drill the wheat back of house & help me on shed bal of day Nate Miller Pay for His seed today*
Thursday 9/10/08:	*Glenn went to Pennyan to Fair I work on shed was a vary hot day 102 degrees at 1:30 p.m.*
Friday 9/11/08:	*hang doors on shed fix bridge down by Peach orchard. Clark wright & Cap Hyna were here today to dinner I ship Tub butter to Cook today I Hire a man today for grapes etc. $1.00 & board was a hot day Glenn & I went to lodge the man*

Saturday 9/19/08:	*took Bal. Concords to Dock that were Picked the Boys Pick Dels (Delawares) Tom went to Pennyan Pay Him what was coming to Him. went to shop had Sadie shod at Halleys chd. then took Jennie Sullivan & Maud Woodhouse home.*
Wednesday 9/23/08:	*Jen Pack Tom & will Pick Glenn & I went to Prattsburg & to the Fair with Grapes & Peaches was a hot & Dusty day we did Fairly well around the village Nothing at the Fair no crowd. Stored 235 Baskets Grapes at Will McConnels tonight for Tomorrow home about dark*
Thursday 9/24/08:	*Glenn Harrold & I went to Prattsburg to Fair with Grapes & Peaches vary hot day sold out Peaches & some Grapes*
Wednesday 10/07/08:	*Glenn will & I Pick catawbas clean out lake cellar & draw in 86 boxes Jennie Pack some grapes in trays to ship to Newark. turn cows on the hill this morning & around by cemetary warm day south wind getting cloudy Jennies time tonight 18 days*
Saturday 10/10/08:	*in foornoon I took 200 Baskets Concords to Dock. then had team shod at Holleys Pay for it. Glenn & will Pick. Afternoon we drew in whatt was Picked & Pick & draw in what more time we had heavy white frost last night & freeze in some places rain at night we Pick the last tree of Peaches when we came up tonight*
Wednesday 10/14/08:	*I drew grapes for Frank Hathaway from vinyard to Roffs Dock Albert Walter & I Glenn & Topsie put up some Grapes in trays to go to Newark & did the washing I used Mid & Ladie Tony is lame got a rusty nail out of His foot tonight*
Thursday 10/29/08:	*in morning we took lumber wagon Drill & Roller over to Boyd Barn then help the Pressers draw their Press over & we Pressed what Hay —— tons from the north mow Deign helped us also one hand from Fossets the Press went away after supper.*
Saturday 11/21/08:	*in foornoon I went to lake got some Grapes Topsie Packed 12 baskets to ship I Cooks made crate & crate them ready for shipment Glenn Plow on His place Afternoon we all went to Prattsburg I drive my team. Mr. Elmer Earley Payed me for His Timothy seed today. I settle Cooks lumber Bills at Ringroses & Clarks also. this afternoon was a lovely day*
Tuesday 12/08/08:	*in foornoon I Plowed snow a little in afternoon help fossett Press Hay after Press came high west wind & cold. Cornwells man came I payed bal in full for Piano warmer tonight*
Monday 12/14/08:	*Mooved baled Hay from floor to west end of bay ladie was sick went to town got some medicine for Her. Harrold help me with Hay after school nice day ladie all right tonight*
Wednesday 12/16/08:	*this morning drew load of Hay from my stack to Boyd barn to feed Glenns Heiffer & my calf. then measure the Hay in my barn found*

	7 tons of undivided hay. Then I went to town & got some Groceries. Afternoon I Plow on Ed McConnels no frost
Saturday 12/19/08:	*get out Bobs draw manure from cow stable to vinyards. bought tank for Kerocine Oil the Boys brought it up from Hardware Glenn & I draw Oil from Bbl. & put it in tank tank cost $5.00 drove to town in foornoon got new stocking Rubbers for Harrold Pay for them bring couch mattrass & morris chair from Hyatts to Boyd house for Glenn. all went to town this eve. fine sleighing Glenn & I went to lodge. Not vary cold.*
Sunday 12/20/08:	*was a beautiful day Topsie Harrold & I went to Church Glenn went to Pennyan in afternoon. Topsie Harrold & I went to Church again this eve. fine sleighing good sermon*
Wednesday 12/23/08:	*Put Plows under shelter in Vinyard Barn moove some of clover hay in Barn & draw in Part of straw stack was a beautiful day*
Sunday 1/31/09:	*Snow & Blow some in foornoon & colder & growing cold Glenn went to Bran chport met Nellie Hulse at car. drove to Pulteney to Elder Browns. The Methodist ministers & they were married then came home to Dinner soon after Dinner Glenn drove & took Nellie up to her Folkses for Her to bid them goodby then I took Glenn & Nellie to Branch Port to car to start for a short trip on their Honeymoon was bright but cold 4 degrees above tonight*
Tuesday 3/02/09:	*Glenn & I weigh out & Divide the vitriol & take His & mine for Cross vinyard over to Glenns barn. then He & Nellie went to their Own home to stay I rig Plank on wide tire. afternoon I clean out cow stable drew two loads to Peach orchard left one on wagon as it was night Harrold went to school.*
Thursday 3/11/09:	*cold snow & blow Dell Grease Harnice & make Planting Pegs I went & drew up ice for the market Boys from the Dock House pretty hard Deal on Horses Glenn Albert & I from this way drew I made 6 trips had my dinner at Ned Cases.*
Thursday 4/08/09:	*wind still Blowing not quite so hard. but has blown vary hard during night blew our Privy down about middle of garden. Dell & I finish Pulling Brush on my vinyard then we drew Privy back & Albert & Charles helped us set it up & then we nailed in the side which was brooke out by the Roll. Harrold & Dell went to town this eve, a few moments*
Friday 4/09/09:	*get wagon ready & Dell draw manure from cow stable & basement to the Peach ground there was about 2 inches snow on ground this morning & snow squalls all day I work in shop some repair some harnice put handle in Dell's maul that He brought up last night from town the handle had to be fitted I drove Daisey with lines Some in afternoon Snow & blow terrible in eve. Glenn was over for His milk.*

Sunday 4/11/09:	*was a nice day Glenn & Nellie were over to dinner with us. all went to Prespiterian Church to easter exercises in eve.*
Saturday 4/17/09:	*rain this morning Dell & I finish cleaning our Oats then I went to Dock & got some Freight from Montgomery Ward & co. Harrold went with me Glenn got His after dinner from the Order Dell tighten wires on my vinyard while I was gone. Afternoon Dell tighten wires Glenn Harrold & I went to Geo. wageners Auction I bought a hammer 55 cents nice Afternoon all went to town this eve.*
Wednesday 5/05/09:	*Ada & Charles Walters Dell & I finish willowing my Vinyard except the 9 rows young vines the wire is not on yet. I run out of willows went up to Nate Millers got 98 lbs willows let Glenn have 70 lbs of them. Ada Northrup did not help today*
Friday 6/18/09:	*Clarrence Dell Harrold & I strawtye on cross vinyard in foornoon cold & strong northwest wind. in afternoon clean out garden. Harrold cultivate Potatoes on hill. Dell & Clarrence sprout clintons after we get done with garden*
Monday 6/21/09:	*I work at Sprayer, in foornoon Dell Sprout Clintons Ada Northrup tie & harrold help her after helping wash: Afternoon Dell & I finish young vines except one Row did not have wire for that. then I help Ada & Harrold Dell sprout. Clarrence Finger work this Afternoon He sprout in Clintons Topsie Harrold Dell & Glenns Boy went to town this eve, to Medicine Show.*
Thursday 7/01/09:	*all tie Vines until noon. Afternoon Mifs Patch & Bertha went back to Fossets Payed Mifs Patch $5.00 in full could not make change with Bertha Dell Harrold & I rake & put up the clover & get Barn ready for Hay on south side Mrs Drew tied vines*
Monday 7/05/09:	*was a fine day Harrold tied vines Dell & I Sprayed on the cook vinyard Dell & I went to Dock this eve. got some Freight Glenn went to town with us had fireworks to Glenns when we got back from Town*
Thursday 7/08/09:	*was a nice day strong wind from North Dell & I drew what Hay we had cocked & cocked what was cut & I cut the Piece by Cemetary Glenn sent man over toward night help Dell cock. Albert came over after Choores helped Dell & I finish cocking Harrold raked drove on Rope also took mare over & drew off for Glenn Harrold & Topsie went to town after supper a few moments Bertha Miller & Grace Patch were over from Fossetts this eve, the man came with the Pictures He had been finishing for us I Payed Him for His work $2.50*
Tuesday 7/27/09:	*at home not able to sit up much had my clothes on all day however. the Dr. came again this eve. said I was doing well. Dell came & worked today He & Harrold drew in wheat rakings & fix some fence go to mill & go & get Basket Handles & bring up some for Glenn also brought 8000 to Glenn & 16000 home. hot day.*

FIGURE O-3. Looking northwest with Owen vineyards in the foreground, the barn and house on the right and the small cemetery so often referred to just visible in the sunlight near the grove of trees on the left.

Wednesday 7/28/09: *am feeling a little better Dell tuck Clintons & help Glenn load some Hay Harrold helped Dell some & went to town with Topsie. Albert Walters was over to see me this afternoon a little while Glenn & Nellie were over this eve. has been a hot day Dell drove ladie to town this eve. was not gone long.*

Friday 8/05/09: *Dell shock Oats for Glenn until near 4 P.M. get through then they draw my Oats up from Ed McConnels & put them in Glenns barn Ed & I divide in the shock this year He drawing His Part home 2 loads apiece from about 7 achors. Harrold & I drove up to James Prossers & engaged some grain for feed in foornoon. smith Dailey was down & Painted some Rooffs the Kitchen east Porch & the shed Roofs. was a hot day. in afternoon toward night Topsie Harrold & I went down & called on Mrs. Cook. then to landers & bought a Basket Handling machine $4.00 did not pay for it: Dell went to town this eve.*

Monday 8/09/09: *Dell helped Glenn draw his oats & a little Rye Sprinkle several times I am better but quite weak. in after noon Dell dug some trench for new building I went to mill with Harrold took some grist for Glenn & me. left some grain to Glenns for Pigs I took some boards to mill for Heds & Handles for Picking Boxes had some irons mended at Deans put Handling machine togather when I get home*

Tuesday 8/10/09: *Dell dug trench Harrold & I go to lake get ladder Paint some Roofing etc. from cotage then Harrold drew stone for trench Dell filled stone in trench then helped Harrold draw I helped level up a little. I clean out shop some & handle some baskets. nice cool day*

Monday 8/16/09:	*A Baby Boy came to Glenns this morning rain during latter Part of night Topsie & Ms. Hulse were to Glenns. Dell & I with Fossett & man worked on main vinyard Road our team & theirs on Road scrapper about ¾ day then Dell went up to Harry Fosgates & got manure Spreader. I went to town for Glenn. Nell & Babie booth smart rain hard tonight some Thunder & lightning*
Tuesday 8/31/09:	*Dan Tyler came & helped put up wall Dell & I helped Him in foornoon Afternoon we helped Foscett thresh Dan & Harrold finish the wall. for shed & Packing wall.*
Wednesday 9/01/09:	*Dell & I help Fossett finish threshing then to my place Foscett & man Glenn Dave Steever Gene Northrup & Lew Hopkins help me had 58 bush Oats one rye 260 bush wheat got done & mooved to Glenns one of threshers stay here all night the Others go home. had my Democrat to drive Harrold went to Branchport & got Glenns new man this afternoon cool day*
Thursday 9/02/09:	*thresh over to Glenns my Oats I raised on Ed McConnels 60 bush. Glenn grain 242 bu Oats Rye 4. Dell help over to Steevers then we draw our straw from Glenns up to my Vinyard barn clean it out & put straw in thare then drew 4 cords wood to Fred Browns man put Binder in barn Dell went to town this eve I went over to Glenns a few moments & Harrold a little warmer today*
Tuesday 9/14/09:	*Dan Tyler came & put cement botom in Packing house Dave Steever help & Dell & I. I worked at siding what time I could get. Dell & I had to go for a load of sifted gravel in afternoon was a vary hot day. Frank Hall came this morning & got His wheat & Payed for it $36.50. Payed Albert Walters $10.00 I borrowed of Him some time ago. went to town this eve payed Ida Osburn $1.33 amt. of my acct. Glenn rode down with us. also sent some money Orders.*
Wednesday 9/22/09:	*Dell & Geo. finish sowing wheat in my vinyard & draging New man (Dick) put wood in woodshed, in foornoon Afternoon pick a few Grapes in my Concords. this afternoon. Nice day. Miss Mapes paste labels today I work at Packing house in foornoon. borrow canvas of Glenn & albert. & put on Packing house Roof. in foornoon after Boys got done sowing & draging*
Thursday 9/23/09:	*Harrold went & got Jennie Sullivan to Pack we began this morning Dell & I went to Peach orchard but found our Peaches had been Picked by some unknown Party. Geo. & Dick Picked on my Vinyard Dell & I drew in load when we got back from vinyard. in afternoon I went to mill & Had Bill & Tony shod at Deans Pay for shoing & grinding rain did not bring grist home rain vary nice & still coming when we go to bed. Pay Mit Hall in full for Cobling Glenn got 4 Pigs from my place 2 I gave Him & 2 He is to pay for at $2.00 each.*

FIGURE O-4. Loading train with grapes in baskets in Penn Yan, New York. Each car could hold 5000 baskets. Collection of the Penn Yan Historical Society.

Friday 9/24/09:	*in morning hook load & take to Dock 698 Baskets concords. from my vinyard Consigned to wise the Boys began Picking Apples when I got gone. brought grist home. Today went to Branchport in afternoon got Roofing for Packing house & got medicine for Topsie from Castilo Pay for it 50 cents quit Packing at noon take Jennie home this afternoon. Harrold went from school with me. Dell & I went to lodge Glenn went with us Miss Mapes went over & stay with Nellie pay will Caryell in full for windows & frames for Packing house $2.00*
Saturday 9/25/09:	*Dell & I put the Roofing on Packing house was nice day Geo. & Dick finish Picking Apples Harrold had to go to town & get some nails & some Roofing as we lacked a little then He & the Boys finished drawing in the apples & put them in west end of North shed about 70 crates did not pick up windfalls went to town this eve. the Boys went payed Geo Dean $7.50 in full for His weeks work let Dick have $2.00 making $3.00 He has had Harrold took Bal. grapes Packed to dock this morning at 6½ cents for Andrew McKay Miss Mapes rode to Dock with Him went to Pennyan*
Tuesday 10/12/09:	*in foornoon Albert & I went to Walter Taylors with our Agawams I took part of His load Harrold went with me as there was no school Dell Geo. & Dick Pick concords on my Vinyard cool day some squalls of snow or sleet. I drew in concords afternoon 111 boxes Jennie Pack*

today Harrold & I go to town this eve get some things Box stationary for Jennie 25 cents Glenns help was all over this eve.

Monday 10/25/09: *Dell & Dick gather apples around house & the garden stuff cut Popcorn &c. then Pick up part of cider Apples Jen began Packing worked on Cook grapes in foornoon. then on mine afternoon on acount of getting wagon in to load. I took grist to mill & got Bill & Tony shod payed for it at Deans. a vary fine day I work in Packing house Handle Baskets &c. after I get home Dick Harrold & I went to town this eve, Dick got new shoes for Himself*

Tuesday 10/26/09: *Dell & Dick dig Potatoes Glenn & I drew my clintons to Fosters Glenns man Oto Crist work in Packing house for me until 4.15 P.M. Harrold & I went on hill & help the Boys finish Picking up potatoes & drew in what was dug 39 Boxes & put them in cellar. rain last night but clear away & was a fine day today. Dell Dick & I hook Baskets this eve. the women (except Hellen) & harrold went over to Glenns Jennie Packed 603 baskets for me today*

Monday 11/29/09: *went to Pierce & Nichols got 50000 handles 4000 covers & 10000 hooks. drew up load of grapes from lake cellar before breakfast then drew another load after returning from town then draw over a few baskets from Alberts Jim & I work in Packing house Jennie Packed. Harrold went to school. I got a few baskets covers & handles from Glenn I drew one more load from lake tonight Jim & I Paste a few labels this eve. Harrold & Jen. went to town*

Monday 12/13/09: *Rain sleet & snow during night & considerable during day Glenn was over & He & Gim grind axes in foornoon I work in shop Afternoon Topsie & I go to the Funeral of Nelson Bennett. rain some & a vary strong wind from south Gim thresh & clean beans on barn floor then help Glenn saw some wood*

Monday 1/24/10: *was a nice day thaw a little Colder toward night Glenn & Gim worked in woods I helped Albert put Ruff on His Barn.*

Thursday 1/27/10: *I take Harrold to school Gim clean out cow stable & draw manure on field then we go to woods cut a few stakes & draw out a load Glenn helped us in afternoon we cut a large white Oak for stakes & Plank &c. snow some drew out one load stakes wind blew hard at night*

Monday 2/21/10: *the Brown Boys came & finish sawing to Glenns & begin on Pile in my woods Broke down about 3-30. Foscett helped up thare Gim & I help to Glenns & in woods also thaw all day. Gim & I draw in one load straw when we get home*

Thursday 2/24/10: *warm up a little Clarks folks went after dinner. The boys came to saw but brook their engine again So did not saw. Gim put in some wood in woodhouse Topsie went with Harrold to Dancing school this eve. My cold some better but cant speak loud yet.*

FIGURE O-5. View from Pulteney toward Keuka (Crooked) Lake toward West Point where the lake branches; left branch goes to Branchport and the right branch goes to Penn Yan.

Monday 3/13/10:	*in foornoon rig hog pen ring hogs put floor in ladies stall & seperate hogs Afternoon Gim do some trimming west side of my Vinyard rough March day I went to Holleys had big team shod all round Gim & I cut some feed after I get home Glenn & I went to lodge this eve.*
Friday 3/25/10:	*in foornoon Gim & I drag out brush from lake Vinyard. Afternoon Start the Plow on the Hill Gim Plow. I take Colts & go to Abe youngs got load of Stakes from His mill change in weather colder some. Glenn & I went to town to lodge this eve*
Friday 4/8/10:	*I drew 2 cords wood to H.B. Nichols Father for Glenn on H.B. Nichols acct. $4.00. Gim Glenn & His Boy Benn Piled wood in woods I help them in afternoon got done about 4 P.M. then Gim Harrold & I went to Abe Youngs got some Ash lumber which youngs had got out on shares for me then went to Goodriches got some Old Tin Ruffing & drew to my vinyard Barn was a fine day but a Cold North & Northwest wind Ed Osborn payed me $5.00 on acct.*
Saturday 4/9/10:	*Gim Harrold & I willow on my Concords until toward night was a nice day but cold north wind toward night tonight Harrold Topsie & I went to town Glenn went to Prattsburg with His Hogs at 9 cents @ lb. live weight He Payed me $10.00 loaned some time ago. Had my Grape wagon to go with*
Monday 4/18/10:	*rain this morning landers man helped me drive stakes on lake vinyard today Gim helped Topsie wash then willow on my vinyard in*

	foornoon. Harrold willow some to lake Gim staple & tighten wires to lake Harrold & Mrs. Geo. Barton willow I went to town this eve. Harrold went with me
Tuesday 4/19/10:	*Mrs. Barton & Harrold willow Gim & I drive stakes staple & tighten wires to lake. rain some in afternoon one vary hard shower. Sow had 10 Pigs this eve I stay with Her until 11 P.M.*
Friday 5/13/10:	*Glenn came for me this morning about 1 A.M. to help Him with His cow Bess. She could not calve alone cold north wind some Frost at daylight—Cow & calf all right now. Harrold Topsie & I went to Pennyan today got some clothes chd at Maxons & shoes &c. for Harrold & Topsie Gim draw stone from Cook vinyard Glenn & I went to lodge this eve.*
Wednesday 5/25/10:	*rain nearly all night & still coming this morning about 8 A.M. Slack & I went over to Brake centers in my vinyard Gim helped Topsie Glenn came & Broke centers about an hour rain Brook us off Gim sprout some Potatoes & clean south cellar. Albert came over in afternoon & Blowed some Stumps for me on my Vinyard Glenns mare had colt last night. He called me about 4 P.M. today the colt was not right we worked until after midnight but the little fellow died*
Friday 6/10/10:	*Gim churn then finish Hoing my Vinyard then Plaster Corn & Sprout in my Clintons I went to dock in morning get some lime & Plaster Chd. to Cook then rig drill for Beans drill Beans for Abbott this afternoon. went to lodge with Glenn this eve. Topsie & Harrold stay with Nellie Cold east wind today*
Thursday 6/30/10:	*the Girls Both tie for me in foornoon afternoon tie for Glenn they worked on lake vinyard for me. Harrold & I finish mowing on hill & some in yard I do cythe mowing on hill Glenn helped us put up our hay this afternoon was a clouddy day & rather warm Mrs. Bardeen went home tonight Harrold took Her.*
Friday 7/01/10:	*in foornoon rig wagon clear out barn get ready to draw hay hot & vary smoky Glenn & Albert helped me draw hay this after noon. Glenn & I went to town this eve. to I.O.O.F. instalation*
Saturday 7/02/10:	*Albert & Glenn help me finish getting in my Hay in foornoon Afternoon Harrold cultivate garden & go to dock get Points &c. shipped from Cook I work in garden do glenns chores &c. hot day I went to Branchport for Cook & children (the Girls & a ladie friend) Harrold & Topsie took Edith to town. Clark Wright & wife & babe came to our house. Glenn Nellie & Babe start for Canadice about 4 P.M.*
Saturday 7/09/10:	*helped Glenn in His Hay was a hot day. Harrold drove on fork Harrold went to town this eve. Pay Glenn $10.00 I borrowed of Him $3.90 Pay Mrs. Bardeen bal in full for Her tying*

Tuesday 7/12/10:	*Harrold mowed Piece of grass around old cemetary for me this foornoon I helped Glenn finish drawing His Hay. Afternoon Harrold raked mine Glenn & Baker cock it I mowed the Orchard then the rain came was a fine Shower some Thunder rain Came about middle of Afternoon Edith helped Topsie wash this foornooon & did not work this afternoon she went up to Hulses toward night*
Thursday 7/14/10:	*Edith tie vines today on lake vinyard hot day take Tedder to shop get some Bolts in it which I lost out then about 10 A.M. Glenn & Baker came over & we shook out hay on side Hill by cemetary then I Run over it with Tedder we begin drawing from Orchard we finish drawing in my hay this afternoon we work until 8:30 P.M. to finish had supper at 6 let Baker take my big team Plow &c. home with Him to Plow some vinyard for Himself*
Monday 7/25/10:	*in morning Glenn & I went with His team & drew two loads wheat in Tomers Barn for Jo. Lee. Jo helped us. then Glenn had mowing machine did some mowing Baker was up & got Billey & Cart. I mowed around stubbles & mowed my Alfalfa then work t binder in afternoon Harrold went to Branchport & got 6 bags feed at Hurlbutts in foornoon. Glenn Nellie & babe were over this eve. Sprinkle a few times today but no rain of consequence*
Tuesday 7/26/10:	*Put bal of binder togather go to shop have irons made for Cultivator Harrold Rakes alfalfa & around the stubbles hot in afternoon*
Thursday 7/28/10:	*was a fine day finish binding Glenns Oats & went on hill & cut about 1½ hours on mine Glenn & Harrold shocked for me used Glenns team & Tony. helped Fosscett thresh wheat in afternoon*
Saturday 7/30/10:	*Harrold went to Pratsburg got threshing Coal for Glenn & me was nice day I plow first round in my Concords. Harrold got home about 1 P.M. wert Caskey Came toward night stay all night we all went to town this eve. I Payed Moon for my strawburries $1.60*
Wednesday 8/1/10:	*was a nice day drew in Glenns Oats & 4 loads of mine Pearl Ingraham helped & Baker helped in afternoon I went to town after dinner & got a new sling and some fixtures for the track in barn rain a little at night Martin Caskey was here to dinner also wert as he had not gone home when His Father came.*
Tuesday 8/2/10:	*after unloading load Oats went down & cut 4 achors Oats & Barley for Pearl Ingrham He furnished one horse & the twine got home at noon. Afternoon drew in the Bal. of my Oats Pearl helped Baker came up & He & Glenn helped got done about 4.P.M. Baker went home Perl & I Sprout in Clintons after we got Horses in &c. Glenn Harrold & I went to school meeting this eve. Harrold went to mill 4 bush wheat of Fosscett 8 bush Oats of Glenn took grist for Glenn also.*
Friday 8/12/10:	*in foornoon went to Holleys had the mares shod all round no new*

	shoes. nice day. took Binder Reel off & drew Binder in after I get home. Afternoon I get Fosscetts rake & rake loose Oats for Geo. Wagener. 2 hours then return rake & work in my Clintons tucking & sprouting Harrold went to lake (Drakes) with some Potatoes in morning. He worked in Vinyard about 2½ hours in afternoon Glenn & I go to town this eve.
Saturday 8/13/10:	*Harrold & Topsie start with the mares & carriage for wayland I help glenn Plow last round in His vinyard with little Plow use His Horses we finish by making last round in His Old Vinyard in foornoon nice Breeze. Afternoon Plow in young Vinyard until about 3 was offle Hot. we quit & go to Ball game to Pulteney*
Tuesday 8/16/10:	*in foornoon Glenn help me sprout & tuck Clintons in afternoon I plow first round & Glenn drew Oats for Geo. Wagener. was a hot day 105° at one P.M.*
Wednesday 8/31/10:	*finish threshing about 11 AM 314. Oats 258. wheat. wind blew hard from south put Oat straw back in Barn. Afternoon Clean up barn Mrs. Hoad wash for Topsie today Harrold went for Her & took her home about 3.P.M Clark wright & mirt. was down today were here to dinner*
Saturday 9/17/10:	*Glenn & Oto with Glenns team & my Big team finished draging my lake vinyard. Co work on wood Pile I went to Walter Taylors & Hammondsport with Fosscett to see about grapes I sold my Dellawares & Clintons & Ives to wissman to be delivered at cellar Hammondsport at $52.00 @ Ton I sold Glenns Dellawares also. same place & Price Glenn & Boys dress one of my Hogs toward night I help them finish when I get home Mrs. Lamphire help Topsie today*
Wednesday 9/28/10:	*here my Diary is Blank until Oct.*
Tuesday 10/18/10:	*finish Picking Catawbas Glenn & Oto help also Mr. Highs Henry Thomas & Ed Galigar Afternoon Ed. Henry Oto & Mr. Hughs Pick in my Clintons Glenn & I finish drawing in the Catawbas was nice day.*
Thursday 10/20/10:	*Henry. Mr. Highs & I finish my Clintons this foornoon Cut my Pop Corn & draw in Corn. in foornoon. afternoon draw in Pop Corn & begin diging Potatoes was a warm day*
Friday, 10/28/10:	*in morning Sent Henry to Branchport with 14 Trays of grapes to ship to Cook He took one tray from williams. to ship for Him (Williams) I went with Glenn down to Mr. Stones to look over a place which Glenn is talking of buying Henry went to the mill for grist after getting back from Branchport. He took the 1000 baskets from my Packinghouse to Glenn Afternoon Egbert Stewart Came to see my Concords did not sell to Him. later in day Weissman Came I sold to Him for $50.00 a ton & 15 cents each for every tray not*

FIGURE O-6. The sidewheel steamer Yates on Lake Keuka.

	returned in good shape as when taken Grapes delivered to Hammondsport. Glenn & I went to lodge this eve. Cold raw day Henry & I drew Jag of Chunks from my woods. then Henry drew 2 cords wood from the Buzz Pile to my place. I payed O. Retan $10.00 Borrowed of Him I Borrow $10.00 of Glenn
Friday 11/04/10:	*finish refilling & slating Concords Glenn drew one load Henry drew ballance to Dock (2 loads). I went to H.Port in afternoon settle for my Concords I Payed Boat charges of 2½ cents a box 511 boxes to H.Port Frank Hathaway went with me Stormy day*
Wednesday 11/30/10:	*in foornoon went up in woods where Glennn & Walter were working then back & husk corn bal of day the Democrat & Chronicle man was here today had dinner with us. was a nice day but hard traveling some snow Flurries*
Monday 12/26/10:	*at home nice day cold some better nail a few Picking Boxes draw some water from lake for washing as Cistern is Dry Harrold went to Prespiterian Church to Christmas tree this eve, Glenn & Walter were over & used Grindstone this P.M.*
Monday 1/02/11:	*was a warm day but colder toward night I trim a few vines on my Concords Glenn & walter dig Skunks in foornoon get 4. Afternoon they work in woods some the snow is nearly gone Harrold went to*

	Branchport for Emma's Sister Cora she came & stayed all night with Emma
Friday 2/10/11:	*was a good march day wind & snow squalls. I made two trips to Drakes with 2 cds. wood each. & put 2 cds. more on sleighs for morning Harrold went to school. Glenn was over this eve a few moments a little Baby Boy came to His house today all right colder tonight wind northwest Mrs. Walters was over today twice Topsie about the same, (I drew the two loads today which Glenn & I put on sleighs yesterday He did not go.)*
Wednesday 2/22/11:	*Harrold & I put up the bal. of Cooks wheat got Alberts Sleighs & went to Branch port with it booth teams fine in foornoon but blustering with heavy snow squalls. afternoon & eve. Colder at night Mr & Mrs. walters were over this eve. Glenn was in a few moments the Dr. Called to see Topsie this P.M. toward night bought 1 bush Pea Beans at the Elivator to eat Payed $2.10 for them*
Monday 5/15/11:	*Harrold & I finish drawing our ashes to Vinyard draw down some wood Saw a few more bolts for Box stats & harrold drew them with the ones to Bakers to the Mill I plant some sweet corn & cultivate some in garden &c. Harrold took buggie to shop to have tires set & ½ rim in one wheel nice day.*
Friday 7/21/11:	*Harrold & I Plow on the Hill in Clintons was vary hot, use single rigs afternoon get ready & start for Rochester Harrold took me to the dock took Boat for Pennyan Ed McConnel Joined me at Keuka we stay in Pennyan tonight as the Boat got in late rain hard in eve, gave Eva Check for Bal of wages this P.M. to tomorrow night ($8.70 check)*
Saturday 7/22/11:	*Ed & I leave Penn. this morning on morning train for Rochester to the Growers & Shippers exchange meeting fine day & vary good time meeting was adjourned for two weeks in same place arrived home on eve. train left Rochester about 5 P.M. Harrold came for us at Branchport He took Eva to Prattsburg this P.M. Harrold Reports a heavy rain last eve. with Just a bit of Hail*
Monday 8/7/11:	*Cut Alberts Oats on Alexanders 5 achors was a vary Hot day use two of my Horses. bring machine home & put it in barn go to town with Harrold this eve.*
Wednesday 8/23/11:	*in foornoon help Baker thresh they mooved to our place set up & have dinner Mrs. walters helped Harrold with the dinner hresh for me 222 bu Oats 43 bu wheat Albert helped me draw in load of straw after they were gone that we shoved off stack nice day to work. Harrold took washing up to Mrs. Hoads to be done this eve. Payed Fosscett $2.00 bal. in full that I owed Him also Payed Baker the money for His Pears I took Branchport (yesterday) Payed this morning*

Thursday 8/24/11:	*do Choores clean up Barn floor then begin raining I do some mending & Patching rain harder toward night Dr. Called to see Topsie*
Tuesday 9/19/11:	*take some crates & ladder to Orchard set Jo to Pick Apples I turn Car in barn Plow some in west Bean lot. then drag use 3 horses was a warm day draw in what Apples Jo Picked at night*
Wednesday 9/27/11:	*Joe Pick Grapes ½ day today Carl & I start for evaporator with Apples Brake axel on Grape wagon down by fred Browns tenant house came back got other wagon & go with load home P.M. feed & get brooken wagon home. Carl & I went up to Glenns this eve. got a man Glenn brought from Pennyan for me fron Corning Gim Haner by name.*
Tuesday 11/28/11:	*Harrold & I went up & help Glenn put Ruff on His wagon house was a strong south wind we took Tony up to Him*
Sunday 12/24/11:	*Topsie some easier today Glenns folks were down a while toward night snow a little in afternoon frooze Just a little last night*
Monday 12/25/11:	*at home was lovely warm day did not freeze but Just little last night sun shone warm & bright most of day today Glenns folks were down had a fine time Considering Jo McConnels man brought a single bed with springs & matrass for Topsie this morning I settle with landers & pay Him for my Baskets $89.60 He took bal of Baskets & Handles home with Him*
Tuesday 12/26/11:	*was a nice warm but Cloudy day after Choores Harrold went to Branchport for a Mrs Grey from Pennyan to Care for Topsie I work in shop in P.M. making Head Rest for Topsie rain at night I went over to help Baker doctor sick Cow Harrold went over & saw Claud a few moments still raining at 11 P.M.*
Saturday 1/6/12:	*was a cold day Abert Came over & help me change my stoves set Coal heater in place of wood heater Cook Came toward noon we settle Harrold took Glenns wagon home & got my Democrat went to Warner Williams & got Mrs. Cook then I took booth Mr. & Mrs. Cook to Prattsburg was 4° below when I got back from Prattsburg this eve Harrold went to town & got some Groceries Fosscett was over a few moments & help me put fosset in my molasses Bbl.*
Wednesday 1/10/12:	*4° above Zero this morning Harrold drove to Francis Crosses for Mrs Cross to come & stay the day with my Wife she did some sewing while here I took washing up to Mrs Hoad after they got back also took Cloth & thred for Mrs. Hoad to make 4 sheets for us. Fosscett was over this eve & gave me His young Calf Harrold & I went over & got it was a Heiffer Calf of the Holstein Breed Mrs. Cross Stayed with us tonight 2° above Zero but the wind has gone down*
Saturday 1/13/12:	*Cold day about Zero & 10° below at night Chores & in house in foornoon. Albert Came over & helped me get down my Dickies out I*

	went to town got some Groceries had Tonies shoes pulled & His Neverslips put on at Deens chd. then I went to Fs. Crosses & got Mrs. Cross to stay with us over sunday.
Monday 1/15/12:	*at home was a Colday Cut up Pork Mrs. Walters Came over & tried out lard for us did not Pay her today we also made sausage I ground it she mixed & packed it for us. wind blows terribly tonight Harrold went to town for Groceries this eve.*
Sunday 1/28/12:	*at home nice but clouddy day Mr. Cross & Son were here to dinner John Hathaway was up for a whip Cracker. Harrold went to Church this eve. the Raizy Boy was here & went with Him. got a Telaphone from Rozell from Livonia betwean 12 & 1 that mother Died last night Burial tuesday but we cannot go Topsie is vary Low today not seeming to notice much Mrs. Walters was over this eve*
Friday 2/2/12:	*was a nice day some snow squalls Harrold drew down load of wood in foornoon I drew down Jag in afternoon & Emptied the ashes. let Albert have 7 bush wheat lacking 2 lbs. and got another sack of Flour of Him (50 lbs.) He payed me the bal between the two sacks I have had & the wheat of $4.60 we weighed Topsie today she weighs 70 lbs. Harrold went to Branchport to dance tonight was a terrible night cold & blustering*
Tuesday 2/6/12:	*was a nice cold day. Albert & I went down & helped put up ice in & fill Landers Ice house to have some ice next summer if we wish it Harrold went up to Glenns help get His ice Boat on lake then come down where we were cutting ice & skated a while. He went to town & got Express Package from laflers & Botle Alcohol I sent to H. Port with lafler for.*

FIGURE O-7. Seneca Owen many years after he wrote these pages. Collection of Louis and Pauline Owen.

CHAPTER XXI

Wells Barns

Improvements in all areas of farming and barn building continued throughout the nineteenth century, but not all changes, particularly in barn design, are obvious to the casual passer-by. For instance, close to 200 barns were built, most in western New York, which except for surface decoration look like most other gambrel-roofed barns on the outside; however the inside reveals a strong difference in the framing. The window caps of these barns resemble a "lazy W" with the arms in a horizontal position, the trademark of the Wells Barn. Figure 21-1 shows a 36 by 98 foot Wells barn, in excellent condition, on the Delbert Thomson/Edward Mulligan Farm, in the town of Avon, Livingston County.

In a quest for a frame that would allow full use of a hay track and still provide maximum strength, John Talcott Wells Sr., of Scottsville, began experimenting with fabricated truss frames when he entered the construction business in 1871. On April 23, 1889, the United States Patent Office awarded Wells patent number 401,870 for his truss design. Figure 21-2 shows one of his patent models.

Although the Wells teams (see figure 21-3) built barns over three generations as far away as the Hudson River Valley, most were located in the Genesee Country. Other buildings constructed using the patented truss included auditoriums on Staten Island and at Silver Lake.

The Wells advertisement in figure 21-4 illustrates some of the major advantages offered by the design: The center of the barn remained open, allowing the use of a hay track; also, the truss design was quite strong and the barn could be built from relatively small dimensioned lumber, either purchased or sawn on the property from existing stands of trees. Toward the end of the nineteenth century it became increasingly difficult to obtain or afford traditional timber framing material. The use of black walnut, cherry, and oak began to give way to softer woods such as pine and poplar, and even then it was difficult to find trees of sufficient girth and length to permit the older post and beam construction. The readily available 2 by 3 inch planks in the Wells design were spiked together to make the strong curving ribs of the truss. Wells barns tend to be taller than other plank barns and frequently they are longer. Most of the Wells barns were built as basement barns with provision for a dairy herd in the basement level. Commonly a granary occupied the space between the large barn doors on the upper level with two large end mows filling the space from floor to peak.

After a lifetime of haying in Montana, Iowa, Maryland, and finally in New York, William Elkins commented in a 1986 letter on his experience with a Wells Barn, "[they have] the most practical and convenient hay mow I had ever seen. I had never seen such huge grapple forks of hay

Opposite: Interior view of the Guthrie Wells barn, Mumford, New York.

FIGURE 21-1. 36 by 98 foot Wells barn on Thomson/Mulligan Farm, east of Route 15 on North Avon Road, Avon, New York. The gable features the identifying Wells window hood.

lifted from the hay wagon, rise to such great heights and easily pass through the extra large track door and race along the hay track until the load was tripped. Mowing hay was always hard work but it was almost a joy pitching the hay to the edges and corners without having to work around the beams and posts as was the case in most barns."[1]

The firm known as J.T. Wells & Sons included John Talcott Wells and three of his five sons: Robert, Nolan, and John Talcott, Jr. After the senior Wells retired in 1914, the sons worked

FIGURE 21-2. Wells patent model showing a detail of the laminated truss. The entire model is approximately five feet tall. Big Springs Museum, Caledonia, New York.

FIGURE 21-3. The Wells barn builders pause for a portrait. This is an early Wells barn without the stiffening inverted end trusses, which were added in the mid 1890s; this feature helps date the photograph in the early 1890s.

together for one year, then John Wells, Jr. assumed sole control of the firm. The last typical Wells truss barn, the Stokoe barn in Scottsville, was constructed in 1942 and the last non-truss structure was completed in 1951.

According to company records, construction of a Wells truss barn for Abram Cameron in Caledonia was begun in May, 1896 and completed before October, when the barn was painted. Elm from the farm was sawn for the trusses, timbers and horse stall plank floors. In addition, two box-car loads of lumber were shipped from Buffalo at a cost of $580.80. Total cost for the barn was $1619.62 including lumber and labor ($1473.69), track and car ($43), painting ($47.93), and hardware ($55).[2] Wages reportedly ranged from $1.15 to $2.15 per day according to the worker's skill. The prices of the Wells barns rose slowly compared to the inflation of the 1970s and '80s. For example, the 32 by 92 foot barn constructed in 1914 for Walter Cox of Scottsville called for a finished price of $2500, an

THE WELLS PATENT BARN TRUSS.

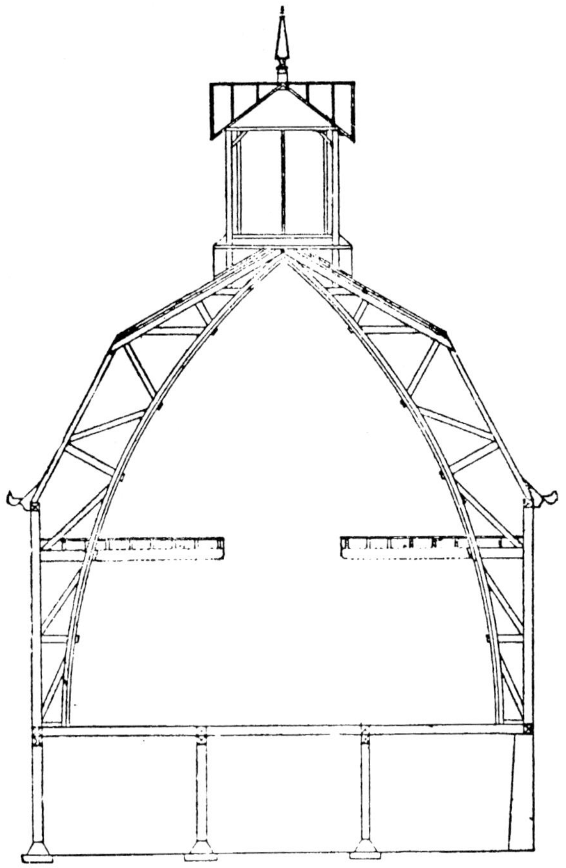

THIS CUT REPRESENTS THE

WELLS PATENT TRUSS

FOR

GRAIN AND HAY BARNS,

and shows a section over the drive floor, with a Scaffold supported by iron rods from above, which will leave the interior of the Barn free from Posts.

The lower story in the cut represents the basement.

The features we wish to call particular attention to is the absence of all Cross Beams, this allows the load to be run into the Mow as soon as it clears the wagon, the advantages arising from this combination are valuable, it does away with the necessity of hoisting the load clear of the track, thus saving time and travel of the horses, the load does not have to drop from the peak of the barn to the mow, thereby packing so hard as to cause mow burning; the sling full of Wheat, Oats or Hay hangs near the mow, and the operator can swing it to any part of the mow.

This Barn Frame can be sawed out on the farm with a Portable Saw Mill, if there is timber suitable, such as Spruce, Ash, Red Elm, Soft Maple, Hemlock or any wood that is tough.

This style of barn can be put up without a Mortice, Tenon or Gain, cut in the Sill or the Plate, thereby saving in the cost of labor, there is no large Timber above the Sills, the End Bents are built after an improved plan, which does away with any large posts or beams.

Correspondence solicited.

Working plans, terms and right of use can be obtained by addressing

J. T. WELLS & SONS,

Scottsville,

Monroe Co., N. Y.

FIGURE 21-4. Wells Patent Barn Truss advertisement, circa 1891. Collection of Jean Melville.

FIGURE 21-5. The 1915 Getzinger/Bailor barn was constructed with two older buildings at right angles to the large barn; in winter doors on tracks could be closed to keep the west and northwest winds out of the barnyard.

increase over the Cameron barn of only $880.38 in eighteen years. In this instance the Wells firm provided all the materials.[3]

The 40 by 110 foot Getzinger/Bailor barn, Geneseo, was built in 1915 on an old foundation (see figure 21-5 and 17-8). The massive size of this barn can be seen in the view of the east facade. It is 55 feet from the barn floor to the peak and the basement adds another 10 feet to the total. The barn is now owned by Peter Bruckel.

The Wells trusses varied depending on the actual size of the barn, but typically the ribs were made of small cross-section sawn pieces of lumber (often a full 2 by 3 inches) nailed together with 3 by 6 inch or 3 by 8 inch braces spiked and bolted to tie the outer chords of the truss together (see figure 21-6). The average distance from the ground to the peak of a Wells barn ranges from 45 to 65 feet. An interior photograph of the 38 by 100 foot James Guthrie barn in Mumford, built about 1923, details the trusses making up two bents, the end trusses that stabilize the west wall, and the close-spaced rafters and roof boards (see figure 21-7). An interesting comparison can be made between the trusses on the 1889 patent model (figure 21-2) and those on the west side of the Getzinger/Bailor barn shown on raising day, July 10, 1915 (see figure 21-8). All the historical Wells raising photographs, many preserved by various members of the Wells family, are from the remarkable collection of Wells historian Jean Melville of Scottsville.

In many instances an existing suitable foundation was used by the Wells company. When Portland cement and concrete came into general use in the 1890s, the Wells company used this material instead of masonry construction for new foundations. The sills (often of hemlock) and the floor joists were installed, then the floor boards loosely placed to provide a working surface for constructing the bents. The bents, each comprising two trusses, were made according to a pattern drawn in chalk on the floor. A pattern was drawn for each one so the finished bent did not have to be moved very far at raising. Because of the gentle curve of the ribs no special difficulty was encountered in bending the 2 by 3 inch planks to the proper curve. The completed bents were stacked

FIGURE 21-6. Detail of Getzinger/Bailor Wells barn trusses showing the laminated ribs and chords plus the spiked and bolted diagonals of the truss form.

FIGURE 21-7. Interior view of Wells trusses in the Guthrie Barn, Mumford, New York. Careful examination of the photograph reveals the bolt and spike pattern used to construct the trusses. Also, the inverted end stiffening trusses are apparent.

in the order of their use until the raising. The last bent produced is the first raised, with the lower portion of the truss being constructed close to the point of insertion in the sill.

Through a series of photographs, the various steps involved in raising a Wells truss barn are shown. Figure 21-9 shows an end bent being raised in the Selden/Stokoe barn, December, 1942. The inverted end trusses were introduced in the middle 1890s to stiffen the barn against both the wind and the outward pressure of the hay. In the Thornell/Bruner Wells barn, (see figure 21-10), built without these end trusses, the end bent posts were a hefty 9¾ by 12 inches in cross section (see figure 21-11).

After the end bent was set and temporarily

FIGURE 21-8. This is the only photograph in this book of an existing Wells barn on raising day. Getzinger/Bailor barn, Reservoir Road, Geneseo, New York. Jean Melville Collection.

FIGURE 21-9. Raising an end bent of the last constructed Wells barn on the Selden/Stokoe farm, South Scottsville, Monroe County, New York, December, 1942. The gin (ginny) pole to which the pulleys are fastened can clearly be seen, along with the inverted end trusses. Jean Melville Collection.

FIGURE 21-10. South end of the 1894 Thornell/Bruner Wells barn, Mendon Center Road, Pittsford, Monroe County, New York. It is 65 feet from the barn floor to the peak in this early Wells barn built for Henry Thornell. The vertical feeling is emphasized by the board-and-batten siding. Box troughs collect rain water to feed the cistern on this hilltop location. A roof of this size, according to the 1878 American Agriculturist, should supply about 2,000 barrels of water per year.

FIGURE 21-11. Circa 1894, 36 by 85 foot Thornell/Bruner Wells barn, Mendon Center Road, Pittsford, New York. The interior reveals the traditional end bent construction. This is one of the earliest extant Wells barns, dating from the period when the design was still being modified; even the traditional lazy "W" window casing is absent. In this barn each arch has six laminations.

braced, the rest of the bents were raised in their turn. The length of a Wells barn was varied by increasing or decreasing the number of bents. The small 30 by 50 foot Richard Updaw Wells barn on Route 15 in Rush has only two trussed bents. The ends are stabilized with the typical inverted trusses. The longest Wells barn discovered, the Hawley Farms in Batavia, is 175 feet long. The power to raise the bents came at first from man, then from oxen or horse power and

FIGURE 21-12. Raising the end bent with steam power on the Frazee barn, December, 1917. The penciled caption under the Wells family photograph identifies as "Uncle Ed" the man helping to run the steam roller.

FIGURE 21-13. Raising the second bent on the 1899 William Trescott barn (now destroyed), Conesus, New York.

eventually from both steam and gasoline-powered machines. Figure 21-12 shows the application of steam power to bent raising. The 1899 raising of the first interior bent on the William Trescott Farm in Conesus is shown in figure 21-13.

Figure 21-14 looks along the long axis of the 1914 Walter Cox barn, Scottsville, New York. The second bent has just been raised and the next bent to be raised can be seen lying flat in the foreground. The bents were laid out and built on the barn floor in reverse order of their raising. As each bent is raised it is tied to the previously raised bent with a girt, purlin or temporary brace. When all the bents are up and stabilized the really hard work of barn building is mostly over (see figure 21-15). The finished shape of a Wells barn appears when the roof rafters are added to the structure in figure 21-16. Increased solidity and

FIGURE 21-14. 1914 raising of the Walter Cox barn, Scottsville, New York. The view is along the long axis of the barn, over the top of the horizontal finished bents.

FIGURE 21-15. All the bents are raised on the Frazee barn, December, 1917. Joy is expressed by John Talcott Wells' son John, who balances on a purlin; the frame is well along.

FIGURE 21-16. *The final shape emerges on the Trescott Farm Wells barn, 36 by 120 feet, 1899.*

FIGURE 21-17. *A Wells barn ready for sheathing. Unidentified.*

FIGURE 21-18. The 1902 Wooster/International Salt Co. Wells barn was an excellent example of the type with its double doors, cupola, slate roof, and typical window. Cuylerville, New York, Photograph by Richard Quick.

form emerge as the studs and girts are added (see figure 21-17).

The sheathing, roof shingles, doors, decorative trim, and cupola (where applied) provided the final carpentry steps in the construction of a barn's exterior. The Wells barn built in 1902 at Cuylerville for John T. Wooster was a completely detailed example of the Wells barn, including all the features listed in the company advertisements (figure 21-18). The lumber was supplied by Wooster's brother in Seattle, Washington, who, not incidentally, was in the lumber business. The total cost for the lumber, including rail shipment, was $3,000.[4] The barn had the typical patented truss system supporting a slate roof crowned with a ventilating cupola. Eventually it was owned and used by the the International Salt Company, which had a nearby shaft, to store feed for the mine mules. Also, because the salt irritated their hooves, the mules were brought periodically to the surface and stabled in the Wells barn. This barn has now been destroyed and the mine shaft closed even though salt is still mined nearby.

Without doubt the Wells Barn was as strong and efficient as the advertisements claimed. However, a Wells barn interior also creates an aesthetic response that quickens the heart of anyone fortunate enough to experience it (figure 21-19).

FIGURE 21-19. *Interior of the Thomson/Mulligan Wells barn. The hay track is suspended immediately below the ridge. East of Route 15, North Avon Road, Avon, Livingston County, New York.*

CHAPTER XXII

Modern Materials in Barn Construction

THE NINETEENTH CENTURY was punctuated by the introduction of "new" building materials as well as by the invention of labor-saving machines of all sorts. At the end of the century the two outstanding new materials were steel for framing and reinforced concrete for walls, floors, arches and so forth. Inevitably the season of the wooden barn would be affected by the use of these new materials and a new era would begin.

During the period of this study the use of iron in barn construction grew slowly, with wrought iron spikes, nails, and hinges the earliest applications. At first, iron was scarce and expensive within the Genesee Country because its weight increased its transportation cost. The need for locally-produced iron stimulated the construction of furnaces for cast iron and triphammers for wrought iron, from which blacksmiths forged nails and spikes. In the Genesee Country a mill for the production of both cast and wrought iron was established before 1810 just south of Avon, at a site now called Triphammer Falls. The small output of this mill precluded the use of large quantities of iron in barn construction.

By mid-century cast iron and wrought iron in large quantities became available from major iron-making centers such as Buffalo and Pittsburgh. The canals and later the railroads brought these products to the Genesee Country at reasonable prices. Furnaces in Rochester also supplied smaller, but essential, quantities of iron. Although cast iron and wrought iron were much used in domestic and commercial building from mid-century to the 1890s, they were little used in barn building. By the Civil War small quantities of steel could be produced by the Bessemer process, but steel was not used for building frames until the 1890s, when it became a marginally economical building material. At first used only as a substitute for wrought iron, steel was beginning to be used with I beams and H columns hot riveted together to form the skeleton frame of tall buildings. However, steel was not widely used for barn framing until the twentieth century.

Portland cement was patented in England in 1824 and imported into the United States during the late 1830s. Non-reinforced concrete, as already seen, was used in barn walls and floors under the names of "mortar" and "grout" beginning in the 1830s. A smooth, hard concrete floor was good for a grain barn because it could be kept clean and seed was not easily lost. Easy cleaning also made concrete a good material for

Opposite: Detail of the "Block Barn" emphasizing the fireproof elements of the concrete blocks and terra cotta roof.

FIGURE 22-1. The 347 by 50 foot fireproof concrete "Block Barn," called McKinney by builder William Simpson, constructed 1907 to 1909. Cuba, Allegany County, New York.

dairy or cattle barns. Reinforced concrete, patented in 1875, was a major improvement because the embedded steel rods provided the tensile strength which concrete itself lacked.

Reinforced concrete was used in barn foundations and floors soon after its introduction. But not all concrete used for foundations was reinforced with iron or steel rods because some thought that the primary stress was compression, which concrete alone could tolerate. The years prior to the First World War were years of limited but spectacular experimentation with concrete and steel in the construction of barn buildings. Not until after the Second World War did steel and reinforced concrete become prominent in new barn construction. The years between the world wars saw a general continuation of wooden barn construction.

McKinney, The "Block Barn"

One of the most often touted characteristics of concrete and steel was their fireproof nature. The search for fireproof building materials had been going on for centuries, but gained momentum in the mid-nineteenth century with the introduction of cast iron. William Simpson learned of the terrors brought by fire when a covered wooden quarter-mile track and training barn he owned near Cuba, New York, burned to the ground following a lightning storm in the early 1900s. Among the colts destroyed in the fire were most of those sired by the renowned trotter

McKinney, which Simpson had purchased for $50,000, a price which astounded the racing world.[1] McKinney's offspring were very highly prized and their loss might have defeated a less tenacious person.

Devastated but determined to continue, Simpson decided to build a new stable on a farm in Cuba, but this time the stable would be fireproof. Concrete seemed to fill the bill; specifically, Simpson was intrigued with the possibilities of concrete blocks. Concrete blocks had been used infrequently in the United States starting in the 1860s but they became popular by the end of the first decade of the twentieth century. The inventor of a not-yet-patented block-making machine was persuaded to bring his invention to Cuba by the new barn's Buffalo architect, John Coxhead. The inventor wasn't sure he could make enough blocks for a barn this big—347 by 50 feet—because his machine made only one block at a time. It required two years but the blocks were eventually produced. The basic blocks measured 12 by 24 by 12 inches. A similar block-making machine is shown in figure 22-2. The concrete blocks were cast at the site and allowed to dry under controlled conditions to prevent the formation of stresses in the blocks (see figure 22-3). Cuba photographer P. H. Kellogg produced a remarkable series of photographs documenting the construction of the "Block Barn," from the time of its initial construction in 1907 through its completion in 1909. In addition to the basic blocks, special moldings and sills had to be made, each with a different mold or form. Concrete blocks were also used for the interior walls, but the floors and arches were poured concrete. Wood was used only when necessary for doors, stairs and such. The roof was of red-orange terra cotta tile produced by the Ludowici Celadon Terra Cotta Company in Alfred, famed for its ceramics.

Considerable experimentation with concrete occurred in the 1890s and 1900s and various advertisements in farm periodicals encouraged

FIGURE 22-2. Concrete block-making machine advertisement. 1907.

use of this material, touting its low cost and durability. All one needed was the proper form and practically any building shape could be produced (see figures 22-4 and 22-5).

Kellogg began recording the construction of the Block Barn with the pouring of the foundation (see figure 22-6). Figure 22-7 shows the

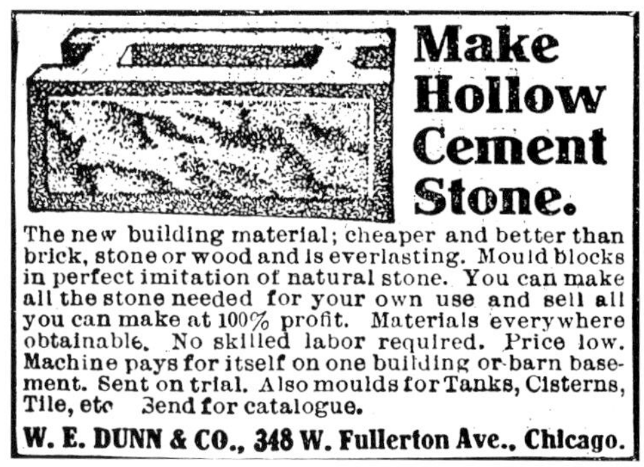

FIGURE 22-5. 1905 advertisement for a cement block mold. Advertisements began to appear about 1905 for block making machines.

FIGURE 22-3. Block drying room at the McKinney barn site, Cuba, New York, circa 1907. Photograph by P. H. Kellogg.

FIGURE 22-4. *Advertisement for the Atlas Cement Machinery Co. The Rural New-Yorker, July, 1905.*

FIGURE 22-6. *Preparing the concrete foundation. Photograph by Kellogg.*

FIGURE 22-7. Prepared forms for arches and roofs. Photograph by Kellogg.

FIGURE 22-8. A horse-powered boom crane assisted with the pouring of the concrete. Anonymous. Postcard from the Dana Kline collection.

FIGURE 22-9. Pouring concrete on the roof. Photograph by Kellogg.

FIGURE 22-10. The barn as it appeared in 1985.

forms for the round arches and the roofs ready for pouring the concrete. The wooden forms were removed after the concrete hardened.

In figure 22-8 the horse-powered boom crane used to lift blocks, mortar, tiles, and the workmen themselves occupies the center of the photograph. The hollow concete blocks lie on the scaffolding to the right, waiting to be laid. Figure 22-9 shows the concrete bucket used to deliver concrete to the wheelbarrows during the pouring of the roof supports. Photograph by Kellogg.

The barn required two years and $200,000 to construct (see figure 22-10). The structure originally contained forty stalls, a foaling room, oak panelled office, a grand, open staircase and dining room. There was also a center aisle long enough to allow winter exercise of the trotters and wide enough to allow the horse and sulky to turn around. The second level, with its rounded concrete arches and high ceilings, was used for fodder storage (see figure 22-11).

The completed barn was named McKinney. In the years between its completion in 1909 and the outset of World War I such notables as the Shah of Persia and Czar Nicholas II sent mares to be foaled at the McKinney Barn (see figure 22-12). The barn stood empty for some years and although there was some damage due to ice heaving of the blocks, the structure did not succumb to fire, and that would have pleased William Simpson. The barn was purchased and restored in 1985 by Mr. and Mrs. Sam Nicholas for raising and training Arabian horses. Johrhar Arabians, as the operation is now known, has some fifty horses, with seventy-five to eighty an expected maximum. Within the Block Barn there are now thirty-five stalls. Besides resetting and

FIGURE 22-11. *Block Barn haymow. Photograph by Kellogg.*

FIGURE 22-12. *Simpson's prize trotter McKinney.*

FIGURE 22-13. A typical terra cotta roofed gambrel barn on Hamilton Hill Road, Alfred Station, Allegany County, New York. Because of the weight of these durable, fire-proof tiles, more substantial roof rafters were required. The small dormer is used to permit loading of the barn with a conveyor.

repointing some of the concrete blocks, workmen replaced more than 2,100 roof tiles. Interestingly, the replacement tiles were found in Alfred, where they had been stored for some fifty years. Because the replacement tiles were already weathered they blended well with the original tiles.

Alfred, New York, is noted for its clay products. The first kiln went into operation in 1889 at the Celodon Terra Cotta Company. The unglazed, low fired clay called Terra Cotta was produced from high quality local clay and used to make many different building components, including roofing tiles such as those used on the Cuba Block Barn. 8,000 tiles per day could be produced at the peak of operation.[2] As a result of the interest of one of the company's founders, the

State College of Ceramics was soon located in Alfred. In addition to Simpson's McKinney Block Barn, there are many wooden barns in the Alfred and Alfred Station area that use similar red terra cotta tiles from the Celodon Terra Cotta Company (see figure 22-13).

Of course pre-cast concrete blocks were used for foundations much as "cinder" blocks are used today, but as has been shown they were also used for entire structures such as barns as well as homes, shops and factory buildings. Although many were skeptical about the strength of concrete blocks, time has shown that they have stood up well, depending like any masonry construction on the original quality of the mortar and its maintenance. Concrete blocks of the sort used on the McKinney barn were distinctive and when they appear the date of their use falls within fairly narrow limits (1907–1925). While a number of examples of entire structures built with concrete blocks are included in this book, it would be a distortion to suggest that this method of construction could be considered popular.

The Everett Doty barn, built about 1912 in Geneseo, was also of pre-cast concrete blocks. Instead of being cast on the spot these blocks were bought from the Nickerson brothers in nearby Lakeville (see figure 22-14). For this barn the concrete was cast in a steel mold to resemble rusticated, quarried stone providing a pleasant surface texture (see figure 22-15). The general form of the gambrel-roofed barn is repeated in the milkhouse to the right. The barn's design required engaged buttresses to support the main walls. The milking parlor has a floor of poured concrete to facilitate cleaning (see figure 22-16). The Doty barn is now owned by Kenneth Book.

The creamery in Lakeville, until recently called the Conesus Milk Producers' Cooperative Association Building and earlier the Genesee Valley Milk Product Company, was constructed about the same time as the Doty/Book barn and utilized

FIGURE 22-14. Doty/Book dairy barn. Circa 1910. Reservoir Rd., Geneseo, Livingston County, New York.

FIGURE 22-15. Concrete block detail from Doty/Book barn.

FIGURE 22-16. Doty/Book milking parlor.

FIGURE 22-17. Concrete block creamery building. Circa 1915. Lakeville, Livingston County, New York. Across the road, next to the lake, stood a wooden ice house which was twice as high as the creamery and four times wider.

the same type of pre-cast concrete blocks (see figure 22-17).

An additional notable example of a structure built almost entirely with the new materials is the huge Riding Hall at Ashantee, south of Avon (see figure 22-18). Martha B. Wadsworth, wife of Herbert Wadsworth, accomplished horsewoman, horse breeder, and riding instructor, was instrumental in the building of the Riding Hall, which opened in 1914. The Genesee Valley Breeders Association, Inc. was established the same year and Martha Wadsworth was its president from 1914 to 1933. Under her direction Genesee Valley hunters gained an international reputation, but she was equally interested in work horses used on the farm.

The 200 by 100 foot structure of reinforced concrete provided space for riding and horse training even in deepest winter; Mrs. Wadsworth alone had some seventy horses to exercise. A steel trussed roof permitted the omission of interior posts, which would have diminished the riding space. The "framing" of the structure is quite obvious in morning sunlight, which also picks out

FIGURE 22-18. 1914 concrete and steel Riding Hall. Ashantee, Livingston County, New York.

the form lines of the concrete walls (see again figure 22-18). An aerial photograph of the adjacent Wadsworth horse barns at Ashantee is shown in the chapter on barn arrangement (see figure 17-6).

Troop "M," First New York National Guard Cavalry, was formed at the Armory at Ashantee (the Armory was destroyed during World War I and the unit moved to Geneseo upon its return); the troop used the Riding Hall frequently for practice as long as it remained a horse cavalry. This unit and its successor, the 27th Reconnaisance Company, served in the Mexican War, World War I, and World War II.

The use of concrete in farm buildings increased tremendously throughout the period from 1890–1915. Cement companies produced booklets designed to familiarize farmers with the pos-

FIGURE 22-19. McClintock/Bruno barn built by the Groton Bridge Company, McClintock Road, Dryden, Tompkins County, New York. The 1907 photograph, from the collection of the Dewitt Historical Society of Tompkins County, shows almost one half of the bank barn. The two main poles of the gins used to raise the prefabricated steel truss sections can clearly be seen. The builders are ready to raise the fifth bent.

sible uses of concrete on the farm, suggesting either poured or blocks.[3] Concrete barns were the largest concrete structures presented in these pamphlets. Among the attributes of concrete was its safety from fire, wind and storms, and also its ability to be rat-proofed, water-proofed and easily cleaned. Other suggested concrete structures included dairy and ice houses, silos, corn cribs and piggeries. While a number of concrete barns were built in this initial period of interest, the use of concrete to construct entire barns waned in the 1920s.

Steel framing is fairly common today in barn construction, but in the first decades of the twentieth century it was too expensive to be practical. An outstanding photograph shows the construction of a steel truss frame on the McClintock/Bruno barn in Dryden, Tompkins County (see figure 22-19). No other steel framed barns built before World War I were found in the region, and although it is a bit east of the Genesee Country this fine example demanded inclusion. The barn was constructed in 1907–08 by the Groton Bridge Company of Groton, New York at a reported cost of $4,000.[4] Although undocumented, local stories suggest that the Groton Bridge Company produced this frame at a very small profit, in the hope that others would wish to adopt the design, thus opening an entirely new area of business for the company. The riveted steel frame was erected on a 45 by 115 foot poured concrete foundation. It is approximately 55 feet to the peak of the barn.

The McClintock/Bruno barn included thirty stanchions and four horse stalls. A hay track was installed and beneath the horizontal truss a granary was located. The barn was sheathed with vertical boards, roofed with decorative, multi-colored slate and, of course, equipped with lightning rods. Willet McClintock stored potatoes there as well as cabbages and some fruit. On a hill above the barn, a windmill-driven pump supplied water to a tank in the barn. The tank was high enough to provide gravity generated water pressure.

The McClintock/Bruno barn has proved to be quite sturdy, requiring only normal maintenance, but the cost of construction seems to have been too high to permit replication. This, indeed, may be the only one constructed by the Groton Bridge Company. A cast date plate was placed on the barn wall by the company marking the completion of the venture.

Previous examples to the contrary, the new building materials were slow to gain acceptance, at least for the construction of a total barn, although many arguments favored concrete and steel. Maintenance costs were said to be low, the materials durable, and many found beauty in the concrete, especially in pre-cast blocks. Although the structures were fireproof, the contents were still flammable, and there still is no way to eliminate that problem. Skepticism concerning the superiority of the new materials and their relatively high cost relegated concrete and steel to a secondary role until after World War II, when experimentation with these materials proved their worth for all sorts of buildings, including the famous Quonset hut.

Sawn timber framed barns continued to be built in large numbers through the 1930s, some of rather traditional design and others incorporating those design changes discussed in Chapters VII, XX, and XXI. While many samples of twentieth century barns could be offered, the series of photographs from the barn built by Albert Schleede of Lyons in the 1920s is particularly instructive. See figures 22-20, 22-21, 22-22, and 22-23. The barn foundation is poured concrete and the framing timbers all sawn. Because a Case gasoline tractor was used for power the construction crew could be small. It is clear that it was no longer necessary to assemble the entire bent on the ground, since the Case tractor working through the pulleys on the gin pole was able to raise the upper purlin bent even after the lower framing was up and boarded. The balloon framing of the roof rafters was typical of the period.

FIGURE 22-20. Building the Albert Schleede barn, Maple Street Road, Lyons, Wayne County, New York, 1920s. The Case gasoline tractor on the left provides the power, working through the pulleys on the gin pole to raise the lower portion of the first end bent. All photographs in this sequence are from the collection of Mrs. Ralph Schleede.

FIGURE 22-21. The lower portions of the sawn timber framed bents are almost in place. Side bracing or trussing provides lateral strength to the barn while leaving the center open, not for a hay fork but for tractor/wagon movement.

FIGURE 22-22. Raising the purlin post portion of the end bent, again using the Case tractor for power, working through the pulleys on the gin pole.

FIGURE 22-23. Raising the balloon-framed rafters.

FIGURE 22-24. The completed gambrel-roofed barn.

In the period between the wars, truss framed barns such as the Wells and the plank truss continued to be built. However, farming changed tremendously following the First World War and of course barns were destined to change as well. In fact, the new barn designs created during the period from World War One to the present are so numerous and complex that they must remain for others to describe.

In many locations throughout the Genesee Country, one can see combinations of new and old barns, some carefully preserved, most serving some useful function. A careful examination of the photograph of the Stewart Farm in Geneseo, Livingston County shows a rather typical farm of the 1980s in the Genesee Country. Barns of a variety of ages range from the wooden board and batten gable-ended barn, to the wooden gambrel-roofed barn close to the silos, to the new metal sheathed, pole construction cow barns (see figure 22-25).

Equipment evolution has brought about the round hay baler, which may well have an important impact on the continued existence of many barns. Hydraulics made round-baled hay a possibility, and because this method of baling, with its heavy, tight bales (600–1,900 lbs.), requires no barn for protection, it now seems possible that the long season of the wooden barn has almost passed (see figure 22-26). Perhaps, however, as in the past new uses or modifications of these magnificent structures will be found to accommodate the changing needs of the farm.

One final example of the reuse of a nineteenth

FIGURE 22-25. Stewart Farm with barns of varying ages, each serving a useful purpose. Route 63, Geneseo-Mount Morris Road, Town of Geneseo, Livingston County, New York.

century barn can be seen in the removal of a barn to a working farm. The construction of the New York State Thruway carved away a large portion of the Stephen McNall farm in West Henrietta. As a result, farming as it had been practiced became impossible. A passing old-order Amish man from East Randolph, New York, in the Conewango Valley near Jamestown, noticed the intact windmill on the McNall farmstead and inquired about it. One thing led to another and soon McNall was asked if he would part with his 36 by 80 foot, 1880s basement barn, (see figure 22-27 for an aerial view of the farm). The barn, once used for dairying and potato storage, was still sound and could be used elsewhere on a working farm. McNall said yes.

Some twenty-five Amish men arrived in July of 1960 to carefully dismantle the barn and in three days it was "unbuilt." Figure 22-28 shows the roof being removed. In figure 22-29 only the rafters remain of the roof structure, the roof boards and shingles having been removed. On

FIGURE 22-26. Round baled hay. Bales weigh close to half a ton.

FIGURE 22-27. Aerial view of the McNall Farm, Route 15, West Henrietta, Monroe County, New York. The barn on the top right is the barn that was carefully taken down to be reused in East Randolph, New York. Anonymous photograph.

FIGURE 22-28. Beginning the dismantling of the McNall barn, July, 1960.

FIGURE 22-29. The timber and other lumber were carefully removed, ropes being used to lower the framing members. Rochester Democrat and Chronicle photograph, July, 1960.

FIGURE 22-30. Shingles, roof boards, rafters and side boards have all been removed; the remaining sawn framing timbers will go next. Anonymous.

the lower level, the side boards have been removed on the west side. Figure 22-30 shows just the foundation with the first level frame remaining in place. The timbers and boards were loaded onto flat bed trailers and removed to East Randolph. Eleven months later the barn had been raised and was useful once more—a creative solution to a problem by both parties.

Barns are still dominant structures in the rural landscape but each passing day seems to produce a new threat to their continued existence. Perhaps as they are better understood and their importance in our lives is more fully recognized, their future will prove to be "a time to keep" and "a time to build up" rather than "a time to break down" and "a time to cast away." If they disappear it will be too late to realize how much they meant to all of us, regardless of our backgrounds, occupations and life style. Part of the rich, varied rural landscape will be no more.

Endnotes

Preface

1. In 190 years in the Genesee Country the total number of barns built was certainly more than 30,000; in 1917 the 1301 square miles of just Livingston and Monroe Counties numbered more than 9,000 individual farms, most with wooden barns; an exact number of barns will never be known. American *Agriculturist Farm Directory and Reference Book, Monroe and Livingston Counties, New York, 1917* (New York: Orange Judd Company, 1917), pp. 19, 25.

Chapter I *The Fertile Land of the Genesee*

1. Edmund B. O'Callaghan, ed., *Documentary History of New York* (Albany, 1849–51), Vol. I, p. 238.
2. Erkuries Beatty, "Journal of Lieutenant Erkuries Beatty," in *Pennsylvania Archives,* (Harrisburg, 1890), 15: 241.
3. Frederick Cook, ed., *Journals of the Military Expedition of Major General John Sullivan* (Auburn, N.Y.: State of New York, 1887), p. 60.
4. ibid., p. 371.
5. The traditional definition of an acre was the amount of land a person and team could plow in a single day. This subjective definition has been refined to 43,560 sq. feet or approximately a parcel of land 200 feet on a side; one square mile contains 640 acres.
6. Orsamus Turner, *History of the Pioneer Settlement of Phelps and Gorham's Purchase, and Morris' Reserve* (Alling, Rochester, 1851; reprint edition, Geneseo, N.Y.: James Brunner, 1976), p. 506.
7. ibid., 505–6.
8. Henry O'Reilly, *Sketches of Rochester with Incidental Notices of Western New York* (Rochester: William Alling, 1838; reprint edition, Geneseo, N.Y.: James Brunner, 1984), p. 383.

Chapter II *Log Barns and Cabins*

1. Henry Clune and Robert Volz, ed., *A Journey through the Genesee Country, Finger Lakes Region and Mohawk Valley from Patrick Campbell's Travels in the Interior Inhabited Parts of North America in the Years 1791 and 1792,* Edinburgh, 1793 (reprint edition, Rochester, N.Y.: The Friends of the University of Rochester Libraries, 1978), n. pag.
2. William Siles, "A Vision of Wealth: Speculators and Settlers In the Genesee Country of New York, 1788–1800" (PhD dissertation. University of Massachusetts, 1978), pp. 211–215.
3. Basil Hall, *Travels in North America in the Years 1827 and 1828* (Edinburgh, 1829; reprint edition, New York: Arno Press, 1974) p. 146.
4. Alex Bealer and John Ellis, *The Log Cabin,* (Barre, Massachusetts: Barre Publishing, 1978), pp. 9–22.
5. Eric Sloane, *A Museum of Early American Tools* (New York: Ballantine Books, 1964), p. 70.
6. Ethan Lanphear, *Scenes and Memories and Travels of Eighty-Two Years* (Plainfield, N.J.: Lanphear, 1900), p. 16.
7. Orsamus Turner, *Pioneer History of the Holland Purchase of Western New York* (Buffalo, Derby and Co. 1850; reprint edition, Geneseo: James Brunner, 1974) p. 470.
8. An Englishwoman (Wright, Frances), *Views of*

Society and Manners in America in a Series of Letters From That Country to a Friend in England During the Years 1818, 1819, and 1820. (New York: Bliss and White, 1821), pp. 142, 146.
9. Henry Conklin, *Through Poverty's Vale* (Syracuse: Syracuse University Press, 1974), p. 45.
10. "The Wintering of Stock," *The Genesee Farmer,* XIV (1853): 372.
11. Turner, *Phelps & Gorham,* p. 535.
12. William Boyd, *History of the Town of Conesus, Livingston Co., N.Y.* (Conesus, N.Y.: Boyd, 1887), p. 60.
13. ibid., p. 51.
14. Patrick Shirreff, *A Tour through North America together with a Comprehensive View of the Canadas and United States* (Edinburgh, 1835; reprint edition, New York: Benjamin Blom, Inc., 1971), p. 87.
15. ibid., pp. 129–135.
16. Turner, *Holland Purchase,* p. 470.
17. Turner and McIntosh, in their various books, generally agree as to early land values.

Chapter III *Experiments in Land Speculation*

1. O'Reilly, *Sketches of Rochester,* p. 169.
2. Lockwood L. Doty, *A History of Livingston County, New York* (1876; reprint edition, Geneseo, N.Y.: Livingston County Historian, 1979), p. 274.
3. ibid., p. 259.
4. Arthur Parker, *Charles Williamson, Builder of the Genesee Country* (Rochester, N.Y.: Rochester Municipal Museum /Rochester Historical Society, 1927), p. 25. See also Helen Cowan, *Charles Williamson, Genesee Promoter—Friend of Anglo-American Rapprochement* (Rochester, N.Y.: The Rochester Historical Society, 1941).
5. Helen Cowan, *William Berczy's Williamsburg Documents,* Foreword by A. J. H. Richardson (Rochester, N.Y.: The Rochester Historical Society, 1942) pp. 141–266.
6. O'Reilly, *Sketches of Rochester,* p. 253.
7. James Wadsworth Ledger, July 19, 1832, Wadsworth Family Papers, State University of New York at Geneseo, Geneseo, N.Y., pp. 9–10.

Chapter IV *Roads, Steamboats, and The Triumph of Canals*

1. Turner, *Phelps and Gorham,* p. 536.
2. O'Reilly, *Sketches of Rochester,* p. 354*.
3. Doty, *A History of Livingston Country, New York,* p. 335.
4. Bob Whittier, *Paddle Wheel Steamers and Their Giant Engines* (Duxbury, Massachusetts: Seamaster Boats, Inc., 1983), pp. 6–10.
5. *Livingston Journal,* Geneseo, New York, July 28, 1824.
6. W.H. McIntosh, *History of Ontario County, New York* (Philadelphia: Everts, Ensign & Everts, 1876), p. 55.
7. ibid., p. 55.
8. An Englishwoman, pp. 170–171.
9. "The Road Law and Shade Trees," *The American Rural Home,* (April 29, 1871): p. 133.
10. An Englishwoman, p. 169.
11. Caroline C. Richards, *Village Life in America: 1852–1872* (1913; reprint edition, Williamstown, MA: Corner House Publishers, p. 124.
12. Turner, *Holland Purchase,* p. 555.
13. O'Reilly, *Sketches of Rochester,* p. 247.
14. ibid., p. 333.
15. Neil McNall, *An Agricultural History of the Genesee Valley* (1952; reprint edition, Westport, CN: Greenwood Press, Publishers, 1976), p. 247.
16. H.C. Blinn, "Diary," 1856, Genesee Valley Historical Collection, State University of New York at Geneseo, Geneseo, N.Y.

Thayer Gauss Diary

1. Thayer Gauss, "Diary," Bloomfield, New York, 1826–1858, East Bloomfield Historical Society Collection.
2. W. H. McIntosh, *History of Ontario County, New York,* p. 209.

Chapter V *Hewn Timber-Framed English Barns*

1. McNall, *An Agricultural History of the Genesee Valley,* pp. 66–69.

2. Doty, *A History of Livingston County, New York,* p. 272.
3. Ulysses Hedrick, *A History of Agriculture in the State of New York* (Albany: New York State Agricultural Society, 1933), p. 139.
4. Turner, *Phelps and Gorham,* p. 432.
5. Turner, *Holland Purchase,* p. 555 and p. 384.
6. Doty, *A History of Livingston County, New York,* p. 578.
7. "Wheat," *The Genesee Farmer:* I (1831): 21.
8. Fraser-Walker 1816 article of agreement, Fraser family collection.
9. Conklin, *Through Poverty's Vale,* p. 196.
10. Turner, *Phelps and Gorham,* p. 203.
11. J. Howard Pratt, *Memories of Life on the Ridge* (Albion, N.Y.: Orleans County Historical Association, 1978), pp. 109–15.

Chapter VI *Bank Barns and Other Enlarged Gable-Roofed Barns*

1. William Buckminster, *The Cultivator's Almanac and Cabinet of Agricultural Knowledge for the Year 1840* (Boston: D.H. Williams, 1840), p. 85.
2. "Side-Hill Barn," *The Genesee Farmer,* XIII (1852): 20.
3. "Hay Harvesting Machinery," *American Agriculturist,* XXXVI (1877): 250.
4. Gardner Weeks, *Grappling Hay Forks and Railway Hay Conveyors,* Syracuse, N.Y., undated [circa 1878], p. 5.
5. ibid., p. 9.
6. Robert F. Halsted, "19th Century Barns Fast Disappearing," *The Livonia Gazette,* (June, 1981): 2.
7. Graeme Quick and Wesley Buchele, *The Grain Harvesters* (St. Joseph, MI: The American Society of Agricultural Engineers, 1978), p. 27.
8. Hussey had a patent for his reaper six months before McCormick's, but McCormick's seems to have been available earlier thus confusing claims for the "first" reaper.
9. McNall, *An Agricultural History of the Genesee Valley,* p. 118.
10. James H. Smith, *History of Livingston County, New York,* (Syracuse, New York: D. Mason & Company, 1881), p. 119.
11. The reel is a device, still used today, which rotates its horizontal paddles in order to stand the grain up into position for cutting. Figure 6-21 shows McCormick's reaper with a reel.
12. Joseph Hall, "McCormick's Patent Virginia Reaper," *The Genesee Farmer,* XII (1851),: 148.
13. Ardrey, R. L. *American Agricultural Implements* (Chicago: Ardrey, 1894, reprint edition, ARNO Press, New York Times, New York: 1972), p. 228.
14. C.H. McCormick, "McCormick's Patent Virginia Reaper," *The Genesee Farmer,* XI (1850): 173.
15. "Editor's Table," Editorial, *The Genesee Farmer,* XIV (1853): 34.
16. Ardrey, *American Agricultural Implements,* p. 229.
17. Paul Knickerbocker, "Knickerbocker Family History," (unpublished), 1984.

Chapter VII *Mills and the Sawn Timber-Framed Barns*

1. *Combination Atlas Map of Genesee County, New York,* (Philadelphia: Everts, Ensign & Everts, 1876).
2. A tub mill utilized a horizontal wheel in a deep tub or barrel (iron bound); a pipe directed the water to the blades. The tub mill did not freeze up in winter and was the precursor of the turbine.
3. Turner, *Phelps and Gorham,* p. 367.
4. Turner, *Holland Purchase,* p. 471.
5. ibid., p. 383.
6. Turner, *Phelps and Gorham,* p. 347.
7. ibid., p. 358.
8. Doty, *History of Livingston County, New York,* p. 492.
9. Turner, *Phelps and Gorham,* p. 433.
10. Turner, *Holland Purchase,* p. 382.
11. Basil Hall, *Travels in North America,* p. 135.
12. O'Reilly, *Sketches of Rochester,* p. 361.
13. McNall, *Agricultural History of the Genesee Valley,* p. 120.
14. Frances Trollope, *Domestic Manners of the Americans* ed. by Donald Smalley (New York: Alfred A. Knopf, 1949), p. 376.

15. Solon Robinson, *Facts for Farmers,* (New York, A. J. Johnson, 1866), pp. 325–330.
16. L.D. Snook, "Barn Flooring," *American Agriculturist* XLIII (1884): 18.
17. The name "Oxbow" is derived from a nearby cutoff meander or oxbow of the Genesee River.

Chapter VIII *Cobblestone Construction*

1. Although anonymous and undated this tintype appears to date from the Civil War. The main house is an upright and wing Greek Revival structure and the main barn is a large English style barn. The newer barn to the right has the board and batten siding which became popular just before the war.

Chapter IX *Greek, Gothic and Other Decorative Barns*

1. Sir Walter Scott, *Ivanhoe* (London, 1830; reprint edition, New York: Dodd Mead & Co., Inc., 1941), p. 154.
2. J. Sheldon Fisher and Lewis F. Fisher, *A Victor Album,* (Fishers: The Victor Historical Society, 1981), p. 55; also, conversation with J. Sheldon Fisher.
3. Clater, Youate, Skinner & Mills, *Farmers' Barn-Book* (Philadelphia: J.B. Lippincott & Co., 1866), p. 278.
4. Andrew Jackson Downing, *Victorian Cottage Residences* (1873, reprint edition, New York, Dover, 1981), pp. 178–9.
5. Calvert Vaux, *Villas and Cottages* (1864, reprint edition, New York: Dover, 1970), p. 244.
6. William P. Wadsworth, *Riding to Hounds in America* (Berryville, VA: The Chronicle of the Horse, 1962), p. 21.

Chapter X *A Crop Sampler*

1. "An American Silo," *American Agriculturist* XXXVI (1877): 336.
2. Hedrick, *A History of Agriculture In New York State,* p. 365.
3. Doty, *History of Livingston County, New York,* p. 578.
4. Turner, *Phelps and Gorham,* p. 431.
5. McIntosh, *History of Monroe County,* p. 36.
6. O'Reilly, *Sketches of Rochester,* p. 34.
7. Thomas Jefferson, *Garden Book* (Philadelphia: Betts., 1944), p. 243.
8. "Agricultural Progress in New York," *The Genesee Farmer* XIV (1853): 309.
9. *Livingston Republican,* 30 July, 1885.
10. Mary R. Root, *History of York, Livingston County, New York* (Caldedonia, N.Y.: Big Springs Historical Society, 1940), pp. 9–76.
11. "Salt on Hay," *The Genesee Farmer* I (1831): 227.
12. J. Cozzens, *Fifth Annual Report of the American Institute, April 20, 1847* (Albany: Benthuysen, 1847), p. 522.
13. "Hay Cured, not Dried," *American Agriculturist* XXXIV (1877): 334.
14. J. Howard Pratt, *Memories of Life on the Ridge,* (Albion, New York: Orleans County Historical Association, 1978), p. 51.
15. Richard Silvernail, "The Agriculture Land Use of the Western Finger Lakes Region of New York," (PhD Dissertation, University of North Carolina, 1960), p. 172.
16. McIntosh, *History of Ontario County,* p. 30.
17. Joseph Kane, Historian, Town of Erwin, Steuben County, New York, personal letter.
18. Jonathan Periam, *The Home and Farm Manual* (Thompson and Co., New York and St. Louis, 1884, reprint edition, Greenwich House, New York, 1984), p. 238.
19. Marion Rhodes, "Tobacco in Big Flats," Big Flats Historical Society Bulletin, Big Flats, N.Y., 1982, p. 1.
20. Joseph Kane, *Tobacco Growing in Erwin,* (Erwin Museum, Painted Post, New York, 1985), p. 1.
21. *The Industries of Rochester,* (Rochester, Elstner Pub., 1888), p. 86.
22. *American Agriculturist* XLIII (1884): 279.
23. "Small Beer," *The Genesee Farmer* I (1831): 332.
24. Lydia Child, *The Frugal Housewife* (1832; reprint edition, Worthington, Ohio: Worthington Historical Society, 1965), p. 86.
25. McIntosh, *History Ontario County,* p. 210.
26. Hedrick, *History of Agriculture In New York State,* p. 156.

Chapter XI *Horticulture*

1. O'Reilly, *Sketches of Rochester,* pp. 43–4.
2. McIntosh, *History of Ontario County,* p. 210.
3. Hedrick, *History of Horticulture In America to 1860* (New York: Oxford University Press, 1950), p. 214.
4. "Bringing Wheat to Market," *The Genesee Farmer* I (1831): 281.
5. Herbert Wisbey, *The Sodus Shaker Community,* (Lyons, New York: Wayne County Historical Society, 1982), pp. 10, 25.
6. William Aeberli and Margaret Becket, "Joseph Harris- Captain of the Rochester Seed Industry," *The University of Rochester Library Bulletin,* 35 (1982):75.
7. "Mr. Hayward's Seed Barn," *The American Rural Home* (Aug, 1871): 253.
8. Hedrick, *A History of Agriculture In New York State,* p. 245.

Chapter XII *Railroads Change the Genesee Country*

1. "Miscellanies," *The Genesee Farmer* I (1831): 200.
2. Harry Pierce, *Railroads of New York,* (Cambridge: Harvard, 1953), p. 27.
3. McIntosh, *History of Monroe County,* p. 45.
4. O'Reilly, *Sketches of Rochester,* p. 347.
5. ibid., p. 344.
6. Oliver Jensen, *Railroads in America* (New York: American Heritage Publishing Co., Inc., 1975), p. 36.
7. William S. Kennedy, *Wonders and Curiosities of the Railway* (Chicago: S.C. Grigg & Co., 1884), p. 61.
8. Frank W. Stevens, *The Beginnings of the New York Central Railroad, A History* (New York: G.P. Putnam's Sons, 1926), pp. 336–8.
9. Edward Mott, *The Story of The Erie* (New York: Collins, 1900), p. 408.
10. Lee Benson, *Merchants, Farmers and Railroads* (New York: Russell & Russell, 1955), p. 95.

Chapter XIII *A Brief Look at Nineteenth Century Agricultural Implement Manufacturing in the Genesee Country*

1. James L. Orr, *Grange Melodies* (Philadelphia: National Grange of Patrons of Husbandry, 1910), pp. 52–3.
2. McIntosh, *The History of Monroe County,* p. 159.
3. ibid., p. 159.
4. Root, *History of the Town of York,* p. 8.
5. Ardrey, *American Agricultural Implements,* pp. 58–61.
6. ibid., p. 28.

Chapter XIV *Gambrel-Roofed Barns*

1. Copy negatives were loaned by Mrs. Herman Facer (formerly Dorothy Wright): the original negatives can no longer be located. Mrs. Facer also vividly recounted the moving events.
2. Fiske Kimball, *Domestic Architecture of the American Colonies and of The Early Republic* (Scribner, New York, 1922, reprint edition, Dover, 1966), p. 47.
3. For example, the October, 1867 *American Agriculturist* contains an early illustrated article on the advantages of the gambrel-roofed barn.
4. The Alabama Town Historian's Collection also includes a group portrait of women who prepared the raising meals.
5. Wilbur Siebert, *The Underground Railroad From Slavery to Freedom* (New York: Arno Press & New York Times, 1968), p. 80.
6. Robert B. Roosevelt, "The New York Fishery Commission," *American Agriculturist* XLIII (1884): 327.
7. *Genesee Country Scrapbook* (Rochester: The Rochester Historical Society, Vol. XV. 1976), p. 53.
8. Roosevelt, "The New York Fishery Commission," p. 327.

Chapter XV *Agricultural Information in Almanacs, Periodicals, Manuals, at Colleges and Fairs*

1. Sherriff made the trip to investigate the

possibilities of farming in North America for his brother. Sherriff, *A Tour Through North America,* p. i.

2. Artel, "Threshing Oats," *The Genesee Farmer* XII (1851): 46.
3. H.L.E., "Threshing Oats," *The Genesee Farmer* XII (1851): 70.
4. Hedrick, *A History of Agriculture In The State of New York,* p. 131.
5. McIntosh, *History of Monroe County,* p. 37.
6. *History of Wyoming County, N.Y.* (New York: F.W. Beers & Co., 1880), p. 75.
7. "Tenth Annual Fair of the N. Y. State Agricultural Society," *The Genesee Farmer* XII (1851): 237.
8. Hedrick, *A History of Agriculture In The State of New York,* pp. 132–3.

Chapter XVI *Livestock*

1. *The Genesee Farmer* IX (1848): 8.
2. Paul Johnson, *Farm Animals in the Making of America* (Des Moines, IA: Wallace Homestead Book Co., 1975), p. 62.
3. James H. Smith, *History of Livingston County, New York,* p. 116.
4. McIntosh, W. H., *History of Ontario County,* p. 27.
5. *Orwine* is now spelled *Erwine* and is a part of Steuben County which was formed from Ontario County.
6. "Town Meeting Ledger," Painted Post, Ontario County, 1794, Collection of Mr. and Mrs. Richard Platt.
7. Hedrick, *History of Agriculture In The State of New York,* pp. 372–3.
8. M.J.P., "Heavy Fleeces," *American Agriculturist* XXXVI (1877): 336.
9. "Dairy Husbandry in New York," *The Genesee Farmer* XIV (1853): 107.
10. Conklin, *Through Poverty's Vale,* p. 45.
11. "Dairy Husbandry in New York," p. 105.
12. ibid., p. 107.
13. Edward Mott, *The Story of the Erie* (New York, John S. Collins, 1900), pp. 406–408.
14. ibid., p. 409.
15. William Gordon, *Stories and History of the Erie Railroad —Rochester Division* (Rochester: Gordon, 1965), p. 24.

Chapter XVII *Barn Arrangements*

1. Thomas C. Hubka, *Big House, Little House, Back House, Barn, The Connected Farm Buildings of New England* (Hanover and London: University Press of New England, 1984), pp. 180–204.
2. Conklin, *Through Poverty's Vale,* p. 165.
3. Broom corn is a variety of sorghum having elongated, stiff, many branched seed stalks.
4. *New York State Inventory* (Albany, 1892).
5. Turner, *Phelps and Gorham,* p. 453.
6. Orson Fowler, *The Octagon House, A Home for All* (1853; reprint edition, New York, Dover, 1973), pp. 174–8.
7. ibid., p. 175.

Chapter XVIII *The Black and White Farm Barn*

1. David Thorp, "Black and White Farm," *Livingston County Agriculture News* (May, 1975): 17–8.

Chapter XIX *Changing Sources of Power on the Farm*

1. P. Barry, "Agricultural Implements at the Great Exhibition," *The Genesee Farmer* XIII (1852): 88.
2. James Disher, *Iron Man Album* (May/June, 1973): 40.
3. Floyd Clymer, *Album of Historical Steam Traction Engines* (New York: Bonanza Books, 1959), p. 10.
4. McIntosh, *History of Monroe County,* p. 263.
5. Robert F. Halsted, "19th Century Barns Fast Disappearing," *The Livonia Gazette* (June, 1981): 2–15.
6. ibid.

Chapter XX *Truss Framed Barns*

1. "Barns Without Beams," *American Agriculturist* (March, 1878): 98.

2. Jennings "Reward" poster copy, Wayne County Historian's Office, Lyons, New York.
3. Jennings Advertisement broadside, 1879. Collection of Daniel La Gasse.
4. ibid.
5. Wayne County Historian's Office, Lyons, New York.
6. Halsted (Edited by Powell), *Barn Plans and Outbuildings* (New York: Orange Judd Company, 1904), p. 10.
7. ibid., p. 9.
8. Tape interview with William P. Wadsworth.

Diary of Seneca Owen, Pulteney, 1908–1912

1. William R. Gordon, *Keuka Lake Memories* (Rochester: Gordon, 1967), p. 206.

Chapter XXI *Wells Barns*

1. William S. Elkins, Letter to Daniel Fink, January 7, 1986 letter, Caledonia, New York.
2. Ella McGinnis, *Pictorial Wheatland,* Vol. 5 (Wheatland, N.Y.: Wheatland Town Historian, 1978), p.29.
3. Wells documents, Jean Melville collection, Scottsville, New York.
4. Account Book, Mrs. John Wooster collection, Perry, New York.

Chapter XXII *Modern Materials in Barn Construction*

1. William R. and Florence Simpson, *Hockshop* (New York: Random House, 1951), pp. 46–9.
2. "The Alfred Sun," November 27, 1969, p. 1.
3. *Concrete Suggestions for the Farm, Fall and Winter–1924* (Allentown, PA: Lehigh Portland Cement Co., 1924).
4. *Dewitt Historical Society Newsletter,* Number 3, Ithaca, New York, 1984; also, "History of Willett's Farm" (unpublished) by Linda Bruno.

APPENDIX I

Transcription of Wadsworth's Broadside

NOTICE TO NEW SETTLERS

Geneseo, Ontario County March 1809

THE SUBSCRIBER offers for sale the following Townships and Tracts of Land, in the Counties of Ontario, Genesee, and Allegheny, in the State of New York.

A tract containing upwards of 60,000 acres, situated within six miles of the Landing in Fall-town, on the west side of the industrious and enterprising settlers, one half of the land consisting of every other three hundred acres throughout the tract, will be sold for wheat, pork and neat cattle; the wheat and pork to be delivered at Fall-town Landing. The very flourishing settlements of West Pulteney, Braddock's Bay and Fairfield are within this tract: The inhabitants in these settlements have been remarkably healthy. Vessels of 200 tons, sail from Lake Ontario up the Genesee River to the lower falls: this place is called Fall-town Landing, and is only six miles from the tract now offered for sale. A barrel of flour can now be sent from Falltown-Landing to Montreal for one dollar, and a barrel of pot-ashes for one dollar and an half; these prices will be reduced, as the business of transportation increases. Most articles of American produce command as high prices at Montreal as at New York.

The intervals and swails in this tract are timbered with elm, butternut, white and black ash, walnut, &c. the up land with sugar maple, beach, basswood, hickory, wild cherry, white oak, black oak, chestnut, &c. There are a number of groves of excellent white pine timber. There are no mountains or ledges, and scarcely one hundred acres of waste land in the tract. Some of, the intervals or flats will produce, if well cultivated, 80 bushels of corn, 800 weight of hemp, or 2000 weight of tobacco on an acre and other crops in proportion.

Also, the Township of Troupton, situated eighteen miles south of the village of Geneseo, and adjoining the village of Dansville. This tract is within twelve miles of Ark Port a landing place on the west branch of the Susquehannah river; a barrel of flour may be transported from Ark Port to Baltimore, for a dollar and a half, and other articles of produce in proportion; the situation of this Township is considered very healthy, the lands are fertile and uncommonly well watered.

Also, the town of Henrietta, being Township No. 12 in the seventh range, on the west side of the Genesee river; this tract is within eight miles of Fall-town landing, and adjoins the flourishing towns of Hartford (now Avon) and Northfield; the lands in Henrietta are excellent and the settlement very flourishing; the lots adjoining the Genesee River containing handsome portions of timbered flats, are put at five dollars per acre, the back lots at four dollars per acre.

Also, a number of Lots in a tract of Land, usually known by the name of Allen's Flats, or the Mount Morris tract, situated in the forks of the Genesee River, fifteen miles south of the great State Road to Niagara, and four miles from the village of Geneseo. The tract contains about 10,000 acres, 5000 acres of which are flats or interval. It has lately been surveyed into lots of convenient size; the village lots contain from one to forty acres, and the farm lots about one hundred acres each. The village lots contain from one to forty acres, and the farm lots about one hundred acres each. The village is situated on elevated ground timbered with white oak, and bids fair to be a very healthy situation. The subscriber will sell the upland and lease the flats, or sell both upland and flats, as applicants prefer.

It is fully ascertained that the flats or interval, on the Genesee river are perfectly adapted to the cultivation of hemp. Mr. Stephen Colton, from Long Meadow raised ten hundred weight of excellent hemp the last season, on one acre of flats in Geneseo. One hundred and six bushels of Indian Corn have been raised on one acre in Allen's flats.

Hemp may be transported by water from the mouth of the Genesee river to Montreal; or it may be sent from Ark Port down the Susquehanna river, in Arks to Baltimore, or it may be sent by land to Albany.

The price at which Lots in the above tracts are put, if from two to five Dollars per acre.—The subscriber usually requires the purchase money to be paid in four equal installments to be made in two, three, four, and five years from the time of purchase, with one year free of interest; in some of the tracts he gives a credit of six and eight years.

Liberal encouragement will be given in the different settlements to Carpenters, Blacksmiths, Shoemakers, Millwrights, and other tradesmen.

The subscriber in order to encourage the settlement of substantial New England Farmers, will exchange a few lots for improved farms.

The tract of Country, in which the above described townships are situated, tho' North of New Jersey, resembles that state in the mildness of its climates. Peaches, Apricots, and Nectarines grow to great perfection on the Genesee River.

A valuable Salt Spring is discovered in Braddock's Bay township. Salt can now be afforded at this spring at one dollar per bushel; when the works are extended salt will probably be afforded at fifty cents per bushel, the same price at which it is now sold at the Onondaga salt works.

A turnpike road is completed from Albany to Canandaigua; and from Canandaigua to Geneseo, and thence to the above mentioned settlements there are excellent waggon roads.

The subscriber has still for sale a number of reserved and other Lots of Land, in the midst of flourishing settlements, in the towns of Geneseo, Hartford, Bloomfield and Pittstown; some of these lots contain handsome improvements.

JAMES WADSWORTH

APPENDIX II

Transcription of Jennings' Broadside, ca 1880

ATTENTION FARMERS!

THE JENNINGS BARN is acknowledged by every one that has seen the frame to be the cheapest, strongest and most convenient frame now in use, without any exceptions, and is endorsed by all leading carpenters throughout the United States. As there are hundreds of them now in use, it sustains the fact that this plan has only to be seen to be appreciated. Remember that there are no cross-ties on the interior, or any obstructions whatever on the floor—the building being clear from one end to the other and from the floor to the peak of the roof, so that a horse-fork can be used to very great advantage.

PATENTED JULY 29, 1879.

In looking at the above cut, fig. 3, a person can readily see that the principle of this construction is such there is no outward pressure to the building after it is finished, but has an inward tendency, if anything; and in order for the building to spread it will not only have to break the truss-braces, purlin and main plates, but will have to lift the whole weight of the roof in moving in either direction. In raising these frames the end-bents are raised first, and staylathed, and then the main-post on the sides are raised one at a time or by pairs, putting in the girts before raising, and when raised stay-lathing them to their place; after the main posts are raised, then comes the purlin arm; bring that around in front of the post where you are going to place it, put in the short girt that goes in the arm, and pin it and fasten a guy to the girt, and hand one end to a man that will climb up on the upper girt, who will hold it while the arm is drawn up by means of a pulley that is fastened to the tenon of the main post; after the arms are raised they should be well stay-lathed; the stay-laths on main post and arms must be left on until the roof is finished, and then they can be taken away with safety, as the roof counteracts all inward pressure. The purlin-arms on end beams are set the same slant as the arms on the interior; the best way to connect the purlin-arms to purlin-plate is to notch the arm around the plate instead of a tenon; the only way they can move is sideways, and they can be spiked, which will prevent and hold them secure. Fig. 2 represents the manner of trussing or tieing main and purlin-plates together; the rafters should be well spiked to main and purlin-plates, as very great strength is derived therefrom. The X or truss- braces are gained into main and purlin-plates so that they may be dropped in after the plates are put in position; the plumb bevel being cut on foot of brace. In building a new barn I generally space off the bents equal disant apart, not exceeding twenty feet; in a twenty feet span it is not necessary to spread the X's over ten feet on plate, that is, find centre of span and measure five feet each way, that will find toe of brace. In tieing main and purlin-plates together I used a rod on every inside bent, but none on the end of building; that is, a building with six bents would require 4 rods on a side, and 4 short belts for foot of purlin-arms.

A very quick and easy way to raise these frames is to raise the end bents first and then form a bent of one side with the main-plates on raising the whole side at one time— the end bents being leaned far enough over and stay-lathed to allow the main plate to pass when the side is raised. It is then stay-lathed and a bent then formed with the purlin-arms, with the purlin-plate placed in position. So it will be seen that these frames can be readily raised in six lifts. The ends of the main-plates are slit or mortised clear out to the end of the plate, allowing the tenon of the corner post to be the full width of the post, so that after the sides are raised the stay-laths can be taken off of the end bents— the frame being then brought together and pinned, thereby dispensing with the old way of climbing up to adjust the main and purlin plates.

Three reasons why this barn is cheaper than the ordinary barns: lst, it takes less timber for the frame; 2d, it is less work to frame it; 3d, it is easier to raise and takes less time. Four reasons why these barns should be built in preference to the old plan; 1st they are stronger and more durable; 2d, they are more roomy and convenient for stowing away grain; 3d, they can be built for less money than the old plan; 4th, they are just the barn a farmer wants for a grain or hay barn. Remember that this plan has been thoroughly tested in every particular, as regards strength, durability and economy. Also bear in mind that old barns may have the beams and cross-ties taken out and this improvement put in at very little expense, especially where the purlin- plates are already in, as is generally the case; but if not, they can be put in very easily. If the purlin plates are already in it will only be necessary to put in the X's and the tie-rods, which can be done without disturbing the roof or siding. I generally fasten the toe or foot of these X's with spikes, as the roof being on prevents their being gained in as they would be in a new frame. It will readily be seen that by putting in the X's first and the tie-rods next, and then tightening up the rods you get the strength of the truss; then by cutting in the arm a square angle at the top, so as to fit around the corner of the purlin-plate, and working the bottom of the purlin arm, as shown in the above cut, and then by putting the arm in position as near as possible right up alongside of the beam and nailng a temporary block on the sill so as to keep the bottom of the arm even with the face of the vertical post on the inside of the barn, also to keep it from slipping while you cut out the beams and cross-ties, you can, when that is done, drive the arms in their right position by hitting them on the side with that what is generally called among carpenters a commander or maul. Thus the job is done without cutting any gains or making any mortises, except where the X's cross one another, and leaves the barn free from obstruction and stronger than it was before.

☞ Every person should without fail see this Barn before building, and should investigate the best Barn plan in existence. They will not only save money, but will be well pleased. Anyone buying a right will be entitled to instructions that will enable any practical carpenter to build it without any trouble.

☞ FARM, TOWN AND STATE RIGHTS FOR SALE Carpenters will do well to secure territory in time.

For further particulars apply in person or address, (enclosing stamp,)

DAVID JENNINGS, Patentee,

[Box 115] LYONS, WAYNE CO., N.Y.

The undersigned have bought the right for the town of phelps, Ontario Co., N.Y. Persons calling on them will be shown these Barns. For further particulars address

JUDSON RAYMER, or BYRON RAMER,

LYONS, WAYNE CO

APPENDIX III

Barn Evolution

ATTEMPTS TO DATE particular barns or identify the builder can be both fascinating and frustrating. Rarely is there a corner date stone, or reliable date of construction carved in a beam, or even a surviving document which indicates the date of construction or the builder's name. Instead of an exact dating an approximate date may have to be arrived at through the consideration of many clues. Identification of a building date for the farm house may provide a starting point for dating the original barn. Houses are more often dated than barns. The many factors to be considered always ensure an interesting puzzle for the barn historian while the frequent presence of conflicting and ambiguous data may frustrate as well. The design of the barn, both interior and exterior, the materials used in its contruction and the building techniques employed, as well as the all-too-often missing documentary evidence, can all be used when available to date a particular barn.

The design of the barn in question is important to the dating. While an *exact* timing of the evolution of different barn designs applicable to *all* parts of the Genesee Country is difficult to discern, a useful chart has been created against which an undated barn may be measured. Many variables and considerable overlap make tidy graphing of the evolution of barn designs difficult but not impossible. For example a new design might or might not have been quickly adopted by neighboring farmers. Adoption of a new design in one location might or might not mean it would be used in a neighboring region. Also, established farmers frequently continued to build and use barn types which suited their needs and which had proven themselves. Skepticism of new, untried designs is to be expected. On the other hand a new design was sometimes embraced immediately and used in most cases whenever a new barn was to be constructed. With these limitations in mind, the historian can note many aspects of the design of a particular barn in his attempts to date it; for example: gable or gambrel roof (other)? type of frame? presence of a basement? relation of the barn to the landscape? original purpose of the barn? relation of the barn to other buildings? If the barn has a hay track, was it built for the hay track or are there signs of alterations to accommodate the track? How are the doors arranged? Is there exterior detailing to match some architectural style and if so, what style? An understanding of when various architectural styles such as Greek or Gothic flourished in the area can be very helpful. These and other design factors can help in establishing a date. No obvious absolute pattern exists if design is considered as the only factor in dating, but if it is considered in conjunction with other key factors, it can be useful.

Additional key factors in the dating are the materials used and the preparation of those materials. Hewn timbers are likely to indicate an earlier dating than sawn timbers in barns of similar design. If the timbers are sawn, the historian should note the saw marks. Figure 7-10 shows the common markings. The method of joining the framing members can also be pertinent. Does the joining use spikes or mortise and tenon? If nails are employed, identification of the type of nails used (hand wrought, machine cut, or wire cut) can also be useful. Figure 7-11 shows the three basic nail types. The use of newer materials such as concrete and steel in the construction is also a telling factor. The historian had best be wary in barn dating. For example, because barns were so important (most considered them to be the most essential structures on the farm) they were commonly enlarged, altered and adapted to suit changing requirements. New materials might be used in a barn modification, but just as readily sound old (used) materials might be incorporated (e.g. old, used, hand wrought nails might be used with new lumber or old beams might even be used in otherwise new construction). New apparatus or hardware might also be applied until today the changes and inclusions make exact dating of the original structure next to impossible.

It is always a pleasure to find that oral or written evidence exists to aid in the dating; present owners as well as former owners and neighbors can be mines of information. Precise dating is frequently difficult, of course, because of the lack of documentation which often exists for the dating of homes. But various documents which might prove to be useful in dating sometimes can be found, including tax records, building contracts, maps, 19th Century County Atlases and diaries. With all the hazards of dating the historian sometimes has the joy of discovering a barn with its date of construction readily available, such as the barn shown in Figure A-1, a small 1891 gambrel barn in Webster's Crossing, Route 15, Livingston County, New York. See detail, figure A-2.

As historians in the field are able to attach dates to particular barns, thus increasing the "pool" of dated barns, the dating of further samples becomes easier. The combined efforts of many groups in a region will lead to more precise dating of barns because of the broadened data base. Dating projects in one region should be useful in dating efforts in regions elswhere. Members of school history classes, historical societies, 4-H clubs, scouts, etc. might have access to sources of dates that a single researcher might never locate (i.e., diaries, attic trunks, that carved date on a beam which no outsider is aware of, etc.). With the above ideas and limitations in mind the following chart is presented in hopes that it can be a useful dating tool:

An Outline of the Evolution of Major Genesee Country Barn Types

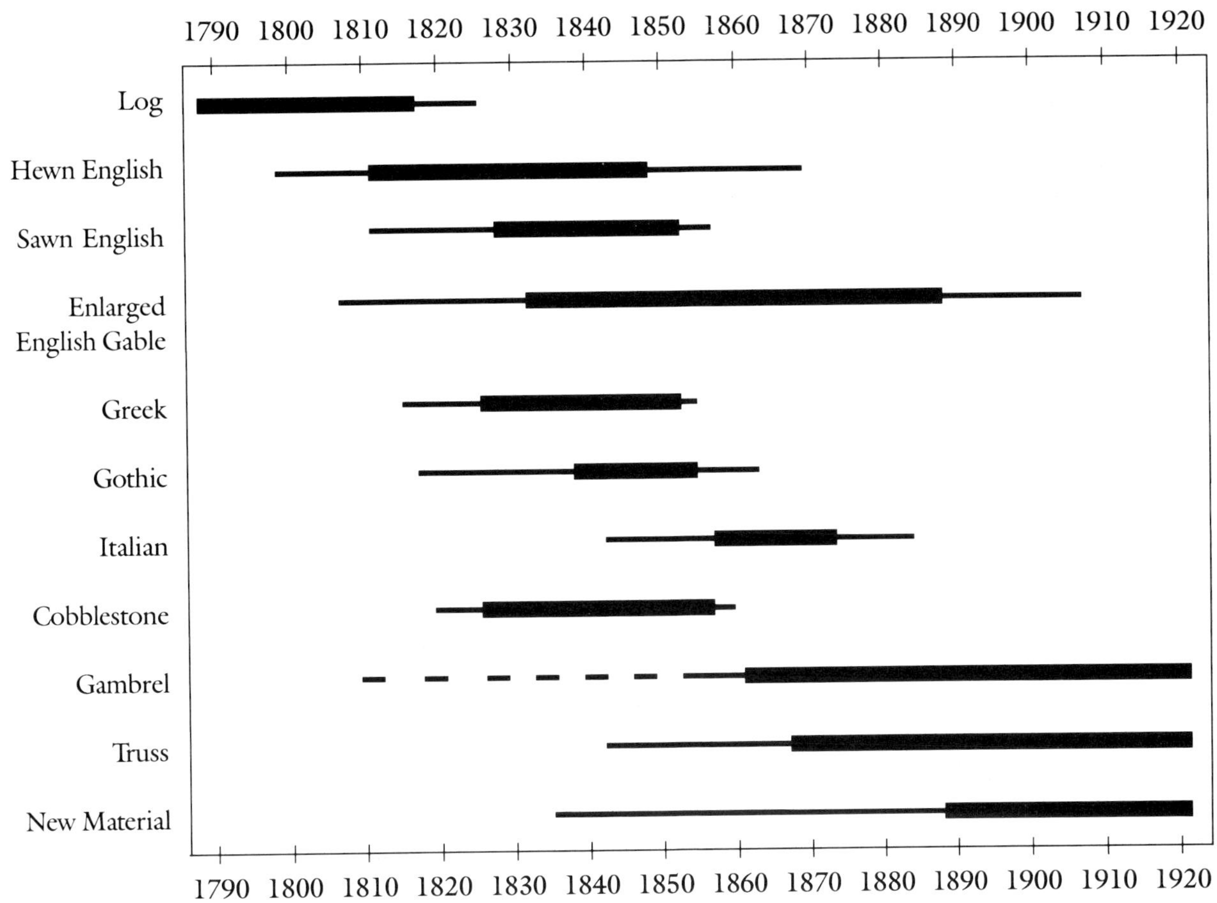

NOTE: no absolute value should be attributed to width of the bar but the wide area signifies the period of most frequent occurrence of a particular barn form.

Dating alone will help create a broader picture of barn construction patterns but understanding who built the barns can also lead to a significant increase in present information. At times the barn builders are known and at other times there are similarities in barns within certain regions that suggest the same builder was at work even if the builder's name might not be known. For example, in certain areas of the Genesee Country horizontal clapboard siding is more frequently seen than in others. One such area is in the Bath & Corning area of Steuben County. The similarity of barns along certain roads suggests that the same builder was involved. The details of the barns indicate clear connections amongst several barns. Horizontal siding appears on barns constructed along County Road 36 between Honeoye and Naples—three are gambrel and 2 gabled. Figure A-3 shows a typical example of these clapboarded barns. In all five examples there are two horizontal framed windows in the gable which provide the final telling resemblance amongst these barns. For most farmers the tightness usually offered by clapboarding was not desired in a hay, grain or cow barn but for barns with other storage requirements such tight boarding was a positive feature. Interpretation of the original intent of the structure can be helped by understanding each and every feature. Also, often a person can use common features such as the siding and windows, doors, foundations, roof pitch, proportions, decorative details, etc. to identify the hand of a particular builder.

As people throughout various regions identify the various builders and also date barns, a clearer picture of the evolution of barns and other structures will certainly evolve.

Figure A-3. Example of horizontal sided barn on County Road 36 between Honeoye and Naples, Ontario County. At least four others in the same area seem to have been built by the same builder; that is, if the physical similarity is an indication. Razed in the Fall of 1986.

APPENDIX IV
Photographing Interiors

PHOTOGRAPHY seems to be one of the most convincing and informative methods of recording the details of a specific barn. With just a bit of help, most who wish to perform this very important photographic documentation can achieve excellent results. Using current photographic equipment most people are capable of photographing barn exteriors by planning for the best light, light that illuminates the most important surfaces. Patience and careful observation of the light falling on a barn are probably the two most important factors in producing successful photographs, if the technical aspects are under control. Some of the photographs in this volume required a four year period before the desired results were obtained. Obstructing foliage, facades facing north, away from the sun, or farm equipment that could not be moved sometimes made repeated attempts necessary.

In this book much emphasis has been placed on the framing of barns, but producing photographs that reveal the framing can be a challenge. Many times the existing barn light is inadequate and supplemental light from a flash unit is required. The ordinary, camera mounted sort of flash often proves to be insufficient for the large spaces involved. Fortunately, the following technique is useful and requires a bare minimum of equipment, much of which is probably already owned by the photographer.

It will be necessary to have a camera with a "time" (T) exposure setting on the shutter or a "bulb" (B) setting. With the T setting the shutter will remain open until it is pressed a second time (in some cameras some other action is required to close the shutter; check the instruction booklet). With the B setting the shutter remains open as long as there is pressure on the shutter release. In newer cameras the B setting is the one most commonly found.

If there is a choice, then a camera with a wide angle lens is desirable. Don't despair, almost any camera with its "normal" lens can do a fine job if no other is available. On a 35mm camera a 50mm lens is considered normal and on a 2¼ inch square camera a 75mm or 80mm lens is considered normal. A camera with interchangeable lenses may already have the proper wide angle lens. On a 35mm camera a lens of 35mm or 28mm is useful. Shorter (wider) lenses can be put to good use as long as one is extra careful of perspective foreshortening and curving of parallel lines in the subject. On a 2¼ inch square camera a 40mm or 50mm lens will be helpful for interior work.

It will be necessary to place the camera on a tripod to steady it during a "time" exposure. The larger the tripod the better. The basic proper exposure is achieved with correct light metering or with an exposure series. In time exposures of barn interiors, a frequent problem is overexposure of the negative in spots where outside light enters the barn through windows and holes. The result is a flare around these bright areas that interferes with the photographic detail in other areas and otherwise degrades the photograph (see Figures 20-10 and 21-11). To compensate in situations where one can't shutter the window or restore a board, additional lighting is needed. Flash of some sort is usually the best or only practical solution. Also, before use the lens must be cleaned with proper lens tissue and lens cleaning fluid to reduce flare and maintain high contrast.

Because underexposure is a most common problem

one should begin with a careful lightmeter reading. In order to establish an exposure base take the reading from a typical wall surface and avoid reading light through a window or open door. By "typical wall surface" is meant that its illumination is representative of the area to be photographed and that it should not be white or black toned but rather a medium gray (color is not important). Then make a series of four exposures with each one twice the exposure of the previous one (e.g. ½ sec, 1 sec, 2 secs, 4 secs). It is less wasteful to achieve the proper exposure in this manner than it is to have to come back to reshoot the photograph.

"Open" flash in which a photoflash is located some place other than the camera site and is triggered independently of the opening of the shutter, is probably the most effective technique to use in the large open spaces of a barn. It can be a flashgun which uses bulbs or an electronic flash (strobe). Some flash units can be used independently of the camera when they have an "open" flash button which trips them at the will of the photographer. An alternate method might be to use a second camera (without film) with a flash attached which can be tripped simply by cocking the shutter and firing the flash at the part of the barn that needs to be illuminated.

To determine exposure in an open flash situation it is perhaps best to work backwords with the flash guide number. Guide numbers are usually supplied by the flash unit manufacturer or the flash bulb maker. Begin by determining the proper "f" stop required to obtain the proper depth of field (sometimes this has to be a bit of a compromise between a "possible, practical" f stop and the desired depth of field). After considering all prior factors such as strength of lighting source, amount of ambient light, and sensitivity of the film, the distance of the flash to the barn wall can be determined: if the f stop is f5.6 for instance, then the guide number can be divided by 5.6 in order to determine the distance from the wall, roof, or bent that the person with the flash must stand for proper illumination of the surface. The indicated guide number* of 380, when divided by the f stop number of 5.6, yields a distance of 67 feet from the surface to be illuminated. However, experience has shown that two thirds of the calculated distance produces the best results, so in this case the distance from the flash to the wall surface should be roughly 44 feet. Because flash illumination obeys the inverse square law of light (flash brightness decreases rapidly with increasing distance from the subject) it is likely that the most powerful flash available will be needed in order to obtain the proper working distances or f stop.

Note: experience has shown ISO films of 400 in black and white or color to be excellent choices for interior photography. These "fast" films provide the desired mix between the light from the flash and natural light exposure. A significant number of the interior photographs in this volume required exposures of 30 seconds to one full minute even with ISO 400 film. In color work where the lighting is extremely contrasty, color print film is more "forgiving" than slide film of exposure problems, and slides can also be made from the resulting negatives if required.

In open flash photography with the camera on B or T there are two sources of illumination. Natural light will illuminate some areas and the flash will fill in those areas that are poorly lighted or marginal. It should be noted that often only one bulb is required at a strategic point to illuminate some critical area. For instance, open barn doors often illuminate the immediate interior well enough but might leave the end of the barn in obscurity. One flash at the appropriate area will rectify the problem and may indeed make the intervening barn framing easier to see and understand. Experience will show you how much coverage the flash reflector produces for a given distance from the subject. From the appropriate distance, first calculate the exposure (time and f stop). Then the person who will fire the flash must find a suitable "hiding" position; at times lying down will be the only possible position. To diminish the likelihood that the flash person will show up on the film, dark clothing should be worn. In a single bulb situation the person cannot stand up (except behind a post or behind some farm equipment or bales) without becoming a silhouette on the film.

One of the fascinating aspects of open flash photography is the fact that multiple flash is not only possible but often desirable. My wife and son often assist in the work and both claim anyone can be quickly taught to operate the flash units. At this point it is a good idea to

* (supplied by the manufacturer for 25B medium peak flashbulbs when used with an ASA/ISO number of 400)

go through a hypothetical multiple flash situation before actually attempting to expose film. First, the camera must be set up on the tripod at the proper location. Next, the firing point(s) must be decided on after a general exposure is determined with the light meter. Suppose the meter indicates f5.6 at 30 seconds. Perhaps there are two dark corners of the barn which require special illumination. Based upon the light meter information for the available light exposure there are 30 seconds in which to set off two flashes in the required locations. There should be little problem setting off the two bulbs within the allotted time although a practice dry run always helps. The flash person goes immediately to the first firing point, which based on earlier calculations, would be about 40 feet from a wall (with ISO 400 film and 25B bulbs in a proper polished reflector). The photographer opens the shutter and gives the flash person the command to "fire"; he fires, and walks immediately to the next firing place. On the way the first bulb should be removed (use a glove or handkerchief, the bulb will be hot) and placed in a cloth bag or jacket pocket; do not drop the hot bulb in the barn lest straw or hay be set on fire, and do not put the hot, spent bulb in with fresh bulbs or they may be discharged. A fresh bulb is inserted and fired at the second location as soon as the bulb is ready. If you stay calm, 30 seconds will be plenty of time for two bulbs. The flash person should not stand still in one place or he will be recorded on the film; he should walk around or leave the barn if possible while the shutter is still open. One additional word of caution: the person firing the flash should take pains to make sure the flash reflector shields the bulb from the camera lens or the flash itself will be recorded.

The filament of flash bulbs burns for a period while the length of an electronic flash burst is about 1/1000 or 1/500 of a second. Sometimes the shortness of the electronic flash duration causes reciprocity failure in the film, which may mean that the flash person will have to stand considerably closer to the surface to be illuminated than calculations suggest. A few experiments will lead to excellent results but make sure to record camera settings, bulb type used and distances to prime lighted surfaces.

This drawing represents a hypothetical situation in which the camera and flash units are separated in order to illuminate the end of the barn effectively. The camera is set up on a tripod with its shutter open while on the bulb setting. Two separate flashes are set off one after another, using the open flash button, at the proper distance from the wall, a distance based on the calculations made using the flash guide number.

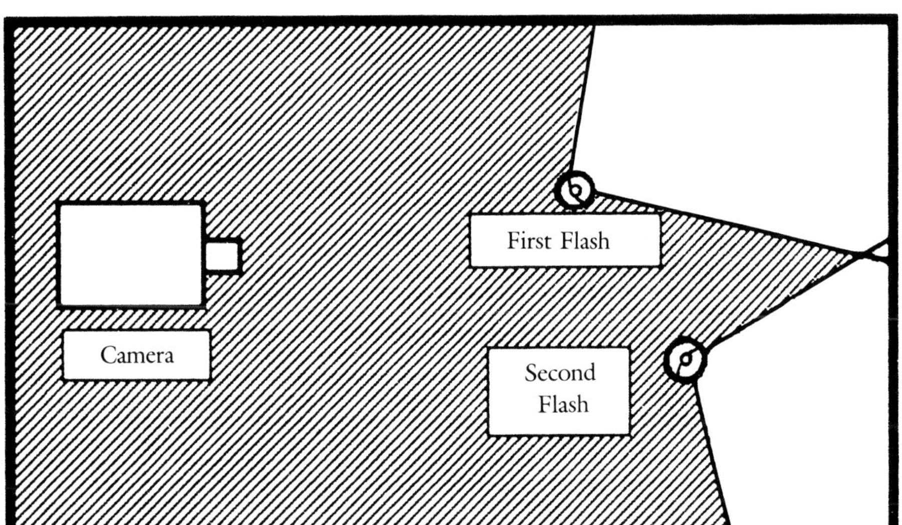

Bibliography

"Agriculture Machine Works of Dow & Fowler, Fowlerville." *Livingston Republican,* September 26, 1867.

Alberts, Robert. *Benjamin West.* Boston: Houghton Mifflin Company, 1978.

Allen, R.L. *Brief Compendium of American Agriculture.* New York: Saxton & Miles, 1846.

American Agriculturist. (Periodical) New York: Orange Judd Company, 1842 to present.

American Farmer and Farm News. (Periodical) Springfield, Ohio: The American Farmer Company, 1881 through 1893.

American Institute. *Fifth Annual Report of the American Institute.* Albany: C. van Benthuysen and Co, Public Printers, 1847.

Anderson, Mildred. *Genesee Echoes.* Castile: New York, Mildred Anderson, 1956.

Andrews, Edward and Andrews, Faith. *Work and Worship.* Greenwich, Connecticut: New York Graphic Society, 1974.

Archer, Robert. *A History of the Lehigh Valley Railroad.* Berkeley, California: Howell-North Books, 1977.

Ardrey, Robert. *American Agricultural Implements.* New York: Arno Press, 1972.

Asher & Adams. *Pictorial Album of American Industry.* New York: Asher & Adams, 1876 (reprint by Rutledge Books, 1976.)

Balla, Wesley. "The Development of The Leasehold System of Land Tenure in the Genesee Valley Under The Wadsworth Family of Geneseo, N.Y." Geneseo: SUNY Geneseo Milne Library, 1977.

Barber, John W., and Howe, Henry. *Early Woodcut Views of New York and New Jersey.* New York: Dover, 1975.

Barbour, James. *Surveys of Proposed Routes of a National Road from the City of Washington to Buffalo.* Washington, D.C.: U.S. Government, 1827.

Barnes, Katherine. *The Genesee Country.* Perry, New York: Barnes, 1972.

Barons, Virginia M. "The Johnston Harvester Co." *The Country Post,* March 9, 1972.

Barry, Patrick. *The Fruit Garden.* Auburn & Rochester: Alden & Beardsley, 1857.

Bartram, John. *Observations . . . Made by Mr. John Bartram in His Travels from Pensilvania to Onondago, Oswego and the Lake Ontario.* London: Whiston & White, 1751 (reprint by Readex Microprint 1966).

Bealer, Alex. *The Tools That Built America.* New York: Bonanza Books, 1976.

Beecher, Mark. *The Tools That Built Niagara.* Lockport, N.Y.: Niagara County Historical Society, 1976.

Beers, F.W. *History of Allegany County, N.Y.* New York: Beers and Company, 1879.

Belluscio, Lynne. *Selected Recipes from The Genesee Farmer, 1831 to 1856.* LeRoy, New York: Lynne Belluscio, 1981.

Benson, Lee. *Merchants, Farmers & Railroads.* New York: Russell & Russell, 1955.

Benson, Tedd and Gruber, James. *Building the Timber Frame House.* New York: Charles Scribner's Sons, 1980.

Betts, Edwin, Annotator. *Thomas Jefferson's Garden Book, 1766–1824.* Philadelphia: The American Philosophical Society, 1944.

Boyd, William. *History of the Town of Conesus.* Conesus, New York: Boyd, 1887.

Brown, William H. *The History of the First Locomotives in America.* New York: D. Appleton and Company, 1874.

Buckminster, William. *The Cultivator's Almanac and Cabinet of Agricultural Knowledge for the Year 1840.* Boston: D. H. Williams, 1840.

Bullock, John. *The American Cottage Builder.* New York: Stringer and Townsend, 1854.

Bunnell, H. O.and Quick, F. N. *Dansville 1789–1902.* Dansville, New York: Instructors Publishing Company, 1902.

Butterfield, E. O. *The Singing Class and Convention Book.* Philadelphia: The J. M. Armstrong Company, 1889.

Canal Museum Staff. *A Canalboat Primer.* Syracuse, N.Y.: The Canal Museum, 1981.

Carleton, Will. *Farm Ballads.* New York: Harper & Brothers, 1873.

Carleton, Will. *Farm Festivals.* New York: Harper & Brothers, 1881.

Carmer, Carl. *The Tavern Lamps are Burning.* New York: David McKay Company, Inc., 1964.

Chazanof, William. *Joseph Ellicott and the Holland Land Company.* Syracuse: Syracuse University Press, 1970.

Child, Hamilton. *Gazetteer and Business Directory of Steuben County, N.Y.* Syracuse: Child, 1868.

Child, Mrs. (Maria). *The American Frugal Housewife.* Boston: Carter, Hendee & Co., 1832, Twelfth Edition (1965 reprint by Worthington Historical Society, Worthington, Ohio).

Cirker, Blanche, editor. *1800 Woodcuts by Thomas Bewick And His School.* New York: Dover Publications, 1962.

Clarkson, Roy. *Tumult on the Mountains.* Parsons, West Virginia: McClain Printing Company, 1964.

Clater, Youate, Skinner and Mills. *Farmers' Barn-Book.* Philadelphia: J. B. Lippincott & Co., 1866.

Clayton, W. W. *History of Steuben County, New York.* Philadelphia: Lewis, Peck and Co., 1879.

Clune, Henry W. *The Genesee Country.* Rochester: Memorial Art Gallery of the University of Rochester, 1975.

Clune, Henry. *Rivers of America: The Genesee.* New York: Holt Rinehart, and Winston, 1963.

Clune, Henry and Volz, Robert. *A Journey Through the Genesee Country, Finger Lakes Region and Mohawk Valley from Patrick Campbell's Travels.* Rochester, New York: The Friends of the University of Rochester Libraries, 1978.

Clymer, Floyd. *Album of Historical Steam Traction Engines and Threshing Equipment.* New York: Bonanza Books, 1949.

Combination Atlas Map of Genesee County, New York. Philadelphia: Everts, Ensign & Everts, 1876.

Conklin, Henry. *Through Poverty's Vale,* Edited by Wendell Tripp. Syracuse: Syracuse University Press, 1974.

Cook, Frederick, Editor. *Journals of the Military Expedition of Major General John Sullivan.* Auburn, New York: State of New York, 1887.

Cotton, J. Randall. "Ornamental Concrete Block Houses," *The Old-House Journal,* XII, (October 1984): 161, 180–183.

The Country Gentleman. (Periodical) Albany, N.Y.: Luther Tucker & Son, 1853–1879+.

Country Life in America. (Periodical) Chicago/New York: Doubleday, Page & Company, 1906 & 1907.

Cowan, Helen, *Charles Williamson, Genesee Promoter—Friend of Anglo-American Rapprochement,* XIX. Rochester, New York: The Rochester Historical Society, 1941.

Cowan, Helen, *William Berczy's Williamsburg Documents,* Foreword by A. J. H. Richardson, XX, part II. Rochester, New York: The Rochester Historical Society, 1942.

The Cultivator. (Periodical) Albany: New York State Agriculture Society, 1830.

Davidson, J. Brownlee. *Agriculture Machinery.* New York: John Wiley & Sons, Inc., 1931.

Davis, Alexander Jackson. *Rural Residences.* New York, Alexander Jackson Davis, 1838 (reprint by Da Capo Press, New York, 1980).

Davis, Kelly. "Dairying in New York State." *The Conservationist.* 36-3, (Nov/Dec 1981): 34–37.

Denton, Daniel. *A Brief Description of New-York.* London: Denton, 1670 (reprint by Readex Microprint 1966).

Doty, Lockwood. *A History of Livingston County, New York.* Geneseo, New York: Edward Doty, 1876 (reprint by Livingston County Historian, Geneseo, New York, 1979).

Downing, Andrew Jackson. *The Architecture of Country Houses,* New York: Appleton and Company, 1850 (reprint by Dover 1969).

Downing, Andrew Jackson. *A Treatise on the Theory and Practice of Landscape Gardening.* New York: A. O. Moore & Co., 1859.

Downing, Andrew Jackson. *Victorian Cottage Resi-*

dences (first printed as *Cottage Residences*). New York: John Wiley & Son, 1873 (reprint by Dover 1981).

Early Arts in the Genesee Valley (Eighteenth & Nineteenth Centuries). Geneseo, New York: Genesee Valley Council on the Arts, #3 in the series, 1974.

Editors. *The Smithsonian Book of Invention.* Washington, D.C.: Smithsonian Exposition Books (Distributed by Norton & Co.), 1978.

Ekblaw, K. J. T. *Farm Structures.* New York: Macmillan, 1917.

Ellis, John, and Bealer, Alex. *The Log Cabin.* Barre, Massachusetts: Barre Publishing, 1978.

Fairchild, Herman. *Geologic Story of the Genesee Valley.* Rochester, New York: Fairchild, 1928.

Farm Directory and Reference Book, Monroe and Livingston Counties, New York, 1917. New York: Orange Judd Company, 1917.

The Farm Journal Illustrated—Rural Directory of Livingston County, New York. Philadelphia: Wilmer Atkinson Co., 1917.

Farm Power, 2nd Ed. Chicago: International Harvester Company of America 1915 (reprint by the American Society of Agricultural Engineers, undated).

Farmers' Library, John Skinner, Editor. New York: Greeley & McElrath, 1846.

Fessenden, Thomas G. *The New American Gardener.* Boston: Russell, 1828.

Fippen, Elmer. *Rural New York.* Port Washington, New York: Kennikat Press, 1971 (reissue).

Fisher, Sheldon, & Fisher, Lewis. *A Victor Album, A Pictorial History of the Town of Victor, New York.* Fishers, N.Y.: The Victor Historical Society, 1981.

Fitch, James. *American Building, The Historical Forces That Shaped It.* New York: Schocken Books, 2nd Edition, 1977.

Fitchen, John. *The New World Dutch Barn.* Syracuse: Syracuse University Press, 1968.

Flint, Charles L. *Milch Cows and Dairy Farming.* Boston: Crosby & Nichols, 1862.

Fowler, Gary. *"The Effects of Land Tenure on Land Management in the Genesee Valley: A Study in Methodology,"* thesis. Syracuse University, 1957.

Fowler, Orson. *The Octagon House, A Home For All,* New Introduction by Madeleine B. Stern. New York: Fowlers and Wells, 1853 (reprint by Dover, New York, 1973).

Fox, Dixon R. "The Old Barn," *New York History,* Burlington: New York State Historical Association, Vol XIX, No. 1, (January 1938): 17–35.

French, J. H. *Gazetteer of the State of New York.* Syracuse: R. Pearsall Smith, 1860 (reprint by Heart of the Lakes Publishing, Interlaken, N.Y, 1980).

Gage, M. H. "From Corduroy Roads to Canals." Rochester: *Upstate New York/Democrat and Chronicle,* (June 25, 1978): 12–19.

Gauss, Thayer. "Diary, 1826–1858." Bloomfield, N.Y.: East Bloomfield Historical Society.

A Genesee Harvest, A Scene in Time 1779. Geneseo, N.Y.: The Genesee Valley Council on the Arts, #7 in series, 1979.

The Genesee Farmer (Periodical). Rochester, N.Y.: Luther Tucker, 1831–1840.

The Genesee Farmer—"second series" (Periodical). Rochester, N.Y.: James Vick, Daniel Lee, Joseph Harris et. al., 1840–1867.

Gillon, Edmund. *Early Illustrations and Views of American Architecture.* New York: Dover Publications, Inc., 1971.

Glassie, Henry. "The Variation of Concepts Within Tradition: Barn Building in Otsego County, New York." *Geoscience and Man,* V (1974): 177–235.

Gordon, William R. *Keuka Lake Memories.* Rochester, New York: Gordon, 1967.

Gordon, William R. *Stories and History of the Erie Railroad-Rochester Division.* Rochester, New York: Gordon, 1965.

Grafton, John. *New York in the Nineteenth Century.* New York: Dover Publications, 2nd edition, 1980.

Gras, Norman. *A History of Agriculture in Europe & America.* New York: F.S. Crofts & Co., Publishers, 2nd edition, 1946.

Hall, Basil. *Travels in North America, in the Years 1827 and 1828.* Edinburgh: Cadell and Co., 1829 (reprint by Arno Press, Inc., New York, 1974).

Halsted, Byron. *Barn Plans & Outbuildings.* New York: Orange Judd, 1881.

Halsted, Byron, revised by Edwin Powell. *Barn Plans & Outbuildings.* New York: Orange Judd, 1904.

Hansen, Hans J., translated by Janet Seligman. *Architecture in Wood.* New York: The Viking Press, 1971.

Harney, George. *Barns, Outbuildings and Fences.* New York: George Woodward, 1870.

Hatch, Alden. *The Wadsworths of the Genesee.* New York: Coward-McCann, Inc., 1959.

Haydon, Roger. *Upstate Travels, British Views of Nineteenth-Century New York.* Syracuse: Syracuse University Press, 1982.

Hedrick, Ulysses. *A History of Agriculture in the State of New York.* Albany, New York: New York State Agricultural Society, 1933.

Hedrick, Ulysses. *A History of Horticulture in America to 1860.* New York: Oxford University Press, 1950.

Hedrick, Ulysses. "What Farmers Read." *New York Historical Association,* Burlington, Vt., XVII, (July 1936): 281–289.

Hiles, Theron. *The Ice Crop.* New York: Orange Judd Co., 1893.

History of Allegany County, N.Y. New York: F. W. Beers & Co., 1879.

History of Wyoming County, N.Y. New York: F. W. Beers & Co., 1880.

Holton, Gladys Reid. *The Genesee Valley Canal.* Brockport, New York: Phillip Maples-Publisher, 1970.

Hornung, Clarence. *Handbook of Early American Advertising Art.* New York: Dover, 1947.

Hubka, Thomas C. *Big House, Little House, Back House, Barn,* Hanover and London: University Press of New England, 1984.

Hubka, Thomas C. "The Connected Farm Buildings of Southwestern Maine" *Pioneer America,* Pioner America Society, Inc. (1977):143–178.

Hurt, R. Douglas. *American Farm Tools From Hand-Power to Steam-Power.* Manhattan, Kansas: Sunflower University Press, 1982.

Jacques, D. H. *The House: A Manual of Rural Architecture: Country Houses and Out-Buildings.* New York: The American News Company, 1866.

Jefferson, Thomas, edited by Edwin Morris. *Garden Book.* Philadelphia: Betts, 1944.

Jensen, Oliver. *Railroads in America.* New York: American Heritage Publishing Co., Inc., 1975.

Johnson, Clifton. *Highways and Byways of the Great Lakes.* New York: The MacMillan Company, 1911.

Johnson, Paul C. *Farm Animals in the Making of America.* Des Moines, Iowa: Wallace-Homestead Book Company, 1975.

Johnson, Paul C. *Farm Inventions in the Making of America.* Des Moines, Iowa: Wallace-Homestead Book Company, 1976.

"The Johnston Harvester Company, 1870." *Western Monroe Historical Society (Annual Report).* Brockport, New York: 1976/1977.

Kennedy, William S. *Wonders and Curiosities of the Railway.* Chicago: S.C. Griggs & Co., 1884.

Kimball, Fiske. *Domestic Architecture of The American Colonies and of The Early Republic.* New York: Charles Scribner's Sons, 1933 (reprint by Dover, 1966).

Klamkin, Charles. *Barns, Their History, Preservation and Restoration.* New York: Bonanza Books, 1979.

Kuhlmann, Charles. *The Development of the Flour Milling Industry in the United States.* Boston & New York: Houghton Mifflin Company, 1929.

Landon, Fred. *Western Ontario and the American Frontier.* Toronto: Ryerson Press, 1941.

Lane, Timothy. "The Dating of Old Barns, An In-Depth Study of Three Barn Groups in the Town of York." SUNY Geneseo Milne Library, April, 1973.

Lanphear, Ethan. *Scenes, Memories and Travels of Eighty-Two Years.* Plainfield, N.J.: Lanphear, 1900.

Lape, Fred. *A Farm and Village Boyhood.* Syracuse: Syracuse University Press, 1980.

Livingston Republican (Newspaper), Edited S. P. Allen to Ray Sherman, 1837 to 1980. Geneseo, New York.

Long, Amos, Jr. *The Pennsylvania German Family Farm.* Breinigsville, Pennsylvania: The Pennsylvania German Society, 1972.

Loudon, John. *An Encyclopaedia of Cottage, Farm and Villa Architecture.* London: Longman, Rees, Orme, Brown, Green & Longman, 1834.

McCulley, Mary (Editor), [by the Officially Appointed Historians of Genesee County]. *History of Genesee County, New York, 1890–1982.* Interlaken, N.Y.: Heart of the Lakes Publishing, 1985.

McGinnis, Ella. *Pictorial Wheatland.* Wheatland, N.Y.: Wheatland Town Historian, V, 1978.

McGinnis, Ella. *Pictorial Wheatland.* Wheatland, N.Y.: Wheatland Town Historian, VII, 1984.

McIlwraith, Thomas F. "The Diamond Cross: An

Enigmatic Sign in the Rural Ontario Landscape." *Pioneer America,* Pioneer America Society, Inc. 13 (March 1981): 27–38.

McIntosh, W.H. *History of Monroe County, New York.* Philadelphia: Everts, Ensign & Everts, Lippincott Press, 1877.

McIntosh, W.H. *History of Ontario County, New York.* Philadelphia: Everts, Ensign & Everts, Lippincott Press, 1876.

McKelvey, Blake. *A Panoramic History of Rochester.* Woodland Hills, California: Windsor Publications, Inc., 1979.

McKelvey, Blake. *Rochester, A Brief History.* Lewiston, New York: The Edwin Mellen Press, 1984.

McKelvey, Blake. *Rochester on the Genesee.* Syracuse: Syracuse University Press, 1973.

McKelvey, Blake. *Rochester, The Water-Power City, 1812–1854.* Cambridge, Massachusetts: Harvard University Press, 1945.

McKinley, Marvin. *Wheels of Farm Progress.* St. Joseph, Michigan: The American Society of Agricultural Engineers, 1980.

McMaster, Guy H. *History of the Settlement of Steuben County, N.Y.* Bath, N.Y.: Underhill & Co., 1853.

McNall, Neil. *An Agricultural History of the Genesee Valley.* Westport, Connecticut: Greenwood Press, Publishers, 1952, 1976.

McNall, Neil. "King Wheat in the Genesee Valley." Cooperstown, New York: *New York State Historical Association,* XXVII (October 1946): 426–443.

Mau, Clayton. *The Development of Central and Western New York.* Dansville, N.Y.: Mau, 1958.

Merrill, Arch. *Land of the Senecas.* Rochester, N.Y.: Creek Books, 1949.

Merrill, Arch. *A River Ramble.* Rochester, N.Y.: Creek Books, 1943.

Merrill, Arch. *The Underground, Freedom's Road and Other Upstate Tales.* Rochester, New York:Seneca Book Bindery Co., 1963.

Merrill, Arch. *The White Woman and Her Valley.* Rochester, New York: Seneca Book Bindery Co., 4th printing, 1968.

Morgan, Lewis. *League of the Ho-De-Sau-Nee or Iroquois.* New York: Dodd, Mead & Co., 1904.

Morrison, Frank. *Feeds and Feeding, Abridged.* Ithaca, New York: The Morrison Publishing Company, 1951.

Mott, Edward H. *Between the Ocean and the Lakes, The Story of Erie.* New York: Collins, 1899.

New Century Atlas of Livingston County, New York. Philadelphia: Century Map Company, 1902.

New York State Divison of Archives and History. *Sullivan-Clinton Campaign in 1779.* Albany: The University of the State of New York, 1929.

O'Connell, John. *Railroad Album.* Chicago: Popular Mechanics Press, 1954.

O'Dea, Jospeh C. *Horses and Hounds of the Genesee,* (Exhibition Catalog). Geneseo, New York: SUNY Geneseo, 1973.

Ogilvie, William E. *Pioneer Agricultural Journalists.* Chicago: Arthur Leonard, 1927.

O'Reilly, Henry. *Sketches of Rochester with Incidental Notices of Western New York.* Rochester: William Alling, 1838 (reprint by James Brunner, Geneseo, New York, 1984).

Orr, James L. *Grange Melodies.* Philadelphia: National Grange of Patrons of Husbandry, 1910.

Parker, Arthur. *Charles Williamson, Builder of the Genesee Country.* Rochester, New York: Rochester Municipal Museum/Rochester Historical Society, 1927.

Parker, Arthur. *The History of the Seneca Indians.* Port Washington, New York: Empire State Historical Publication/Ira J. Friedman, Inc., 1967.

Perelman, N.J. and Rovere, Richard (Introduction). *1897 Sears, Roebuck Catalogue.* New York: Chelsea House Publishers, 1968.

Periam, Jonathan. *The Home & Farm Manual.* New York & St. Louis: N.D. Thompson & Co., 1884.

Peterson, Charles (Annotation and Introduction). *The Rules of Work of the Carpenters' Company of the City and County of Philadelphia, 1786.* Philadelphia: The Carpenters' Company of the City & County of Philadelphia, 1786 (reprint by Bell Publishing Company, Philadelphia, 1974).

Pierce, Harry. *Railroads of New York.* Cambridge: Harvard University Press, 1953.

The Plough Boy (Periodical), Albany: Printed by John O. Cole, 1819+.

Porter, Daniel. "Making Hay Only When the Sun Shines." *Heritage,* Cooperstown: New York State Historical Association, 1 (July/August 1985).

The Practical Farmer (Periodical). Philadelphia: The Farmer Company, 1855–1900+.

Pratt, J. Howard. *Memories of Life on the Ridge*. Albion, N.Y.: Orleans County Historical Association, 1978.

Pratt, J. Howard. *Saga of the Ridge*. Albion, N.Y.: Pratt, 1983.

Preston, Marie. *Avon, Heart of the Genesee Country.* Avon, New York: 1958.

Quick, Graeme, and Buchele, Wesley. *The Grain Harvesters*. St. Joseph, Michigan: The American Society of Agricultural Engineers, 1978.

Rafter, George. *Genesee River Storage Surveys*. Albany, New York: State of New York, Wynkoop Hallenback Crawford Co., printers, 1897.

Rasmussen, Wayne (Editor). *Readings in the History of American Agriculture*. Urbana, Illinois: University of Illinois Press, 1960.

Rawson, Richard. *Old Barn Plans*. New York: Main Street Press/Mayflower Books, Inc., 1979.

Rempel, John. *Building with Wood*. Toronto & Buffalo: University of Toronto Press, 1967.

Report of the Board of Railroad Commissioners for 1856. Albany: Van Benthuysen, 1857.

Reynolds, Helen W. *Dutch Houses in the Hudson Valley Before 1776*. New York: Payson and Clarke for the Holland Society of New York, 1929 (reprint by Dover, 1965).

Richards, Caroline Cowles (Introduction by Margaret Sangster). *Village Life in America: 1852–1872*. Williamstown, Massachusetts: Corner House Publishers, 1972.

Rifkind, Carole. *A Field Guide to American Architecture*. New York: The New American Library, Inc., 1980.

Robinson, Solon, (Editor). *Facts For Farmers; Also for the Family Circle,* Volume I. New York: A. J. Johnson, 1866.

Robinson, W.L. *The Grange, 1867–1967*. Washington, D.C.: The National Grange, 1966.

The University of Rochester Library Bulletin, Rochester, New York: University of Rochester Library, XXXV, 1982.

Rooney, John F. Jr., Zelinsky, Wilbur, and Louder, Dean R., (Editors). *This Remarkable Continent*. College Station, Texas: Texas A & M University Press, 1982.

Root, Mary R. *History of the Town of York, Livingston County, New York*. Caledonia, New York: Big Springs Historical Society, 1940.

Rose, Albert. *Historic American Roads*. New York: Crown Publishers, 1976.

Rural New-Yorker (Periodical). Rochester, N.Y.: D.D.T. Moore, 1850–1857+.

Sadler, Jacquelin, D.J. *American Stables*. Boston: New York Graphic Society, 1981.

Sanford & Co. *Historical Album of Orleans County, New York*. New York: Sanford & Co., 1879.

Saylor, Henry H. *Dictionary of Architecture*. New York: John Wiley & Sons, Inc., 1952.

Schmidt, Carl. *Cobblestone Masonry.* Scottsville, New York: Schmidt, 1966.

Schmidt, Carl. *History of the Town of Wheatland*. Rochester, New York: Schmidt, 1953.

Schmidt, Carl, and Parr, Philip. *More about Octagons*. Scottsville, New York: Schmidt, 1976.

Schmidt, Carl. *The Octagon Fad*. Scottsville, New York: Schmidt, 1958.

Scientific American (Periodical). New York: Munn and Company, 1845–Present.

Sculley, Francis X. "One Big Barn." *Upstate New York/Democrat and Chronicle,* Rochester, (October 11, 1981): 16.

Seebohm, M. E. *The Evolution of The English Barn*. Cambridge: Harvard University Press, 1927.

Shank, William H. *Towpaths to Tugboats*. York, Pennsylvania: The American Canal and Transportation Center, 1982.

Shaw, Robert B. "A Case of Railway Mania." *Trains,* (May, 1977): 22–27.

Shirreff, Patrick. *A Tour through North America*. Edinburgh: Oliver and Boyd, 1835 (reprint by Benjamin Blom, Inc., New York, 1971).

Siebert, Wilbur. *The Underground Railroad from Slavery to Freedom*. New York: Arno Press and The New York Times, 1968.

Siles, William H. "A Vision of Wealth: Speculators and Settlers In the Genesee Country of New York, 1788–1800." PhD dissertation, University of Massachusetts, 1978.

Silvernail, Richard. "The Agricultural Land Use of The Western Finger Lakes Region of New York: 1600–1959." PhD dissertation, The University of North Carolina, 1960.

Simpson, Wm. P. & Florence K. *Hockshop*. New York: Random House, 1951.

Skinner, J.S. *Memoirs of the Pennsylvania. Agriculture Society.* Philadelphia: Skinner, 1824.

Sloane, Eric. *An Age of Barns*. New York: Ballantine Books, 5th printing, 1976.

Sloane, Eric. *Diary of an Early American Boy (Noah Blake—1805)* New York: Ballantine Books, Funk & Wagnalls, Inc., 1974.

Sloane, Eric. *Eric Sloane's America*. New York: Galahad Books, 1982.

Sloane, Eric. *A Museum of Early American Tools*. New York: Ballantine Books, 4th printing, 1974.

Sloane, Eric. *A Reverence for Wood*. New York: Ballantine Books, 1965.

Smith, James H. *History of Livingston County, New York*. Syracuse, New York: D. Mason & Company, 1881.

State Engineer & Surveyor. *Engravings of Plans, Profiles and Maps, Illustrating the Standard Models From Which Are Built the Important Structures on the New York State Canals, Accompanying the Annual Report, 1859*. Albany: Charles Van Benthuysen, Printer for the State, 1860.

Staufer, Alvin F. *New York Central's Early Power, 1831 to 1916*. Medina, Ohio: Staufer, 1967.

Stefferud, Alfred (Editor). *Power to Produce, The Yearbook of Agriculture, 1960*. Washington, D.C.: The United States Department of Agriculture, 1960.

Stevens, Frank W. *The Beginnings of the New York Central Railroad, A History*. New York: G.P. Putnam's Sons, 1926.

Still, William. *The Underground Railroad*. New York: Arno Press and The New York Times, 1968.

Swartout, Barbara. *Ontario County Cobblestones*. Canandaigua, Geneva: Ontario County & Geneva Historical Societies, 1981.

Taber, Thomas T. *The Delaware, Lackawanna & Western Railroad in the Nineteenth Century*. Muncy, Pennsylvania: Thomas Taber III, 1977.

Thomas, John. *Farm Implements*. New York: Harper & Bros., 1854.

Thomas, J.J. *The Illustrated Annual Register of Rural Affairs and Cultivator Almanac*. Albany, N.Y.: Luther Tucker and Son, 1854+.

Thomas, Robert B. *The Farmer's Almanack*. Boston: Richardson and Lord, 1792+.

Thompson, Katherine Wilcox. *Penfield's Past*. Penfield, N.Y.: The Town of Penfield, 1976.

Threshermen's Review and Power Farming. St. Joseph, Michigan: The Threshermen's Review Company, Inc., 1914+.

Tooker, Elisabeth. *The Iroquois Ceremonial of Midwinter*. Syracuse: Syracuse University Press, 1970.

Tresemer, David. *The Scythe Book*. Brattleboro, Vermont: Hand and Foot, Ltd., 1981.

Turner, Chipman. *The Pioneer Period of Western New York*. Buffalo: Bigelo Brothers, Printers, 1888.

Turner, Orsamus. *History of the Pioneer Settlement of Phelps and Gorham's Purchase, and Morris' Reserve*. Rochester, New York: William Alling, 1851 (reprint by James Brunner, Geneseo, New York, 1976).

Turner, Orsamus. *Pioneer History of the Holland Purchase of Western New York*. Buffalo: Derby & Co., 1850 (reprint by James Brunner, Geneseo, New York, 1974).

Tuttle, Donald. "When They Grew Hops in New York." *The Conservationist* (Sept.–Oct. 1981): 35–38.

United States Department of Agriculture. *Farmers' Bulletin*. (No. 1088, 1512, 1626, 1754, 1817, 1820), Washington, D.C., Government Publication Office, 1926–42.

Up & Down The River. Geneseo, N.Y.: The Genesee Valley Council on the Arts, # 6, 1977.

Van Diver, Bradford. *(Geology) Field Guide—Upstate New York*. Dubuque, Iowa: Kendall/Hunt, 1980.

Van Ravenswaay, Charles (Introduction). *The Compleat Farmer*. Clinton, New Jersey: The Main Street Press, (3rd printing), 1976.

Vaux, Calvert. *Villas and Cottages*, 2nd Ed. New York: Harper and Brothers, 1864 (reprint by Dover, 1970).

Voorhees, Theodore, et. al. *The American Railway*. New York: Charles Scribner's Sons, 1889.

Wacker, Peter O. *The Musconetcong Valley of New Jersey*. New Brunswick, New Jersey: Rutgers University Press, 1968.

Wadsworth, W. Austin. *The Hunting Diaries of W. Austin Wadsworth, M.F.H., Genesee Valley Hunt, 1876–1909*. Geneseo, New York: Genesee Valley Hunt, 1984.

Wadsworth, William. *Riding to Hounds in America*. Berryville, Virginia: The Chronicle of the Horse, (6th printing), 1962.

Waugh, F. W. *Iroquois Foods and Food Preparation*. Ottawa: Government Printing Bureau, 1916.

Weale, John. *Rudimentary Dictionary of Architecture Terms*. London: Weale, 1849.

Welling, William. *Collectors' Guide to Nineteenth Century Photographs*. New York: Collier Books, 1976.

Westing, Frederick. *Erie Power.* Medina, Ohio: Alvin Staufer, 1970.

Whitaker, James. *Agricultural Buildings and Structures*. Reston, Virginia: Reston Publishing Co., 1979.

White, John. *A History of the American Locomotive*. New York: Dover, Johns Hopkins Press, 1968.

Wiard, Henry L. *History of the Wiard Plow Company.* (speech delivered in Batavia, New York, 1962).

Wigginton, Eliot, (editor). *The Foxfire Book*. Vols. 1 to 6, Garden City, New York: Doubleday & Company, Inc., 1972 to 1980.

Wilkins, Lois N. *The Mormon Barn of Lakeville, New York*. Livonia, N.Y.: Wilkins, 1978.

Wing, Henry H. *Milk and Its Products*. New York: The Macmillan Company, 1905.

Wisbey, Herbert, A. *The Sodus Shaker Community.* Lyons: Wayne County Historical Society, 1982

Woodbury, I. B. *The Million's Gleebook or New York Melodeon*. New York: Cornish, Lamport, & Co., 1851.

Woodward, George, & Thompson, Edward. *National Architect*. New York: George Woodward, 1869, (Reprint by Da Capo press, New York, 1975).

Wright, Frances (by an Englishwoman). *Views of Society and Manners In America; in a Series of Letters From That Country To a Friend in England During the Years 1818, 1819, and 1820*. New York: Bliss & White, 1821.

Yarrington, James. *Wyoming County New York, An Architectural Tour.* Wyoming County: Arts Council for Wyoming County, 1984.

Zimiles, Martha and Murray. *Early American Mills*. New York: Bramhall House, 1973.

Index

Editor's Note: Words and numbers in **Bold Type** indicate illustrations

Abutment, see ramp
Addison Tobacco barn, 257, **Port, 232, 256**
Advertisements from the 1820s, **58–59**
Adze, gutter, used in hewing, 51
Aerial views of farms, **97, 158, 363–365, 384, 487**
Agricultural Education 334–335
Agricultural implement manufacturing, 289–301
Agricultural information, 331–339
Allan, Ebenezer "Indian", pioneer, mill constructed at Rochester, 43, 167, 171
Allegany County Fair, **330**, 335, **336**
Allen, Lewis, first large orchards, Grand Island, N.Y., 264
Almanacs, importance to farmers, 331
American Agriculturist, 139, 209, 233, 250, 252, 257, 260, 327, 331, 334, 347, 423
American Farmer and Farm News, 334
American Rural Home, 246, 272, 343
Amish, reuse of barn, 485–488, **487, 488**
Anonymous Diary, 67–68
Anthony, E. and H.T., photographic suppliers, 381
Anti-Friction Hay Carrier, 146, **146**
Appert, Nicholas, invents vacuum canning, 267
Apple drying, 264–265, **266–267**
Appleby, J., develops binder knot mechanism, 299
Apples, 70, 264–266; apple jack, 265; cider, 265
Aqueducts, Erie Canal, Rochester, 85–86, **84**, Genesee Valley Canal, 86, **89**
Architect designed barns, 427–428, **427–429, 468–476**
Arkport, a northern terminal for wheat shipping, 77
Arks, importance in shipping farm products, 77
Ashantee Farm, Avon, barn arrangement, 361, **365**, concrete riding hall, 479–480, **480**
Auburn and Rochester Railroad, 279; advertisement, **279**
Aultman Buckeye Twine Binder, 300, **300**
Avon, N.Y., site of first bridge over Genesee River, 80
Ayrault, Allen, 338; twin steers, 340–341, **342**

Bacchus & Fitch Company, Brockport manufacturers of McCormick reapers 295
Baker, A.D., Company, Swanton, Ohio, manufacturer of separators, 396, **398**
Balcom Greek Revival farm house, prime example of, 203, **204**
Baldwin, Samuel, 41
Balloon framing, 176–177, **176**, 183; with gambrel roof, 303, 310, 313
Bank barns, 137–164
Barber, Aaron, see also Mulligan, Gambrel barns, 183–184, 203, **Port, 184, 185**; Greek revival barns, 203, **205**; arrangement, 361, **364**
Barge boards/Verge boards, 216, **219**
Barn alteration, 127, **128–130**, 131, **132**, 157–164, **159–164**,
Barn arrangement, 361–368, **362–366, 368, 384**
Barn bridge (ramp to second level), 135
Barn cleaning, 436, 443; manure trolley, **388**
Barn dances, 313
Barn dating, 96–97, 177, 501–504, **502, 503**
Barn enlargement, 157–164, **157–164**; 303–304, 310–311, **304–305, 309**
Barn/farmhouse style matching, 203–228; Barber/Mulligan, **205**, John Coe, **216**, Cole/Gagnier, **226**, L. M. Loss, **225**, John Maltman, **227**, John Shields, **215**, Sardis Simmons, **207**, Wadsworth/Hartford House Farm, **228**
Barn fires, xvi, **xvii**, 74, 248–250, 303, 327
Barn flooring, proper installation of, 178–179
Barn frame models, English three bay, 113, **113; Wells, 451, 452**; Jennings barn, 416, **418**
Barn framing 74, 95–132
Barn moving, Knickerbocker Farm, 162, **164**; Wright Farm, 303, 305–309, **306–309**; Black & White Farm, 384, 387, **384**; described by Caroline Richards, 387; McNall Farm 485–488, **487–488**
BARN PHOTOGRAPHS (listed by county)
- Allegany: Block (McKinney) barn exterior, **Port, 466, 468–469, 474**, construction, **470–473**, interior mow, **475**; Church/Bromeley Farm nine-sided carriage house exterior, **369**, interior, **370**; Tile-roofed gambrel, **476**
- Cayuga: Horse fork in operation, **149**
- Chemung: U.S. Department of Agriculture Big Flats Soil Conservation Tobacco barn, **256**
- Genesee: Brown Farm, **345**; Clark Farm, **348**; Pafk Farm barn raising, **Dust Jacket, 322**; Taylor gambrel barns, **322**;
- Livingston: Ashantee Farm barn arrangement, **365**; Jones Bridge Farm barn arrangement, **362**; Big Tree Farm enlarged gable barn, **65**, aerial view of the Gen-

esee River Flats, **158**, exterior, **159**, interior, **159, 160**; Black and White barn, exterior, **Port, 380, 383**, aerial view, **384**, construction **382, 383**, interior, **385, 386, 387, 388, 389**; Barn in Genesee River flood waters, **Port**, Chanler basement barn, exterior, **180**, interior **181**; Coe Gothic barn, detail, **Port, 202, 216, 217, 218, 219**; Coe Road Octagon barn, **Port, 371**; Conesus Milk Producer's Coop. Association Building, concrete block creamery building, **479**; Crossett Farm barn arrangement, **365**, exterior, **427**, interior, **429**, blueprint, **428**; Culvert Farm side-hill barn, **Port, 138, 139**, interior, **142**; Doty/Book concrete block barn, exterior, **477**, interior and detail, **478**; Enlarged side-hill, **161**; Fall Brook Farm gambrel barn, **325**; Fish Hatchery barn, **329**; Gambrel-roofed barns, **304, 323**; Getzinger/Bailor Wells barn exterior, **455**, arrangement, **366, 455**, truss detail, **456**, raising, **458**; Hartford House barns, **228, 229**; Lakeville side-hill barn, **138**; Maxwell/Estes barn interior, **311**; McClellan Farm arrangement, **362**; McClellan Farm gambrel, **316, 317**; McNinch barn, exterior, **411**, interior, **412**; Merrimac English barn, enlarged, **124, 125, 126**; Merrimac/International Salt Company truss barn, exterior, **424**, interior, **425, 426**; Mormon barn, **133**; Mulligan Farm barn arrangement, **364**; Mulligan gambrel barn, exterior, **Port, 184**, interior, **185**; Mulligan Greek Revival barns, **205**; Mulligan/Thomson Wells barn exterior, **452**, interior, **465**, distinctive window cap, **Port**, Nations Farm English barn, aerial view, **97**, elevation, **98, 108**, general view, **229**; interior, **94, 118–119, 120, 121, 122**; exterior, **116, 117**; Oxbow basement barn, sawn frame, exterior, **182**, interior, **182, 183**; Riding Hall, **480**; Roof comparison, **115**; Small English barn, **114**; Shaker barns, Groveland Community, **368**; Stapley/Merrimac arrangement, aerial view, **363**; Stewart Farm barn complex, **486**; Trescott Wells barn raising, **461, 463**; Wadsworth Homestead, cow barn, exterior, **222, 223**, interior, **224**, horse barns, **328**; Westerly exterior, **Port**, interior, **Port**; Westerly Gothic stable, **210, 212**, interior, **213, 214**; Wade English barn, **124**; Wade gambrel barn, **312**, interior, **314, 315**; Webster's Crossing barn, **502**; Wingate Farm truss barn, exterior, **430**, interior, **430**; Wooster/International Salt Co. Wells barn, **464**; Yowell barn interior with tobacco, **252**

Monroe: Brown Wells barn fire, **xvii**; Cobblestone barn foundation, **192, 193, 194**; Cole/Gagnier Italian Villa barn, **226**; Cox Wells barn raising, **462**; Ellwanger & Barry Nursery barns, **271, 273**; Enlarged basement barn, **158**; Estes barn, converted to gambrel, **311**; Genesee Country Museum log barn, **Port, 46, 50, 52, 53**; Genesee Country Museum English barn, **131**; Genesee Country Museum hop house/Flannigan hop barn, **259, 260**, interior, **261**; Guthrie Wells barn interior, **Port, 450, 457**; Hayward seed barn, **274**; Henrietta side-hill exposed frame, **141**; Knickerbocker enlarged side-hill, **164**; McNall Farm barn complex, **487**, dismantling main barn, **487, 488**; Pittsford gambrel-roofed barns, **Port, 316**; Richardson/Michel gambrel barn (underground railroad station), **326**; Seldon/Stokoe Wells barn raising, **458**; Shearing/Krenzer raising, **Dust Jacket, 302, 319**, finished barn, **319**; Shields Gothic barn/house, **215**; Stacking hay near Hubbard Farm barns, **247**; Thornell/Bruner Wells barn exterior, **459**, interior, **460**; Vick Flower Farm barns, **270**

Ontario: Boarden/Moody truss barn interior, **422**; East Bloomfield raising, **112**; Gauss interior framing, **68**, shouldering, **123**; Greek revival, **207**; Horizontal sided barn, **504**; Maltman Italian Villa barn, **227**; Moody Farm with windmill on barn, **407**; Padelford Farm gambrel barn, **324**; Side-hill three bay with shed, **Port, 132**; Smith Geneva Nursery barns, **275**; Swift-Pulvino cobblestone barn, exterior, **188, 195, 196**

Orleans: Jones Farm barns, **268**; L. M. Loss Farm Italian Villa barn, **225**; Sumner Farm barns, **264**

Otsego: Farmer's museum, barn door with flail, **99**; Horse collars in barn, **413**

Seneca: Cosad/Brown Ramp barn, **378, 379**

Steuben: Addison Tobacco barn (horizontal doors), **Port, 232, 256**; Altered frame for hay track, **144**; Dzybon Tobacco barn, **255**; Gable converted to Gambrel-roofed barn, **305**; Greek Revival, **207**; McCarthy Tobacco barn interior, **254**; Owen barn, **Port, 433, 438**; Pratt barn, exterior, **127**, interior, **128, 129**; scarf joint, **130**

Tompkins: McClintock/Bruno steel framed barn construction, **481**

Wayne: Burns apple dry house, **267**; Barn moving (Wright Farm), **306–309**, conversion of gable to gambrel roof **309**; Cobblestone barn, Greek revival, **196, 197**; Ennis enlarged barn exterior, **Port, 162**, interior, **Port, 163**; Galbraith Jennings barn interior, **Port, 414, 418, 419, 420, 421**, exterior, **418**; Herbst Farm raising, **186, 187**; Jenkins/Weeks Octagon barn, **373**; Raymer Jennings barn exterior, **416**; Schleede barn raising, **483, 484, 485**

Wyoming: Asel Clark Farm, **348**; Sherman barn arrangement, **363**; Senft Greek Revival barn, **206**; Thomson/Smith Farm Ramp barn, exterior, **374**, interior, **360 376, 377**, plan and elevation, **375**

Yates: Gothic stable, **221**

Barn raisings, in Gauss Diary, 69, 73; detailed account of, 110, 112–113; Alva Paddock, **109, 111**; East Bloomfield, **112**; Herbst Farm, 186, **186, 187**; Shearing Krenzer, 313, **302, 319, 321**; Black & White, 370, 381–389, **Port, 380, 382–389**; Wells Barns, 455–464, **453, 458, 461–463**; McKinney, **471–473**; McClintock/Bruno Steel Frame barn, 482, **481**; Schleede, 482, **483–485**

Barn site selection, 103
Barn storage, 135, 436
Barn, timber framed, 134, 135
BARNS, Log, 47–59; English, 95–133; Side-Hill (basement), 137–141; Enlarged Gable, 157–164; Sawn Frame, 167–187; Cobblestone, 189–198; Greek Revival, 203–206; Gothic Revival, 208–223; Italian Villa, 224–228; Arrangements, 361–378, Gambrel, 303–329; Truss framed, 415–430; Wells Barns, 451–465; Modern material barns, 467–489; (All of the above barns are illustrated)
Barry, Charles, designer of Palace of Westminster, 208, **209**
Barry, Patrick, horticulturist/author of *The Fruit Garden*, 264; 391; influence on grafted trees, 264; Ellwanger & Barry Nurseries, 270–272, **271**; Italian Villa, 271–272, **271, 273**; varieties of pears, 338
Barter, as means of exchange, 77
Basement barns, dairy use, 351
Battle, Captain, Genesee River steamboat captain, 79
Bealer, Alex, author of *The Log Cabin*, 51
Beans, production of, and Civil War influence, 246
Beatty, Erkuries, 1779 description of Genesee Valley, 36
Beer, as a substitute for drinking water; also recipe and commercial production, 258
Belke, August, barn builder, 186
Belvedere (see cupola)
Bierstadt, Charles, photographer, 86, **88**
Big Tree Farm, Geneseo enlarged barn, example of, 157, 161, **158–160**
Bigelow Semi-Portable steam engine, **404**
Binders (harvesters), development and manufacturing in Genesee Country, 299–300, **299–300**; advertisement, **300**
Black & White Farm barn, Groveland 370, 381–389; **Port, 380, 382–389**; silo detail, **389,** name, 384
Blinn, H. C., (Shaker) describes Genesee Valley Canal trip, 91, 93
Block Barn (McKinney), Cuba, N.Y., 468–477, **Port, 466, 468–475**; the trotter, 468–469, **475**; block making, 469, **470**; size of blocks, 469, Shah of Persia, 474; Czar Nicholas II, 474
Board and batten sheathing, 211, **211, 212,** 215
Boarden/Moody Rushville Farm truss barn, 404, 418, **407, 422**
Boiler explosion, description of 394
Boughton, Enos, (Town of Victor pioneer), framed barn, 96, 170
Boughton, Jared, (Town of Victor pioneer), 170, 243
Boxing (shouldering), definition of, 116, **123**
Brewster, Henry, (pioneer of Riga), describes 1805 marketing of wheat, 41, 43–44
Bridge, first over Genesee, 80; Tonawanda RR at Rochester, **278**, Portage, Erie RR Wooden Trestle, 282, **276, 281**, Portage Iron Bridge, 282, **282**; Portage Steel Bridge, 282; Genesee Valley Canal Aqueduct, 86, **89**
Brinkerhoff's Self Raking Reaper, 296–297
Broad axe, used in hewing, 103, **104**
Brown Farm, West Bethany, hog drive, **345**
Brown, Thomas, Wells barn, Scottsville, fire, xvi, **xvii**
Buckwheat, milling of, 170
Buffalo, Rochester & Pittsburgh Railroad, 284
Burns, Samuel, apple dry house, **267**
Bush, William, builds early sawmill, 169
Butler's Rangers, **37**

Cabbage, introduction of, 257
Cache, corn, 34
Campbell, Patrick, description of Genesee Country, 47, 54, **54**
Campfield, Dr. Jabez, Revolutionary War Soldier, 36
Canada, as a market for wheat, 44
Canal boats, 85–93; advertisement, **90** *Harvey Tucker,* a Genesee Valley Canal boat, 93, **93**
Canals, 83–93, **84–93**; maps of, **87,** 89, **283**
Canandaigua, N.Y., first land office in U.S., 37
Canning, vacuum, development of, 266–267; commercial, 267–268
Carroll, Charles, land speculator, 171
Cartier, Jacques, 33
Casaday Sulky plow, 294, **295**
Case agricultural engine, **401**
Case horse powered threshers, 391, **392, 393**
Cathedral of Notre Dame, Paris, 208, **209**
Cayugas, on Wyoming Valley raid, 35
Cazenove, Theophile, partner in Holland Land Purchase, 40
Chandler, Porter, Piffard, Gothic stables, **Port,** 209–211, **210, 212, 214**
Chanler barn, Geneseo, 179–180, **180, 181**
Chapin orchard, (pioneer orchard,) developed Northern Spy apple, 264
Cherry Valley, N.Y., raided by Senecas, 35
Chewing Tobacco, 251, **252,** 253
Child, Lydia, author of *The Frugal Housewife*, recipe for Table Beer, 258
Church/Bromeley polygonal (nine sided) brick barn, Belevedere, N.Y., 370, **369–370**
Cigarettes, 253, production of, 255; Peerless Tobacco Factory, 255; brands, 255
Cigars, 253–255, 313, **255**
Circular saw, first patented, 174; saw marks, **177**
Civil War influences on bean production, 246
Clark, Asel, (Town of Conesus pioneer) raises log cabin, 53
Clark Farm, Covington, **348**
Clermont, Fulton's steamboat, 78
Clearing, forests, 50, 54–57, **49**
Clinton, DeWitt, visits Genesee Valley, 83–85
Clinton, George (Governor), 35
Clinton, James (General), importance to 1779 Sullivan Expedition, 35
Clinton Line Barge, **85**
Clover, seed in Philo Hampton Diary, 200; planting of, 247; **247**

Cobblestone Construction, 189–198, **188, 190–198**; distribution, in the Genesee Country, 189
Coe, John, Gothic Revival barn, South Livonia, **Port, 202,** 216–221, **216–219**
Coe Road Octagon barn, South Livonia, 371, **Port, 371**
Cole/Gagnier, Italian Villa barn, Mendon, 225–227, **226**
Combine, twenty horse, **409**
Concrete, (mortar, grout), 467; reinforced, 467–482; attributes of, 482
Concrete blocks, 469–470; advertisement for machines and molds, **470, 471**; barns made from, 467–482; Doty/Book barn, 352–353, 356, 477, **477, 478**; Conesus Milk Producers' Cooperative Association Building, 477, **479**; Ashantee Riding Hall, 479–480, **480**; McKinney "Block" barn, 468–477, **Port, 466, 468–475**
Cone, Joseph, builder of Black & White Farm barn, 381
Conesus Lake, 33, **38–39**
Conesus Milk Producers Cooperative Association Building, 477, **479**
Conklin, Henry, author of *Through Poverty's Vale*, 51, 56, 109, 110, 349, 367
Connection of farm buildings, reasons for and against, 361
Cooling milk, methods of, 351–356, benefits of, 354
Cooperstown, site of early county fair, 335; Farmers Museum, flail, **99**, horse harness, **413**
Corduroy road, 80–81
Corn, 34–35, 70; storage of, 233–238; husking (shelling) of, 237–238; harvester advertisements, **238**; husker advertisments, **239**; sheller advertisements, **240**
Corn husk mask, 35, **35**
Corn sheller 238, **240**
Cornell, Ezra, 335
Cornell University, established, 334–335; College of Agriculture, 335
Cosad, Charles, Ramp barn, Town of Junius, Seneca County, 377–378, **378, 379**
Country Gentleman, The, 331, 334
Cows, milk, 341, 349–356; breeds, 353; beef, 341, means of shipping, 341–342
Cox, Walter, Wells barn, Scottsville, 462, **462**
Coxhead, John, architect, McKinney "Block" barn, 469
Cradling, definition, 151, **152**; Gauss Diary comments on, 70
Craft, Rev. David, Revolutionary War, soldier describes Seneca Village, 36
Crops, 233–261
Crossett Farm, Geneseo, arrangement of barns, 361, 366; description of, 424–425, 427–428, **427–429**
Cuba, N.Y., Genesee Valley Canal, 86, McKinney "Block" barn, 468–477, **Port, 466, 468–475**; Allegany County Fair, 331, 335, **336**
Cultivator, The, A Consolidation of Buel's Cultivator and The Genesee Farmer, 331
Culvert Farm, Geneseo, example of English side-hill barn, 137, 140–141, **Port, 138–139, 142**
Cummings Patent Thrashing Machine, described, 391, **392**
Cupolas, 161, **162**, 183, **184**, 211, **212**, 215, 216, **217**, 224–227; on tobacco barns, 252, **255**; on decorated gambrel roofed barns 327, **329**; with windmills, 404, **407**
Curtis, Glenn, 432

Dairy farming, shipping milk, 284; 349–356; herd size at mid-century, 349; advertisements relating to, **352**, 356
Dansville & Rochester Railroad, 277
Dating barns, 96–97, 177, **177**, 501–504, **502, 503**
Dederick, P.K., Hay Press Manufactory, Albany, N.Y., 408
Deere, John, obtains patent for plow, 293
Delaware, Lackawanna & Western Railroad, 283–284
Demise of virgin timber, second-growth woodlots, 415
Denonville, Marquis, hostilities towards Seneca Indians, 33
DIARIES:
  Anonymous, 67–68
  Thayer Gauss, 69–75
  Sheffield Peabody, 134–135
  L.E., 165
  Philo Hampton, 199–201
  Jennie Mather, 357–359
  Seneca Owen, 431–449
Dingee, W., horse powered threshing, sweep power mechanism, 391, **393**
Disher, James, boiler explosion, description of, 394
Distilleries, 171, 243
Domestic Horticultural Society, formation of, 268
Doors, 97, 100, 115; hinged 117, 321; rollers, 323
Doty/Book Farm, Geneseo, concrete construction, 352–353, 477, **477–478**, milk house, 356
Doty, Lockwood L., historian, 241, 243
Douglas, Frederick, 327
Dow & Fowler Agricultural-Machine Works, Fowlerville, 297, **298**
Downing, A. J., *The Architecture of Country Houses*, 208; *Cottage Residences*, 220, **220, 225**; *The Fruits and Fruit Trees of America*, 268
Downing, Charles, architect, brother of A.J., 268
Driveway (drivethrough), definition of, 97, **118–119**
Dry foundation, 108, **108**
Dry houses, (evaporators), used in fruit processing, 265, **266**; Samuel Burns apple dry house, **267**
Durham boats, use in river transportation, description of, 77
Durham breed, reason for their raising, 341
Dzybon, Andrew, tobacco barn, Corning, 257, **255**

E., L., diary fragment, 165
Ear Marks, sheep, use of, 346, **346**
Eastman, George, 381
Economy of building materials in barn construction, 415
Education, agriculture, 334–335
Ekblaw, K.J.T., author of *Farm Structures*, proponent of truss frame construction, 423
Ellicott, Mr., explains need for mills, 167
Ellwanger and Barry, Nursery, Rochester, 270–272, **271,**

Ellwanger (continued)
273; advertisement for **272**; varieties of pears, 338
Empire Grain Drill Company, Shortsville, Ontario County, 301, **301**
English barns (also known as, three-bay, Connecticut, hay, grain, common), 95–132, **96–98, 108–109, 111–133**
English barn size, 95; converting to gambrel, 303–309; enlarging, 157–164, **158, 161–164**
Ennis, Robert, side-hill basement barn, Alloway, N.Y., 161, **Port, 162–163**
Entablature, 203; identified, **204**
Enterprise, (steamboat), description of, 78
Erie Canal, 44, 61, 83, 84, 85, 86, 189, 259, 284, **84, 85, 87**
Erie Canal (steamboat), 79
Erie Canal aquaduct, Rochester, N.Y., 85–86, **84**
Erie Railroad, 282; Portage Bridge, **276**, 282, **281–282**; milk trains, 352–353, 433; at Avon, N.Y., 282, **280, 285**
Eshbaugh, Paul, hewing, 103–107, **104–107**
Estes Brothers Farm, Caldonia, barn conversion, 310–311, **311**; plowing demonstration, walking plow, **293**, sulky plow, **294**
Evaporators, apple, see Dry houses

Facts For Farmers, 177, 251
Fairs, 62, 335–338, 435; **330, 336**; New York State Fair, Rochester, 338, **339**
Fall Brook Farm, open gambrel barn, 323–324, **325**
Fanning, in Thayer Gauss Diary, 71–73
Fanning Mills, use in threshing, 101 103, **103**
Farmers' Barn-Book, 211–215
Farmer's Museum, Cooperstown, N.Y., three bay log barn, 51–52; flail, **99**, horse harness, **413**
Farnum, Mrs., Honeoye pioneer describes barn raising, 110
Farquhar Vibrating Separator, 396, **396**
Fences, 343–344; stump fences, 55–56, **56**, 343–344; laws relating to sheep pens, 345
Fertilizers, use of, 246; misuse of 351, **350**
Finger Lakes, 33
Fireproof barn construction, 468
Fires, see barn fires
Fish Hatchery barn, Caledonia, decorative gambrel, 327–328, **329**
Fitzhugh, William, early land speculator, 171
Flail, **99**
Flailing, 75, 98, **99, 100**
Flannigan Farm, Greece, N.Y., hop house, 260, **259–261**
Flat boats, see also arks, 77
Flooding, of the Genesee River, 33; provides rich topsoil, 241, **Port, 243;**
Flooring (threshing), 67, 71, 72, 98, 100, **101**
Flour City, (Rochester) derivation of term, 172
Ford, Henry, Museum, 297
Foster, Jessup & Brown Co., grain drills manufactured in Palmyra, 301
Foundations, dry, 108, **108**; mortared, 135, 189; cobblestone, 193, **192–194**
Fowler, Orson, author of *A Home For All*, 370, 372; octagon barns, 371, 372, **Port, 371, 373**
Fox, Weston and Brownson sawmill, Gang Mills, N.Y., 175
Foxfire, describes modern log contruction, 51
Frame alteration, 127, **129**, 141, **142**
Framing, barn, 74, 95–132
Frazee, Wells barn, 462, **461, 462**
Fremin, Father, describes early Seneca farming, 33
Frick Company, manufacturer of separator outfits, 396, **399**
Fruit trees, planting, in Thayer Gauss diary, 72; Wadsworth plants, 65; conducive to Genesee Country, 263–265, **262, 268**; grafting, 264–265; advertisement, **272**
Fulton, Robert, importance to steam navigation, 78, 79
Furnaces, cast iron, wrought iron 467;

Gable roofed barns, see English barn;
Gable to Gambrel conversions, 303–304, 310–311, **304–305, 309**
Galbraith, Richard, Jennings barn, Lyons, **414,** 418, **418–421**
Gambrel–roofed barns, 303–328, **302–329**; conversion from gable, 303, 305, 310–311, **304–305, 309**; increased capacity, 303; origin of term, 309, **309–310**; gable/gambrel proportions, 310; comparison with gable, 312, 313; construction photographs, **313–321**; varied pitch of roof surfaces, 321–323, **322–323**; Dutch/ "overshoot", 321, 323, **323**; box gutters, 323, **459**
Gang Mills, see Fox, Weston & Brownson
Ganson, Captain John, mill, 167–168
Garland, William, claim of first steam farm engine in Monroe County, 394, **394**
Gasoline tractors and engines, first use of, 405, 408, **408, 409**
Gauss, Benjamin, Revolutionary War soldier, pioneer of Bloomfield, father of diarist Thayer Gauss, 69
Gauss, Thayer, diary of, 69–75, Gauss barn interior, **68, 123**
Gelatin, use of in photographic process, as food, 344
Gelser/Schumaker Farm, see Schumaker
Genesee (steamboat), 79
Genesee Corn Sheller, 238, **240**
Genesee Country description, 36
Genesee Country Museum, log barn 52, 54, **Port, 46, 50, 52–53**; winnowing basket, **102**; English three bay barn, 131, **131**; hop barn, 260, **259–261**
Genesee Country, title to, 36–40
Genesee Farmer (And Gardener's Journal), 139–140, **140**, 143, 152–153, 154, 241, 250, 258, 266, 277, 327, 331–334, **332**, 338, 341, 349, 350, 369, 391; Luther Tucker, publisher, 331; second series *Genesee Farmer*, Joseph Harris publisher, 331
Genesee Pike, description of, 80, also known as Seneca Turnpike
Genesee River, navigation, 77–80; flooding 241, **Port, 243,** Letchworth gorge, **32, 284**

Genesee roads, 1809 map of, **81**
Genesee Valley Canal, 80, 86, **88;** map of, **89**; Portageville aqueduct, 86, **89**; Dansville Branch, 86; southern terminus, 86; locks **88**; loading GVC boat, **93**; decline in wheat shipments, 88; schedule, 91; trip by H.C. Blinn, 91, 93
Genesee Valley Canal Railroad, also known as Rochester & Genesee Valley Canal Railroad, 283, **284**
Genesee Valley Hunt, 137, **138**, 230, **229, 231**; importance of to barn conservation, 230, **229, 231**
Genesee Valley Manufacturing Company, Mt. Morris, grain & fertilizer drills, 301
Genesee Valley Stock Farm, (Westerly Farm, Piffard) 210–211, **Port, 212–214**
Genesee Valley, views of, **Title Page, Port, 62, 243**
Getzinger/Bailor Farm, Geneseo, Wells truss barn, arrangement of, 366, **366**; details of construction, 455, **455–456, 458**
Gin, poles, use in log barn construction 54; illustrated **109, 111, 458, 472–473, 481,** 482, **483, 484**
Gorham, Nathaniel, invests in Genesee Country land, 36–37
Gothic Revival barn architecture, 208–223; Westerly stables, **Port**, 209–211, **210, 212–214**; Shields Farm, 215, **215**; Coe Farm, 216, 219, 221, **Port, 202, 216–219;** A.J. Downing design, 220–221, **220**; Wadsworth Homestead cow barn, 222, **222–224**, Ellwanger & Barry Nursery, 271, **271**
Grafting in fruit trees, 264
Grain drills, 301; Empire Grain Drill Works, 301, **301**; Genesee Valley Manufacturing Co., 301
Granary, location and construction, 103, 180, 395, 451; problems of, 200
Granary of the Country, wheat growing, 241–246
Grange, the (The Patrons of Husbandry), 339
Grapes, packing, 441, shipped via lake, 440; shipped via train, **440**
Greek Revival barn architecture, 203–206, **205–207**
Greek temple, 203, **204**
Green and McAuliffe sawmill, Fairport, 175, **175**
Green Corn Festival (Seneca), importance of, 35
Green, Seth, first superintendent of New York State Fisheries, 327–328, **329**
Groton Bridge Company, steel frame barn, erection of, 482, **481**
Groveland Shaker Community, 367–368, **368**
Guthrie Wells barn, Mumford, 455, **Port, 450, 457**
Gypsum (plaster), use of as fertilizer, 246

Hall, Capt. Basil, comments on pioneering 47–48; on stumps 55; describes lumbering, 171
Hall, Joseph, compares reapers, 152–153
Halladay windmills, advantages of ownership, 404, **407**
Halsted, Byron, author of *Barn Plans and Outbuildings*, on apple evaporators, 265, **266**
Halsted, Robert F., extensive description of hay track use, 149–150; describes gambrel barn, 395–396
Hamilton, Alexander, association with Church/Bromeley Farm, Belevedere, N.Y., 370
Hampton, Philo, diary of, 199–201
Hancock, Massachusetts, Shaker round barn, 369–370
Harmon, Capt., raises barn, Honeoye, 110
Harris, Joseph, Editor, *The Genesee Farmer*, on raising wheat, 241; founder of Harris Seed, 270; on cabbage, 257
Harrison, John, Groveland pioneer, 63
Hartford House, Geneseo, Italian Villa style buildings, 227, **228, 229**
Hart–Parr gasoline tractors, 405, **409**
Harvey Tucker, Genesee Valley Canal Boat, 93
Hawthorne, Nathaniel, 91
Hay, 70, 247–250, 435; storage, 247–250; curing, 250; salting, 250; cocking, 437; hay fork/track, 444; press, 408, **410**, 435; round baled, 485, **486**
Hay Fork, Harpoon, invention of, 143, **143**, **145**; Gardner Weeks, 147, **147–148**; advertisements for, **146–147**
Hay Press, 408, **410**; 435
Hay Track (also referred to as a horse fork), **142**, 143, 145–150, **144, 148, 149**, 179–180, 184, 324; Nellis Patent Cable, 143, **145**; to outside, 149; hood, 324; alternation of barns to accommodate, **142**; changes barn design, 143–150
Hayward, E.S., seed barn, Brighton, N.Y., 272, 274, **274**
Heath, John, patent twine binder, 299
Hedrick, Ulysses P., (agricultural historian), silos, 233; hop production, 260; pear production 266; Rochester nurseries, 270; agricultural periodicals, 331
Hendrick Hay Press, **410**
Herbst Farm barn raising, Lyons, 186, photographed by Herman Wright, **186–187**
Hessian Fly, 245, **245**
Hewing, art of, 103–107, **104–107**
Hogs, 344; advertisements, **344**; hog drive, **345**
Holland Land Company, 40, 61; buys 3.3 million acres, 40, **41**, 61
Holmes, E. B., Member of Congress, promotes McCormick reaper manufacturing in Brockport, 153
Hops, 258–261, **258, 260**; hop houses, 259, 260, **258–261**; fungus, 260
Horizontal siding, 503, **504**
Horses, in Gauss diary, 69, 73; feed, 247; proportion to oxen, 341; versus tractors, 408, trotter McKinney, 468–469
Horse Fork, Tilting, use in hay storage, 143, **143, 145, 149**
Horse powered sweep, used in threshing, 391, models of, **392, 393**
Horticulture, 263–275
Hubbard Farm, North Chili, stacking hay, 248, **247**
Humphrey, Wolcott, carriage house, Warsaw, designed by A. J. Warner, 428
Husking (also shelling) corn, methods of, 238, **238–240**
Hussey, Obed, reaper, 152–153

Hutchinson's Patent Cider Mill, 265, **265**
Hydraulics, use in round baled hay, 485, **486**

Ice, use of 351, 352, used in milk production, 353–354; harvesting, 354, **353–355**, 436; clothes washing, Mather diary, 359; storage of, 354;
Icehouses, construction of, 354, **355**
Implement Manufacturing, 289–301
International Harvester Co., 290; on cost of plowing, 408; gas engine, **408**
International Salt Co., see Merrimac Farm, also Wooster/International Salt Co. Wells barn
Iron bridge, Portage, 282, **282**
Iron, cast and wrought, production of, 170, 467; used for nails, 178, **178**
Iroquois, 33–36, 40
Iroquois trail, becomes New York State Road, 80
Italian Villa style barns, 224–227; A.J. Downing design, **225**; Loss Farm, 225, **225**; Cole/Gagnier Farm, 225–226, **226**; Maltman Farm, 226–227, **227**; Hartford House, 227, **228–229**

Jell-O, invented by Pearl Wait, LeRoy, also commercial success 344
Jenkins, Burgess, Octagon barn, Red Creek, 372, **373**
Jennings Barns, truss frame, 415–416, 418, 423, **Port, 414, 416–421**
Jennings, David, patents truss frame, 415–418; advertising broadside, 416, **417**, 499–500
Johnson & Company, reaper manufacture, Geneseo, 296–297
Johnston Harvester Company, Brockport, 295, **296**; twine binder, 300, **300**
Jones, Horatio, frame barn, 96–97; pioneer wheat planter, 241–243
Jones, John, Groveland pioneer, 63
Jones, Seth, orchard, Kendall, N.Y., 265, **268**

Kellogg, P.H., Cuba, N.Y., barn photographer, records McKinney "Block" barn, Cuba, N.Y., **468–475**
Kelly Brothers Nursery, Dansville, 274
Kelly, Daniel, Groveland pioneer, 63
Kelly, Oliver, founder of the Grange, 339
Kelsey, Jehiel, pioneer, 96; on the building of saw mills, 170
Ketchum's Combined Harvester, 152, **153**
Keuka Lake, **432, 433, 442**; steamboat on, **446**
Kimball, William S., & Co. Peerless Tobacco Factory, Rochester, 255
Knee braces, used in barn alterations, 408, 411, **412**
Knickerbocker, George, Pittsford, side-hill barn enlargement, 161–162, 164, **164**

Land Advertisements, 40, **42**, **43**, 57, **58, 59**, 497–498,
Land Grant Colleges, establishment of, 334–335
Land leases, 64
Land Office, Canandaigua first, in U.S., Phelps & Gorham, 37; Holland Land Company, 40, **41**
Land speculators, need for economic development, 167
Land value, 61, 63; means of increasing, 57
Lanphear, Ethan, Wellsville, describes log cabin building, 49
Leased farms, upkeep of, 65–66; leasing arrangements, 66
Leases, land, 64
Leavenworth Corn Husker, 238, **239**
Lee, Mother Ann, founder of Shakers, 367
Lehigh Valley Railroad, 284
Letchworth Gorge, (Genesee River), **32, 284**
Lincoln, Abraham, Cyrus McCormick/Seymour & Morgan litigation, 296
Little Beard's Town (Cuylerville), described, 36
Livestock, 341–356; advertisement for strays, 342
Livingston Agricultural Works, see Dow & Fowler
Livingston County Agricultural Society, 336
Livingston Republican, 234, 280
Livingston, Robert, gains steamboat monopoly, 78; first importer of Merino sheep, 347
Livonia Gazette, see Halstead, Robert
Log barns, construction of, 47–54, **Port, 46, 50, 52, 53**; notches, 51, 52, **52**
Log cabins, 47–51, **49, 50**; Frances Wright description of, 50
Loss, L.M. Farm, Albion, Italian Villa style buildings, 225, **225**
Ludowici Celadon Terra Cotta Company, Alfred, 469, 476, **476**

Magee, Samuel, describes Williamson's 1796 barn, 62
Maltman, John, Farm, Canandaigua, Italian Villa style barns, 226–227, **227**
Manure, problem with, 350, 351, **350**; barn trolley, **388**
Map of Genesee Country Railroads and Canals, **87, 89, 283**
Maple sugaring in Thayer Gauss diary, 72
Markets, early agricultural, 41–44, 57, 77–78, 82, 84, 86–89
Marsh, Charles, importance in the development of binders, 299
Martin, John, early use of grafts in apple production, 264
Mason jar, revolutionizes the home canning industry, 267
Massachusetts, Commonwealth of, preemption of land in New York State, 36
Massey-Harris/Massey-Ferguson, farm implements, 290, 293, 296, **293**
Mather, Jennie, diary, 357–359
Maxwell/Estes gambrel barn, Caledonia, 310–312, **311**
McCarthy Tobacco barn, Addison, 253, **254**
McClellan, Robert, Farm, Geneseo, converted gable to gambrel, 313 **316, 317**; arrangement, 361, **362**
McClintock/Bruno Farm, Dryden, N.Y., steel framed barn, 482, **481**
McCormick, Cyrus, reapers, 151–154, 295–296; Virginia Reaper, 151, **154**; first binder, 299; folding binder, **299**
McCormick, Robert, father of Cyrus, 151
McIntosh, W.H., historian on Genesee Country wheat, 243;

describes hop raising in East Bloomfield, 259
McKay, Hector, Town of Conesus pioneer, raises log cabin, 53
McKay, John, Caledonia pioneer and miller, 170, 171
McKinney, the "Block" Barn, Cuba, N.Y., 468–477, **Port, 466, 468–475**; the trotter, 468–469, **475**; block making, 469; **470**
McNall, Neill, author of *An Agricultural History of the Genesee Valley*, on wheat shipments, 86, 88; milling sites, 172, 173; livestock population, 341
McNall, Stephen, Farm, Henrietta, barn removal and reuse, 485–488, **487–488**
McNinch Farm, Conesus, milk house, 356, **411**; barn modified, 411, **411, 412**
Melville, Jean, Wells historian, 455
Merino sheep, 347–348, **347**
Merrimac Farm, Groveland, 127; English barn, 117, 127, **124–126**; barn arrangement, 361, **363**
Merrimac/International Salt Co. plank barn, 423–424, **424–426**
Mid-Winter ceremenony (Seneca), 35, **351;**
Miles, Dr. Manly, early proponent of silos, 233
Military Tract, described, **34**, 61
Milk house, 356; Doty/Book Farm, 477, **477**; McNinch Farm, 356, **411**
Milk production, 351; transportation to market, 352; first bottled, 353
Milk trains, 352
Mill Grove, southern terminus Genesee Valley Canal, 86, **89**
Mills, flour (grist), 44, 167–173; located, **38, 39**, 95, 96, **171–172**
Mills, saw, **38, 39**, 40, 44, 74, 95, 96, 115, 134, 167–186; located, **38, 39, 168, 169, 171, 173, 174, 175, 176**
Missouri Grain & Fertilizer Drill, see Genesee Valley Manufacturing Co.
Monroe Agricultural Society, 335
Moody, Robert, Rushville Farm, windmill, 404, **407**; truss barn, 418, **422**
Mormon Barn, Geneseo, 131–132, **133**
Morrill Act establishes Land Grant Colleges, 334; Cornell University, 334–335
Morris, Robert, land speculator, 37, 61; Morris Reserve, 37; buys land west of Phelps and Gorham Purchase (Holland Purchase), 40
Mortise, 105–107, **105**
Mount Morris, flood control dam, 33
Mow, 97, 98, **98**, 140, 180, 200, 203
Mulligan, Edward, Barber Road Farm, Avon, gambrel barn, 183–184, **184–185, Port**; Greek Revival barns, 203, **205**; arrangement, 361, **364**; North Avon Rd. Farm Wells barn, 451, **Port, 452,** 464, **465**
Music, 200, 289

Nails, blacksmith made, 178, **178**; machine cut, 176, 178, **178**; wire nails, **178**; retrieval of, 178; nailmaking machine, 178, **179**
Nations Farm, Geneseo, 97, **97**; English barn, elevation, **98, 108**; description, 115–117; general view, **97**, **229**; interior, 118–121; exterior, **97, 116–117**
Newbold, Charles, patents cast iron plow, 290
Newspaper advertisements from the 1820s, **58–59**
Newtown, N.Y., Battle of, 36
New York & Erie Railroad, 278, 280; see also Erie RR
New York Central Railroad (and Hudson River Railroad), 279, 287, **287**
New York Farmer and Horticultural Repository, 331
New York State Agricultural Station at Geneva, 335
New York State Board of Agriculture, 335
New York State Fair, Rochester, 338, **339**; McCormick/Ketchum competition, 338
New York State Horticultural Society, 268
Nichols & Shepard Company, separator, 396, **397**; engine, **400**
Notch, (log), saddle, 51, **52**, sharp, 51; dovetail, 52, **52**; square, 52
Nurseries, 269–274, **270–275**
Nye/Hopkins barn, Pittsford, largest swing beam in book, 116

Octagon barns, 370–374; Coe, 371, **Port, 371**; Jenkins, 372, **373**
Octagon houses, 370
Oetzel, John, Jennings barn model, 416, **418**
Oliver, James, plow patents, 293, Oliver Chilled Plow Works, 294
One Hundred Acre Tract, Rochester, 167, 171
Ontario Steamboat Landing, **78**
O'Reilly, Henry, historian, describes marketing of wheat, 62; comments on War of 1812, 64; on Rochester exports, 77; on wheat production, 244; compares travel by canal/RR, 278;
Osborne Co., corn harvesting, 238, **238**
Oswaldt Brothers, custom threshers, 396, Herman Wright photograph of, **400**
Owen, Seneca, diary of, 431–449; farm, **Port, 433, 438**; photograph of, **449**
Oxbow Farm, Geneseo, 180–183, **182–183**
Oxen, 67, 69, 70, 73, 82

Packet Boat (Genesee Valley Canal), schedule of 91, **91**
Padelford, H., gambrel barn, Canandaigua, 323, **324**
Pafk, Fred, barn raising Alabama, N.Y., 321, **322, Dust Jacket**
Painting, barns, reasons for not, 113; methods/styles of, 211
Palace of Westminster, example of Gothic Revival Architecture, 208, **209**
Patent model, Wells barn, 451, **452**
Patterson, Alexander, Conesus, barn raising, 53
Peabody, Sheffield Winslow, diary of, 134–135, photograph of, **134**
Peaches, 264, 266
Pears, 266

Peerless Tobacco Factory, Rochester, N.Y., 255
People's College, Havana, N.Y., 334
Peppermint barns, board & batten, Attica/Java area, 211
Periam, Jonathan, *Home and Farm Manual*, 251, 351; illus. from, **350**
Phelps & Gorham Purchase, **34**, 36–37, **38, 39**, 61, dealings with Indian Allen, 167
Phelps, Oliver, invests in Genesee Country land, 36–37
Philips Spiral Corn Husker, 238, **239**
Photographing barn interiors, 505–507
Photographs by Herman Wright, **186–187, 306–309, 400**
Photography, improvements, 381
Pickering Treaty, 40
Pit sawing, 48, **48**, mechanization of, 173, **173**
Pitts Premium Separator (thresher), 391, **392**
Plank frame barns, 423–427; bents, 423, **423–426**
Plank road, 75, 82
Plaster (gypsum), used as fertilizer, 246
Plough Boy, The, 331
Plows, manufacturing of, 289–294; walking, 290–293, **290–293**, 395, **395**; cast-iron, 290; sulky, 294, **294–295**; advertisements of, **292–293**
Plums, 266
Polygonal barn, general information, 369, 374; Church/Bromeley nine-sided barn, Belevedere, N.Y., 370, **369–370**,
Port of Rochester, **78**
Portage, Erie RR wooden bridge, 282, **276, 281**; iron bridge, 282, **282**; steel bridge, 282
Portageville aqueduct, Genesee Valley Canal, 86, **89**
Portraits of livestock, 343, **343**
Potash/pearlash, manufacture and use of, 57
Potatoes, 251, 486
Powell plank frame, description of, 423, **423**
Power on the farm, 391–411, **390–411**
Practical Farmer, The, 334
Pratt, Joel, Prattsburg pioneer builds black walnut timber framed barn, 127, **127–130**
Pratt, J. Howard, author/historian writes *Memories of Life on the Ridge* with a detailed account of a barn raising, 110, 112–113; on tobacco farming in Orleans Country, 251
Pratt, John Henry, father of J. Howard, 110
Preemption line, defined, 36, **38**
Prince, William, pomological author, 268
Pugin, A. W. N., designer of Palace of Westminster, 208, **209**
Pulteney, estate, description of, 40, **39**
Pulteney, Sir William, 40, 61
Pulver Gothic stable, Branchport, 221, **221**
Purlins, definition of, 116, **121**, 161; canted posts, 416, **419**; large cross section, **420–421**
Putnam, Jay, barnwright/carpenter, J. Howard Pratt barn raising, 110, 112–113

Races, horse, 62–63, 335; advertisement, Williamsburgh races, **63**
Rafters, 54, 74, 127
Railroad bridges, Erie RR, Portage, 282, **281, 282**; Tonawanda (Tonnewanta) RR, Rochester, 278, **278**
Railroads, 277–287, **276, 278–287**; effect on farmland price, access to markets, 280–282; milk trains, 284, 352; freight rates, 287; map of, **283**; timetable, 285
Raisings, barn, in Gauss Diary, 69, 73; detailed account of, 110, 112–113; Alva Paddock, **109, 111**; East Bloomfield, **112**; Herbst Farm, 186, **186, 187**; Shearing Krenzer, 313, **302, 319, 321**; Black & White, 370, 381–389, **Port, 380, 382–389**; Wells Barns, 455–464, **453, 458, 461–463**; McKinney, **471–473**; McClintock/Bruno Steel Frame barn, 482, **481**; Schleede, 483–485
Ramp barns, 374–378; Thompson/Smith Ramp barn, Warsaw, 374, 377, **360, 374–377**; Cosad Ramp barn, Town of Junius, Seneca County, 377–378, **378–379**
Ramps, (exterior, including bridge, wharf, abutment), **237**, 323, **324, 423**
Random–width boards, use of, 115, 175
Raymer Brothers, barn builders, 415, 416, **416**
Reapers, invention of, 150–152, early manufacturers, 152–154, **153–155**; Genesee Valley manufacturers, 295–298, **296–298, 300**; Seymour & Morgan, 295–296, **297**; Johnston, 295–296, **296**; Brinkerhof, 296; Dow & Fowler, 297, **298**; Wyckoff, Tuttle & Olin, 298, **298**
Reinforced concrete, use of in barn construction, 468–482
Reisinger, Andrew, Pulteney, N.Y., establishes first commercial vineyard, 431
Revolutionary War, Genesee Valley campaign, 35–36
Reynolds and Bateham, first Rochester nursery, 270
Reynolds, P.C., apple grafting, 264–265
Richards, Caroline, author *Village Life in America*, 387
Richardson, David, underground railroad barn, Henrietta, 325–326, **326**
Richmond, Van Rensselaer, barn, Lyons, builds architect designed barn, 428
Ridge pole, 113
Ridge Road, 80
Riding Hall, Ashantee, 479–480, **480**
Riga, Town of, early description, 44
Roads, 80–83; 1809 map of, **81**
Robinson, Solon, author *Facts For Farmers*, 177, 251
Robinson, William Henry, artist, 1868 view of Rochester, 286, **286**
Rochester & Genesee Valley Canal Railroad Company, 283, **283**
Rochester & Genesee Valley Railroad, 278
Rochester, Lockport & Niagara Falls Railroad, 279–280
Rochester, Nathaniel, 171
Rogers Locomotive, 278, **278**
Root, Mary, York historian, on barn fires, 248, 250
Ross, Craig and Mortimore, Black and White Farm name, 384
Round barns, Shaker, 369–370
Rowe, Jerry, demonstrates hewing, 103, **104–106**

Rumely OilPull Tractor, 405, **410**
Rural New Yorker, 334
Rush, Will, Lyons, barn mover, 303, 305, **306–309**

Sash saw mill, "up and down", 173, **173**
Saltbox barn, **Port**, 131, **132**
Saw marks, illustrated, used in dating barns, 177, **177**, 501
Sawmills, 40, 44, 74, 95, 96, 115, 134, 167–170, **166, 168–169**, 171, 173–175, **173–176**; located, **38, 39**
Sawn-Frame barns, 167–187
Scaffold, 116, **128**
Scarf joint, definition of, 127, **130**
Schedules, railroad, 278, 279, **285**
Schleede, Albert, Farm, Lyons, stave silo construction, 234, **235**; constructs sawn timber framed gambrel-roofed barn, 482, **483–485**
Schumaker Farm Dalton, stump fence, 55, **56**
Scott, Gardner, Wadsworth overseer, 65, 263
Scott, Sir Walter, 208
Seed barn, Brighton, N.Y., 272, 274, **274**
Seed production, 269–274; Seed barn, 272, 274, **274**, Shaker seeds, 269, Shaker seed packets, **269**
Selden/Stokoe Wells barn, Scottsville, 457, 460, **458**
Seneca crops, 33; corn storage, 34–35
Senecas, as inhabitors of the Genesee Country, 33; extinguishment of land rights, 36; Wyoming Valley raid, 35, village described 36
Seneca Turnpike, first major East/West road, 80; also known as the Genesee Pike
Senft, Alfred, Greek Revival barn, Peoria, N.Y., 203, 206, **206**
Seymour & Morgan, reaper manufacturers, Brockport, 295–296; Triumph Reaper, 153, 296; "New York" Reaper, 154, **155**, 296, **297**; Triumph Number 8 Binder, 300
Shaeffer, Peter, framed barn, Avon, 96
Shakers, 367–368; Sodus, 367; Groveland (Sonyea), barns, 367–368, **368**; corn, 237; broom corn, 367; flat roofs, **368**, 368; Round barn, Hancock, MA, 369–370; H.C. Blinn on Genesee Valley Canal, 91, 93; seed packets, **269**
Shearing/Krenzer gambrel barn, construction of, **Dust Jacket, 302, 319**; described, 313; celebration, **321**
Shed additions, saltbox profile, **Port**, 131, **132**
Sheds, cattle, 51, **50**
Sheep, 345–349; early problems with wolves, 345–346, 370; ear marks, 346, **346**; sheep pens, laws pertaining to, 345; Merino advertisements, **348**; ram, **347**; Clark Farm, **348**
Sherman, E.C. Farm, Middlebury, N.Y., 361, **363**
Shields, John, Gothic barn, North Bergen, 215, **215**
Shingles, wooden, 113–115, **114, 115**
Shipping cattle, by railroad, 341–343
Shirreff, Patrick, Scottish farmer, on stumps, 54–55, on agricultural periodicals, 331
Shocks, grain, **150**; drying corn, 233; oats, **402**
Shorthorn Herefords, 210,
Shouldering, used in barn framing, 116, **123**
Sibley, Hiram, & Co., seeds, 270, **270**
Sibley, Joseph, pioneer describes wintering of cattle, 51; on bartering, 77
Side-Hill barns, 137–141, **136, 138–141**
Siles, William, historian, remarks on Genesee Country settlement costs, 47
Silk growing in the Genesee Country, 263
Silos, 233–237, **234–237**; square, 234, **234, 235, 237**; hexagonal, 387, **388, 389**; horizontally sheathed, **Port, 236**, stave, 234–235, **235–236**; advertisements for, **237**
Simmons, Colonel Sardis, Greek Revival barn, Town of Richmond, 206, **207**
Single-cylinder gasoline engines, 405, **405**
Slate roofing, as architectural decoration, 216, **217**
Sloan, John, architect, author of *The Model Architect*, 208
Smith, W. & T., Geneva Nursery, 274, **275**
Smokehouses, Genesee Country Museum log smoke house/ dovetail notching, 52, **52**; Greek revival cobblestone, West Bloomfield, 197, **198**
Snook, L. D., on the proper installation of barn flooring, 178–179
Snowfall, annual, 361
Spenser, Major William, builds Westerly Farm, Piffard, 210, **Port, 210, 212–214**
Stage lines, 82, 83; advertisements, **83**
Standard rail gauge, lack of, 281–282, **280**; established officially, 282; conversion to, 282
Stapley-Merrimac Farm, Geneseo, barn arrangement, 361, **363**
State Fair, New York, Rochester, 338, **339**; McCormick/ Ketchum competiton, 338
Steamboats, 78–80, **78, 79**
Steam powered traction engine, 391, 394–405, **390, 394–405**; plowing described, 395, **395**; threshing 401, 404, **400–403**; advertisement, **405**
Steel, 467; Bessemer process, 467
Steel bridge, Erie RR, Portage, 282
Steel frame barn, Dryden, N.Y., 482, **481**
Sternwheeler, 79
Stewart, Harrold, Farm, Geneseo, 485, **486**
Straw stack, in Mather diary, 357, **400**
Stump, fence, described, 55–56, **56**, 343–344; in Philo Hampton Diary, 200; stump pullers, 55, 57; blowing (blasting), 443; removal, 33–34, 54–56, **54–55, 57**, 199, 200
Sulky plow, 294, **294–295**
Sullivan Campaign (also Clinton–Sullivan), 35–36, 263
Sullivan, John (General), 35–36
Summer-fallowing, 246
Sumner Farm, orchards, Shelby, Orleans, County, 263, **264**
Sweetheimer, James, barn model, 113, **113**
Swift-Pulvino Farm, cobblestone barns, Phelps, largest complex in New York State, 195, **188. 195, 196**

Swine, see hogs
Swing beam, reason for, 116; illustrated, **94, 108, 118–122,** largest encountered, 116
Sy, Dan, barn builder, builds Pafk barn, Alabama, N.Y., **Dust Jacket**, 321, **322**

Table Beer, recipe for, 258
Tamboro, volcano, influence on wheat production, 65
Taylor Farm, gambrel-roofed barns, Elba, N.Y., roof variety, 321, **322**
Taylor, winery, 431
Tenant farming, reasons for and method of, 64
Tenon, 105, 106, **105, 106;** weakness, 423
Thomas, John, pomological author, 268
Thomas, Robert, *Farmer's Almanack*, 1826 dairy description, 351
Thomson/Mulligan Farm, Wells barn, Avon, 451, **Port, 452**, 464, **465**
Thomson/Smith Ramp barn, Warsaw, 374, 377, **366, 374–377**
Thornell/Bruner Wells barn, Pittsford, **Port, 316,** 457, **459–460**
Thorpe, David, 387
Threshing, 71, 72, 73, 75, 98, 200, 201, 359. 395–396, 401–404, **409**, 439, 444; costs of, 332–334; flooring, 67, 71–72, 98, 100, **101**
Threshing machines, 391, 394–396, 398, 401, 404, **390–403**; advertisements, 392, 396, 397, 398, 399
Tie rods, 141,184; in use, **185, 317**
Tilting Horse Fork, use in hay storage, 143, **143, 145, 149**
Tin roofing, advantages, 227
Tobacco, 251–257; areas where grown, 251; chewing, 253, **252**
Tobacco barns, 251–257; **Port, 252, 254, 255, 256**
Tonawanda (Tonnewanta) Railroad, 278, **278**; bridge, **278**
Tornado Windmill Company, Elba, N.Y., 404, **406**
Tractors vs. horses, 408
Trains, milk, 352–353
Traxler stump puller, 55, advertisement for, **57**
Treaty of Big Tree, 40
Treaty of Paris, ends Revolutionary War, 40
Treenail, in timber framing, 106; types of wood used for, 106
Triphammer, Conesus outlet, 170
Trollope, Frances, 175
Trumbull, John, artist, land speculator, 44
Truss framed barns, 179–180, **180–181**, 415–428, **414, 416–430**
Tucker, Harvey, Genesee Valley Canal boat, **93**
Tucker, Luther, 269
Turner, Orsamus, 170, 243
Turnpike, Seneca, 80
Tyler, President John, 210; death of daughter Julia at Westerly Farm, 210–211

Underground Railroad, 324–327; barn, **326**
Union barns, see also, Valentine, Samuel, 211
U.S. Dept. of Agriculture Big Flats Soil Conservation Service Tobacco barn, 257, **256**
Updaw, Richard Wells barn, Rush, N.Y., smallest encountered, 460

Vacuum canning, 367
Valentine, Samuel, Fishers, N.Y., "Union Barns", Valentown Hall Museum, 211
Vassar, Matthew, Gothic stable, A.J. Downing designed, 220–221
Vaux, Calvert, architect/author of *Villas and Cottages*, 222
Ventilation in Italian Villa style barns, 211
Vick, James, establishes Rochester seed firm, 269–270; publishes *Genesee Farmer*, 269; Flower farm, **270**
Victoria Tower, Palace of Westminster, **209**
Vineyards, in Seneca Owen diary, 431–449, **433, 442**; willowing, 437, 442, 443; straw tying, 437
Virgin Timber & second growth woodlots, 415
Volcano, Tamboro, influence on wheat production, the year without a summer, 65

Wade Farm, English barn, Avon, 117, **124**; gable and gambrel barns, 312–313, **312, 314–315**
Wadsworth broadside to new settlers, 40, transcription of, 497, **42**
Wadsworth Culvert Farm, 140–141, **Port, 138–139, 142**
Wadsworth flour, quality of, 243–244
Wadsworth, Harry, steam threshing, **400**
Wadsworth, Herbert, sawmill, Ashantee, 169–170, **169**; Ashantee Farm, **360**, 361, 479–480, **365, 480**; Silo Farm, 234; builds future Wade Farm, 312–313, **312, 314–315**
Wadsworth Homestead, cow barn, 222, **222–224**; horse barns, 327, **328**; portraits of livestock, **337, 343**
Wadsworth, James (pioneer), 40, 44, 64, 65
Wadsworth, James S., builds Hartford House and farm, 227, **228–229**; marries Mary Wharton, 227; instructions to his outdoor clerk, 65–66
Wadsworth, Jeremiah, Commisary General of the Continental Army, 40; Uncle of pioneers James and William, 40; investor, 40
Wadsworth, Major W. Austin, founder of Genesee Valley Hunt, 230
Wadsworth, Martha B., Riding Hall at Ashantee, 479–480, **480**
Wadsworth Oxbow Farm, cattle barn, 180–183, **182, 183**
Wadsworth, W. Austin, present Master of Fox Hounds, 230
Wadsworth, William P., Nations Farm, 97; on Homestead Farm, 222, **222–224**; Master of Fox Hounds, 230; **231**, horse barns, 327, **328**
Wadsworth, William W. (pioneer), first arrives in Genesee Valley, 40, 64; reaper design, 152; early proponent of Merino sheep, 347
Wait, Pearl, inventor of Jell–O, 344
Walking Plows, 290–293, **290–293**, 395, **395**

War of 1812, influence on Genesee Country settlement, 61, 64
Warner, A.J., architect, designs Warsaw, N.Y., carriage house, 428
Washington, George (General), authorizes Sullivan expedition, 35
Watson, Elkanah, establishes model for county fairs, 335
Weeks, Gardner, hay fork, railway hay conveyor, 147, **146, 147, 148**
Wells barns, 451–464; patent model, 451, **452**; raisings, 455–464, **458, 461–463**, distinctive window caps, 451, **Port, 452**; outside of Genesee Country, 451; mowing of hay, 451–452; advertisement, 451, **454**; location of granary, 451; Abram Cameron Wells barn, cost of, 453, 455; Walter Cox Wells barn, cost of, 453, 462, 465, **462**; Getzinger/Bailor Wells barn, 336, **366**, 455, **455, 456, 458**; James Guthrie Wells barn, 455, **Port, 450, 457**; Jean Melville, Wells Historian, 455; Selden/Stokoe Wells barn, 453, 457, 460, **458**; Thornell/Bruner Wells barn, 457, **459–460**; Richard Updaw Wells barn, 460; Hawley Farms Wells barn, 460; William Trescott Wells barn, 462, **461, 463**; Frazee Wells barn, 462, **461, 462**; Wooster/International Salt Company Wells barn, 464, **464**; Thomson/Mulligan Wells barn, 451, **Port, 452**, 464, **465**
Wells, John Talcott, Jr., 452–453, **462**
Wells, John Talcott, Sr., patents truss barn design, 451, **452, 454**
Wells Patent Model, 451, **452**
West, Benjamin, artist/investor in Genesee Country land, 44
Westerly Farm, Piffard, 209–211, **Port, 210, 212–214**
Wharton, Mary C., marries James S. Wadsworth, 227
Wheat, 241–246; pioneer planting, 243; 1838 comments on by O'Reilly, 244; 1877 comments on by W.H. McIntosh, 243; production, 244, **242**; collapse of production, 173; Hessian Fly & Wheat Midge, 245, **245**
Wheeler, Gervase, architect, 272
Wiard, Thomas, Wiard Plow Works, East Avon, 290, 293, **292**; plow, **293**
Williamsburg(h), 62–63
Williamson, Charles, 61–63; barn, 62; fairs, 335
Williamson Road, 62, 80
Windmills, 404, **406–407**
Wingate, George, stave silo, 234–235, **236**; truss plank barn, 427, **430**
Winnowing, 98, 100–103; basket, **102**
Wood, Jethro, cast iron moldboard patent, 290, **292**
Wood, Taber & Morris steam engine, 394, 395, **394**
Wood, types used in construction, 177, 178
Wood, Walter, introduces interchangeable reaper parts, 154
Woodbury, Daniel, modifies horse powered sweep mechanism, 391, **393**
Woodward, George, on balloon framing, 177
Woodward, Orator T., producer of Jell-O, 344
Wool production, 347–348; decline, 349
Wooster/International Salt Co., Cuylerville, Wells barn, 464, **464**
Wright, Frances, Englishwoman describes log structures, 50
Wright, Herman, farmer/photographer, Herbst Farm barn raising, 186, **186–187**; barn moving, 303, 305, **306–309**; threshing rig, 396, **400**
Wyckoff, Tuttle & Olin Company, Perry, N.Y., 298, **298**
Wyoming County Agricultural Society, first fair in county, 335
Wyoming, Pennsylvania, Iroquois raids prompt Sullivan expedition, 35

Year without a summer, reasons for, 65
Young, Mrs. John, Batavia pioneer, on the building of her home, 49–50; on the clearing of land, 56–57
Yowell, barn, South Lima, chewing tobacco, curing of, 251, **252**

ADDENDUM

The James Way Barn

Author's note: After the type had been set and the layout finished, additional information became available about the James Way dairy barn, another plank truss frame system.[1] Earlier information about this plank truss frame barn had simply been too sketchy to include.

The James Way truss dairy barn was an outgrowth of the James dairy stalls, first available in 1906 and designed to save labor and improve sanitation. From limited manufacture of these stalls in a single structure on the William James Farm, in Wales, Wisconsin, the business quickly grew to include cow, bull, calf and hog pens, stanchions, steel stall fittings, watering buckets, litter and feed carriers, carrier tracks, composition columns and ventilators. Soon the company was offering advice on how to build new or adapt existing barns to take the best advantage of the James equipment. By 1916 the James company architecture department had helped design thousands of barns across the United States, some half dozen of them in the Genesee Country.

The James architects considered the plank truss barn best for dairy farming. Their reasons, published in a 1916 pamphlet were essentially the same as those presented by Edwin Powell in 1903 (see chapter XX): the plank truss frame was more economical in its use of lumber, provided more storage space for both machinery and hay and was more easily built than any other type.[2] Figure AD-1 shows a James Way blueprint from the 1916 publication *The James Way.* Reflecting the concern for good ventilation James Way barns are often identifiable from the outside by their use of dormers along the break in the gambrel roof and the galvanized sheet metal ventilators along the peak.

[1]*The James Way* (Fort Atkinson, Wisconsin: The James Manufacturing Co.), 1916.

[2]Ibid., 23.

FIGURE AD-1. The 1916 blueprint of a typical James Way plank truss-framed barn (there were variations in design depending on barn size and other factors). The foundation, stable floor and half of the walls are of concrete. On careful examination of the print the ventilating outtake flues are visible.